Austral Biogeography

Austral Biogeography

Editors: P.Y. Ladiges, C.J. Humphries and L.W. Martinelli

CSIRO
AUSTRALIA

National Library of Australia Cataloguing-in-Publication Data

Austral biogeography.

Bibliography.
ISBN 0 643 05228 3.

1. Biogeography – Southern Hemisphere. 2. Cladistic analysis. I. Ladiges, Pauline Y. II. Humphries, C. J. (Christopher John), 1947– . III. Martinelli, L. W. (Laurence Walter), 1941– . IV. CSIRO.

574.909814

Reprinted from *Australian Systematic Botany,*
Volume 4 Number 1 (1991)

Cover design by Julie Fitzpatrick,
CSIRO Editorial Services Section, Melbourne

Printed by Brown Prior Anderson, Melbourne

Dr Pauline Ladiges is Reader in Botany at the University of Melbourne, and Chairman of the Editorial Advisory Committee of *Australian Systematic Botany*

Dr Christopher Humphries is at the Botany Department, Natural History Museum, London

Dr Laurie Martinelli is the Managing Editor of *Australian Systematic Botany*

Contents

Contributors

R. Berry	Geology Department, University of Tasmania, Box 252C, Hobart, Tas. 7001, Australia.
C. Burrett	Geology Department, University of Tasmania, Box 252C, Hobart, Tas. 7001, Australia.
M. M. Cigliano	Divisi6n Entomología, Laboratorio de Sistemática y Biología Evolutiva (LASBE), Museo de La Plata, Paseo del Bosque, 1900 La Plata, Argentina.
Joel Cracraft	Departments of Anatomy and Cell Biology and of Biological Sciences, University of Illinois, P.O. Box 6998, Chicago, Illinois 60680, U.S.A.
J. V. Crisci	Divisi6n Plantas Vasculares, Laboratorio de Sistemática y Biología Evolutiva (LASBE), Museo de La Plata, Paseo del Bosque, 1900 La Plata, Argentina.
J. M. Dowd	Department of Botany, University of Adelaide, Box 498, G.P.O., Adelaide, S.A. 5001, Australia.
N. Duhig	Geology Department, University of Tasmania, Box 252C, Hobart, Tas. 7001, Australia.
J. R. Grehan	Entomology Research Laboratory, University of Vermont, 655b Spear Street, South Burlington, Vermont 05403, U.S.A.
Charles E. Griswold	Department of Entomology, NHB 164, National Museum of Natural History, Smithsonian Institution, Washington, D. C. 20560, U.S.A.
Ian M. Henderson	Department of Botany and Zoology, Massey University, Palmerston North, New Zealand.
Pauline Ladiges	School of Botany, The University of Melbourne, Parkville, Vic. 3052, Australia.
P. G. Martin	Department of Botany, University of Adelaide, Box 498, G.P.O., Adelaide, S.A. 5001, Australia.
K. G. McKenzie	Geology Department, University of Melbourne, Parkville, Vic. 3052, Australia.
B. Michaux	Private Bag, Kaukapakapa, New Zealand.
N. Møller Andersen	Zoological Museum, University of Copenhagen, Universitetsparken 15, DK-2100 Copenhagen, Denmark.
J. J. Morrone	Divisi6n Entomología, Laboratorio de Sistemática y Biología Evolutiva (LASBE), Museo de La Plata, Paseo del Bosque, 1900 La Plata, Argentina.
J. Muona	Oulanka Biological Station, University of Oulu, SF-90570, Oulu, Finland.
Gareth Nelson	American Museum of Natural History, New York, N.Y. 10024, U.S.A.
John M. Pandolfi	Australian Institute of Marine Science, PMB No. 3, Townsville MC, Qld 4810, Australia.
Lynne R. Parenti	Division of Fishes, National Museum of Natural History, Smithsonian Institution, Washington, D.C. 20560, U.S.A.
S. Roig-Juñent	Divisi6n Entomología, Laboratorio de Sistemática y Biología Evolutiva (LASBE), Museo de La Plata, Paseo del Bosque, 1900 La Plata, Argentina.
O. Seberg	Botanical Laboratory, Copenhagen University, Gothersgade 140, DK-1123 Copenhagen K, Denmark.
R. I. Vane-Wright	Biodiversity Programme, Department of Entomology, The Natural History Museum, Cromwell Road, London SW7 5BD, U.K.
R. Varne	Geology Department, University of Tasmania, Box 252C, Hobart, Tas. 7001, Australia.
J. J. Veevers	School of Earth Sciences, Macquarie University, North Ryde, N.S.W. 2109, Australia.
Carden C. Wallace	Museum of Tropical Queensland, 70–84 Flinders Street, Townsville, Qld 4810, Australia.
Jacqueline Wolstenholme	Museum of Tropical Queensland, 70–84 Flinders Street, Townsville, Qld 4810, Australia.
Amanda Young	Museum of Tropical Queensland, 70–84 Flinders Street, Townsville, Qld 4810, Australia.

Preface

A symposium entitled 'Austral Biogeography' was held as part of the IXth annual meeting of the Willi Hennig Society in Canberra, Australia on 24-27 August 1990. The goal of the symposium was to bring together different schools of biogeographers — cladistic biogeographers and panbiogeographers, to seek common ground. By inviting biologists with interests in Austral biota, together with geologists, there was hope that some generalities on biogeographic patterns would materialise. The Austral biota is one of the most distinctive in the world, and disjunct patterns are perhaps easier to see in the Southern Hemisphere than the more diffuse patterns of the Northern Hemisphere. Methods of cladistic biogeography and panbiogeography have been applied with some vigour to a range of taxa with trans-Pacific distributions, but acceptable explanations of Pacific areas, as determined from individual biotic cladograms, is still debatable (see Nelson and Platnick 1981; Humphries and Parenti 1986).

Following the Hennig meeting, the Australian Systematic Botany Society held their annual meeting with a special session entitled 'Pacific Biogeography at the Crossroads', and a number of those contributions on the same general topic, but with special emphasis on Indo-Pacific biogeography, are included here. One additional paper submitted to *Australian Systematic Botany* (by J. Grehan) is included to complete the volume.

During the last 15 years or so, there has been a revival of interest in biogeography, brought about by new methodological and empirical developments in cladistic (or vicariance) biogeography (see Nelson and Platnick 1981) and panbiogeography (see Craw 1989). Myers and Giller (1988) summed up the present situation:

> 'In our view, the rigorous cladistic core of cladistic vicariance biogeography and the important primary spatial base of panbiogeography form an essential ground plan for an interpretation of the space–time evolutionary framework of historical biogeography.'

Cladistics has promoted a revival in the understanding of the principles of comparative biology. Considerable rigour is applied to the analysis of relationships of organisms; homology has been equated with synapomorphy (Patterson 1982), and the criterion of choice between competing hypotheses of relationship is parsimony.

At the same time, many biogeographers have come to accept Croizat's main ideas on the relationships of space, time and form (Croizat 1964). Under the general title of panbiogeography, Croizat's ideas have been developed by New Zealand biologists (see Matthews 1989 for accounts by some of the leading figures), who have laid emphasis on the relationship of space, time and form with geology, particularly for the New Zealand biota (see Page 1989), bringing the ideas of Croizat into a modern synthesis.

Cladistic biogeographers accept Croizat's general dictum, that the earth and its biota evolved together, as a theory requiring explanation, but at the same time consider historical biogeography as a problem of classification. In their view, many analyses are plagued by accounts of para- and polyphyletic groups, and theories of area relationships are immune from criticism throughout dispersal narratives. In cladistic biogeography, analysis of the relationships of areas is based on an explicit statement of the relationships of monophyletic taxa. There has been a pursuit of methods (see Page 1988; and papers in Hovenkamp 1987; Myers and Giller 1988;

Systematic Zoology **37**, 3, 4) that allow the principles of cladistics to be applied through the analysis of different groups of organisms that occur in seemingly identical areas.

In this volume, most of the papers come from cladists, and as it turned out only a few panbiogeographers were able to attend the meeting. The biological papers cover methodological issues and studies of a rich array of organisms, from flowering plants to spiders, fishes and corals; the geological papers provide explanations of continental breakup, terranes and the complex geology underlying some of these biotic patterns.

Pauline Y. Ladiges and Christopher J. Humphries

References

Craw, R. (1989). Foreword. In 'Panbiogeography — Special Issue.' *New Zealand Journal of Zoology* **16**, i–iv.

Croizat, L. (1964). 'Space, Time, Form: The Biological Synthesis.' (Caracas: published by the author.)

Hovenkamp, P. (Ed.) (1987). 'Systematics and Evolution: a Matter of Diversity.' (Institute of Systematic Botany: Utrecht University.)

Humphries, C.J., and Parenti, L.R. (1986). 'Cladistic Biogeography.' (Clarendon Press: Oxford.)

Matthews, C. (Ed.) (1989). 'Panbiogeography — Special Issue.' *New Zealand Journal of Zoology* **16**, i–iv, 471–814.

Myers, A., and Giller, P. (Eds) (1988). 'Analytical Biogeography: An Integrated Approach to the Study of Animal and Plant Distributions.' (Chapman and Hall: London.)

Nelson, G., and Platnick, N.I. (1981). 'Systematics and Biogeography: Cladistics and Vicariance.' (Columbia University Press: New York.)

Page, R.D.M. (1988). Quantitative cladistic biogeography: constructing and comparing area cladograms. *Systematic Zoology* **37**, 260–74.

Page R.D.M. (1989). New Zealand and the New Biogeography. *New Zealand Journal of Zoology* **16**, 471–83.

Patterson, C. (1982). Morphological characters and homology. In: 'Problems of Phylogenetic Reconstruction.' (Eds K.A. Joysey and A.E. Friday) (Academic Press: London.)

Various authors (1989). *Systematic Zoology* **37** (3, 4).

On Areas of Endemism

Norman I. Platnick
Department of Entomology, American Museum of Natural History,
Central Park West at 79th Street, New York, NY 10024, USA.

The range of papers presented in this symposium — stretching from the truly sublime to the largely subliminal — is too broad to be encapsulated effectively in just a few paragraphs. Some of these contributions clearly pose a serious challenge for biogeographers: if biogeography without areas, and biogeography without cladograms, have now been added to our list of accomplishments, must not biogeography without taxa, and biogeography without organisms at all, be next on the docket? Absurdities aside, one topic that recurs frequently is the problem of defining areas of endemism appropriate to studies of historical biogeography.

At the minimum, it would seem that an area of endemism can be defined by the congruent distributional limits of two or more species. Obviously 'congruent' in this context does not demand complete agreement on those limits, at all possible scales of mapping, but relatively extensive sympatry at some scale must surely be the fundamental requirement.

Although areas of this sort have been treated by some biogeographers, most notably in Donn Rosen's classic studies of Middle American fishes, many recent biogeographic studies utilise areas that may be too large, whereas a few utilise areas that may be too small. To some extent, at least the first syndrome is easy to understand. By the time we have done the field work to sample populations, collected the data to discriminate species, cladistically analysed their interrelationships, and mapped their distributions, the group of organisms in question gains a heavy burden of anticipation for biogeographically decisive resolution.

In systematics, there is no *a priori* reason to expect that a given character system (be it morphological, physiological, behavioural, or molecular) will provide data crucial to the resolution of cladistic relationships at some particular hierarchical level. Is there an *a priori* reason to expect that the distributions and interrelationships of a given group of taxa will provide data crucial to the resolution of area relationships in some particular part of the world?

For example, Henderson (this volume) cites some genera of Trichoptera which each have a few species in New Zealand, and shows that their distributions are often relatively large and overlapping. Such genera are simply not the ideal starting point for analysing area relationships within New Zealand, despite all the taxonomic work already invested in them. There are many other arthropod genera, for example among endemic New Zealand spiders and opilionids, which include 15 or 20 allopatric species each. It is the cladistic relationships within those genera that will be most informative for New Zealand biogeography; compared with them, the trichopteran taxa are relatively widespread and therefore relatively ambiguous.

In other words, biogeographers do have an advantage not enjoyed by taxonomists. In systematics, we have no way of knowing, in advance, what character system will be most likely to resolve a particular set of interrelationships. In biogeography, however, we can always prefer to initiate our studies with those taxa that are maximally endemic — those which include the largest number of species, with the smallest ranges, in the area of interest.

In many studies, however, including several presented here, the areas analysed bear no demonstrated relationship to these minimal areas of endemism. At one extreme, even smaller areas may be defined — on a purely geographical basis (e.g. different river basins) — and treated as separate areas, even though no taxa in the analysis are endemic to them. This can only be interpreted as the triumph of hope over evidence. At the other extreme, large chunks of the earth's surface may be treated as areas (Australia, New Zealand, southern South America, etc.). Such treatments presume that these geographically, rather than biologically, defined regions represent natural groupings of areas of endemism, a presumption that is probably false in most cases.

Charles Griswold's presentation (this volume) on the vicariance biogeography of afromontane spiders provides a perfect example of why such presumptions should be avoided. Griswold's analyses of taxa confined to the highly disjunct montane forests of Africa and Madagascar shows clearly that anyone analysing the problem at a gross level (i.e. treating 'Africa' and 'Madagascar' as if they were areas of endemism, or natural groups of such areas) will be working largely with artefacts, for 'Africa' in this sense is at best a paraphyletic assemblage of areas of endemism, not the sister group of 'Madagascar'.

In short, gross-level studies (such as the classic work of Brundin a quarter-century ago) treating present-day continental masses or similarly large territories as if they must be natural areas provided a tremendous impetus for the study of cladistic biogeography. Nevertheless, such gross-level studies are unlikely to provide an accurate picture of the complex history of most biotas. Any energy invested in them today could be better spent in identifying those taxa that are maximally endemic in, and therefore (at least potentially) maximally informative about, the regions of interest.

3 September 1991 *Aust. Syst. Bot.*, **4**, 1–11

Phanerozoic Australia in the Changing Configuration of Proto-Pangea Through Gondwanaland and Pangea to the Present Dispersed Continents

J. J. Veevers

School of Earth Sciences, Macquarie University, North Ryde, N.S.W. 2109, Australia.

Abstract

The neotectonic supercycle of earth history started 320 million years ago with the initial coalescence of the continents in Pangea. Final coalescence took place 230 Ma ago at the same time as rift valleys induced incipient breakup that became actual from 160 Ma with the start of seafloor spreading in the Atlantic and Indian Oceans. The current phase of seafloor spreading is marked by the growth lines of magnetic anomalies, from which reconstructions of the continents during the past 160 Ma are accurately determinable by eliminating the dated parts of the seafloor. Many small terranes or fragments are not so well constrained. Palaeolatitude is less precisely determined by continental indicators of magnetic inclination. All this physical evidence provides a unique solution for continental reconstruction since 320 Ma. Less definite evidence provided by biota and geological facies has to be accommodated within this physical framework. Before the coalescence of Pangea (> 320 Ma) the constraints are reversed. This paleotectonic phase lacks preserved seafloor spreading so that continental palaeomagnetism, biota and geological facies are the only indicators.

The changing configuration of Australia and its neighbours in the eastern Gondwanaland province of Pangea — India, Antarctica, Lord Howe Rise–New Zealand Plateau — is detailed through seven stages from the mid-Jurassic breakup of Pangea.

Introduction

The Phanerozoic earth underwent a tectonic-climate supercycle (Fischer 1984; Veevers 1990*b*) from a single continent (Pangea) and ocean (Panthalassa) to dispersed continents (Fig. 1) by the growth of the rift (Atlantic and Indian) oceans between the Pangean fragments at the expense of the Panthalassan relic of the Pacific Ocean. Within this supercycle, Australia lay (in modern coordinates) at the eastern end of Gondwanaland/Pangea and at the western edge of the Paleo-Pacific/Panthalassan Ocean (Fig. 2) and finally at the western edge of the Pacific Ocean (Fig. 3). Westward subduction of Panthalassa/Pacific alternated from Chilean-type, producing a magmatic arc along the continental margin, to Mariana-type subduction, with the arc separated from the continent by a back-arc basin (Veevers 1990*a*). By this process, oceanic terranes, mainly Mariana-type volcanic arcs, and continental terranes became accreted along the eastern margin of Australia (Coney 1990). The north and north-western parts of Australia faced an arm of Panthalassa called Tethys and, from the mid-Carboniferous at the latest, calved off continental slivers or terranes that crossed Tethys to collide with Asia (Veevers 1988*a*). The donation of terranes from Tethyan Gondwanaland, India being the largest, was countered by the accretion of Pacific terranes on the east.

1030-1887/91/010001$05.00

Fossils play a vital part in unravelling Phanerozoic earth history. They not only provide a time frame but also indicate the depositional environment, as wide as that of the continents and oceans to as narrow as that of a pond. In establishing time and place, fossils are crucial in reconstructuring palaeogeography but, as noted by Talent (1984) in his account of Australian biogeography past and present, they have specific strengths and weaknesses. The following description of Phanerozoic palaeogeography is preceded therefore by an appreciation of the changing application and precision of fossils in continental reconstruction compared with physical methods.

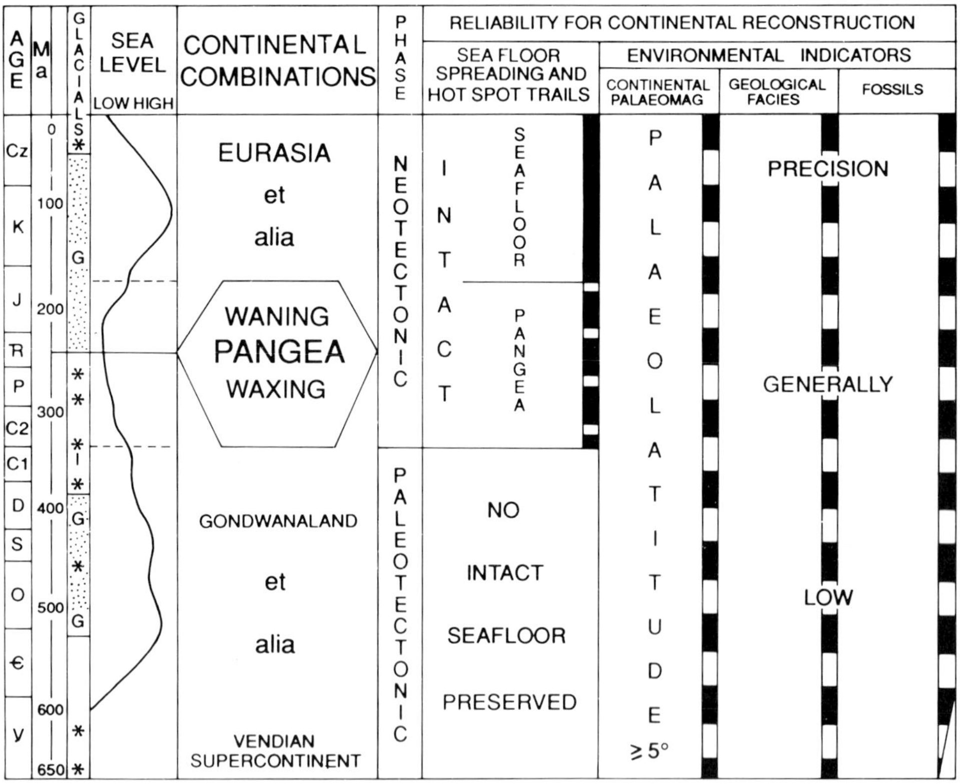

Fig. 1. Reliability of physical and biological evidence for continental and oceanic reconstruction during the Phanerozoic tectonic–climatic supercycle (Fischer 1984; Veevers 1990*b*). * Glacials; I, icehouse state; G, greenhouse state.

Methods of Continental Reconstruction

The pioneers of global palaeogeography (Wegener 1915; Du Toit 1937) based their case for continental drift on two arguments: the jigsaw fit of the conjugate continental margins and the correlation of distinctive geological structures and sedimentary sequences on the continents, including the confinement to the fragments of Gondwanaland of the Gondwana sedimentary sequences and enclosed *Glossopteris* flora. All this was done with virtually no knowledge of the structure, composition and age of the 70% of the surface of the solid earth made up of the deep seafloor. The discovery in the 1960s that the oceanic lithosphere was the working fluid of the plate-tectonic engine, with flow lines indicated by fracture zones and growth lines by magnetic anomalies, led to the confirmation of continental drift as a corollary of plate tectonics. Deep-sea drilling calibrated the age of the magnetic anomalies, and showed that the oldest intact seafloor in the central Atlantic, north-east Indian and western Pacific Oceans is no older than about 160 Ma, or mid-Jurassic. Older seafloor has been resorbed into the mantle at subduction zones, except for rare slivers thrust onto the continents. On the one hand, plate tectonics revealed in accurate detail the neotectonic earth history of the past 160 Ma; on the other, it obscured all of the earlier, paleotectonic part.

Accurate quantitative reconstructions of the continents and oceans back to 160 Ma are implicit in the magnetic growth lines and fracture-zone flow lines of the deep seafloor (see Scotese and Sager 1988 for a recent synthesis). Pangea remained intact from its 320 Ma inauguration to initial breakup 160 Ma ago, though with the possibility of internal shearing. During this neotectonic phase since 320 Ma, the relative position of the continents and intervening oceans can therefore be read from the structure of the present seafloor; their position within an absolute frame can be determined by appeal to the theory of the fixed source in the deep mantle of hotspot trails (Morgan 1981), such as the Emperor Seamounts–Hawaiian Islands chain; and their palaeolatitude and azimuth can be determined, though with low resolution only, by magnetic measurements on the continents.

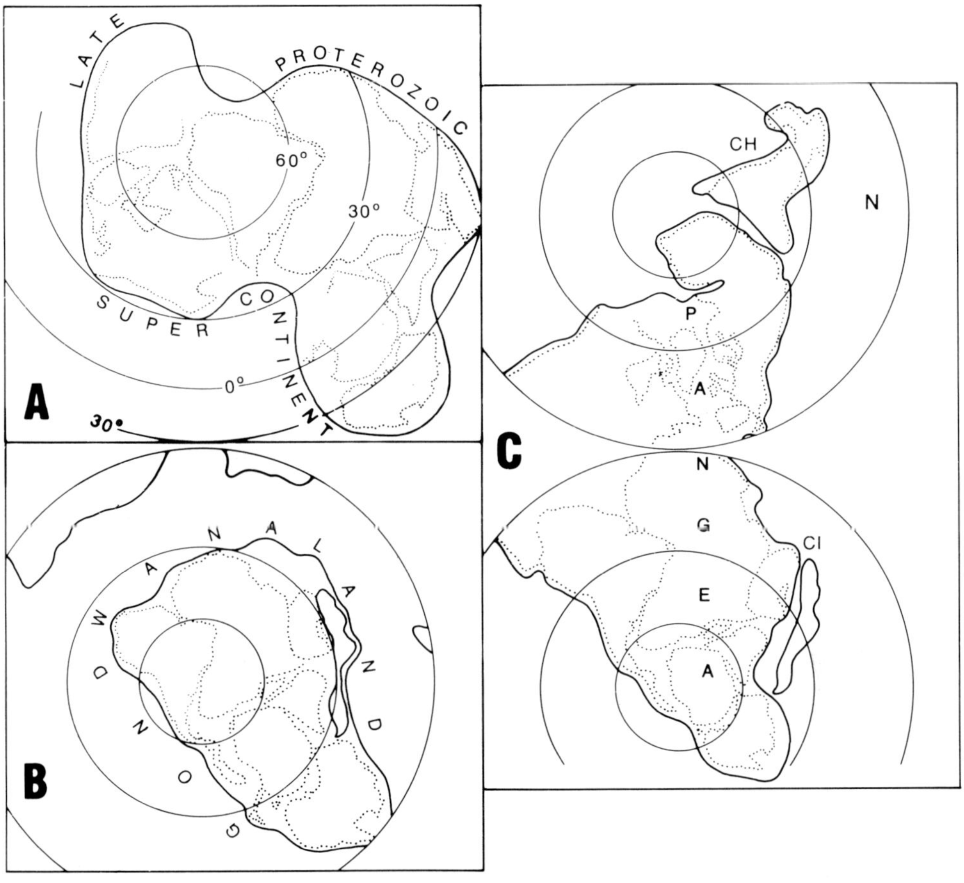

Fig. 2. Global reconstructions for the first half of the Phanerozoic. *A*, Latest Proterozoic (600 Ma) Proto-Pangea on a south polar Lambert equal-area projection (Bond *et al.* 1984). *B*, Mid-Palaeozoic (400–350 Ma) Gondwanaland and other dispersed continents on a south polar stereographic projection (Scotese *et al.* 1979). *C*, Late Palaeozoic Pangea (300 Ma) on south and north (N) polar stereographic projections (Scotese *et al.* 1979). CI, Cimmeria, CH, China (after Veevers 1988*b*).

Direct physical evidence of continental dispersion available in the neotectonic phase overshadows indirect evidence provided by fossils and facies, and anomalies in the distribution of fossils are regarded as problems for biogeography (Fig. 1). Before 320 Ma is a different matter. Information provided by older seafloor (none of which is preserved intact: it was subducted except for a few scraps caught in the continental accretionary wedge) is negligible, and the indirect evidence of continental palaeomagnetism and biogeography/facies is all that is available. Still earlier, before the advent of metazoan fossils about 650 Ma, only the physical methods of continental palaeomagnetism and facies apply. Attempted absolute reconstructions of the Palaeozoic continents and oceans (Zonenshayn *et al.* 1987) have to rely on anchoring palaeomagnetic apparent polar wander paths to isolated intraplate continental magmatic belts interpreted as hotspot trails.

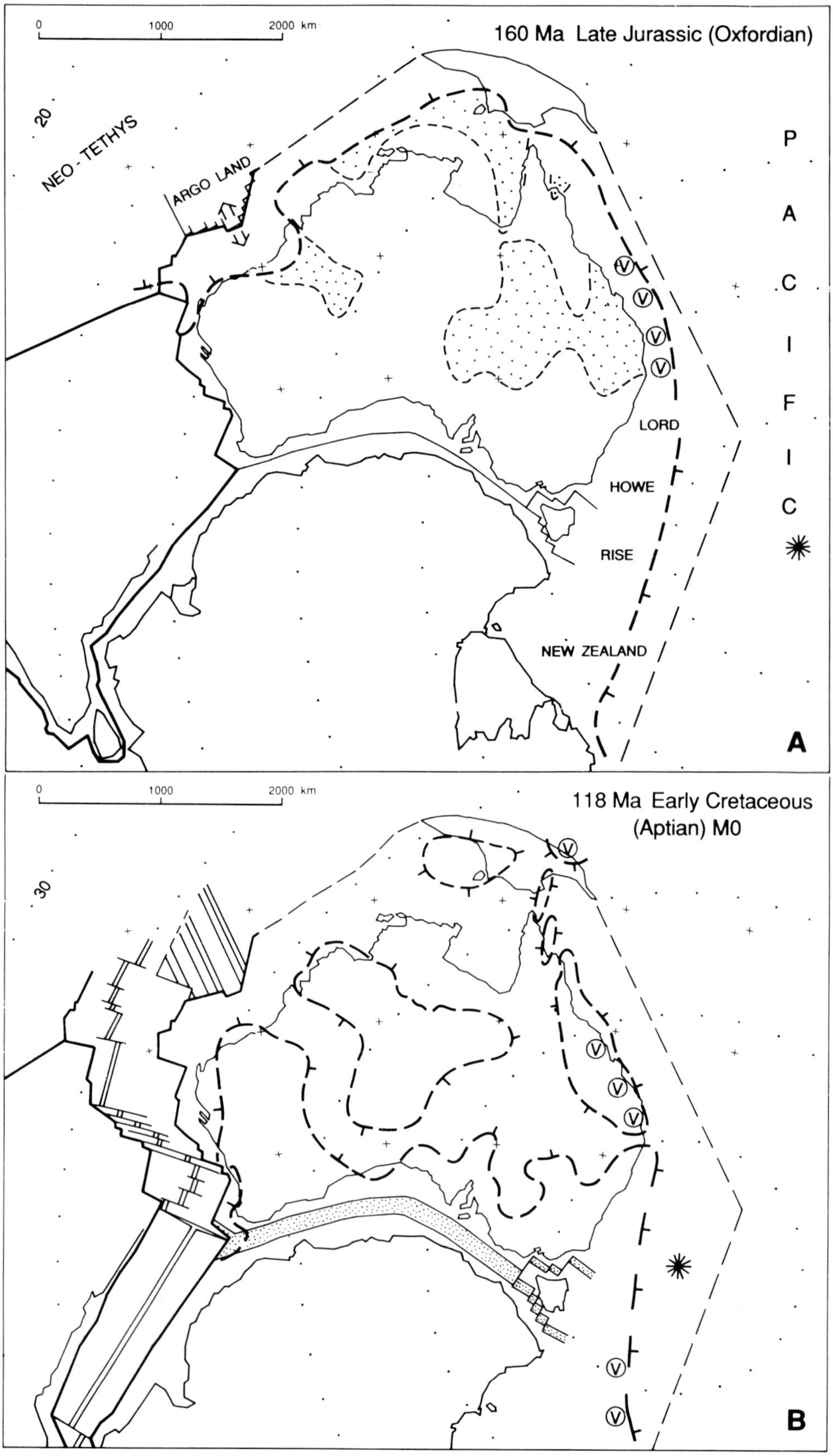
0
1000
2000 km
160 Ma Late Jurassic (Oxfordian)
20
NEO-TETHYS
ARGO LAND
P
A
C
I
F
I
C
LORD
HOWE
RISE
NEW ZEALAND
A
0
1000
2000 km
118 Ma Early Cretaceous
(Aptian) M0
30
B

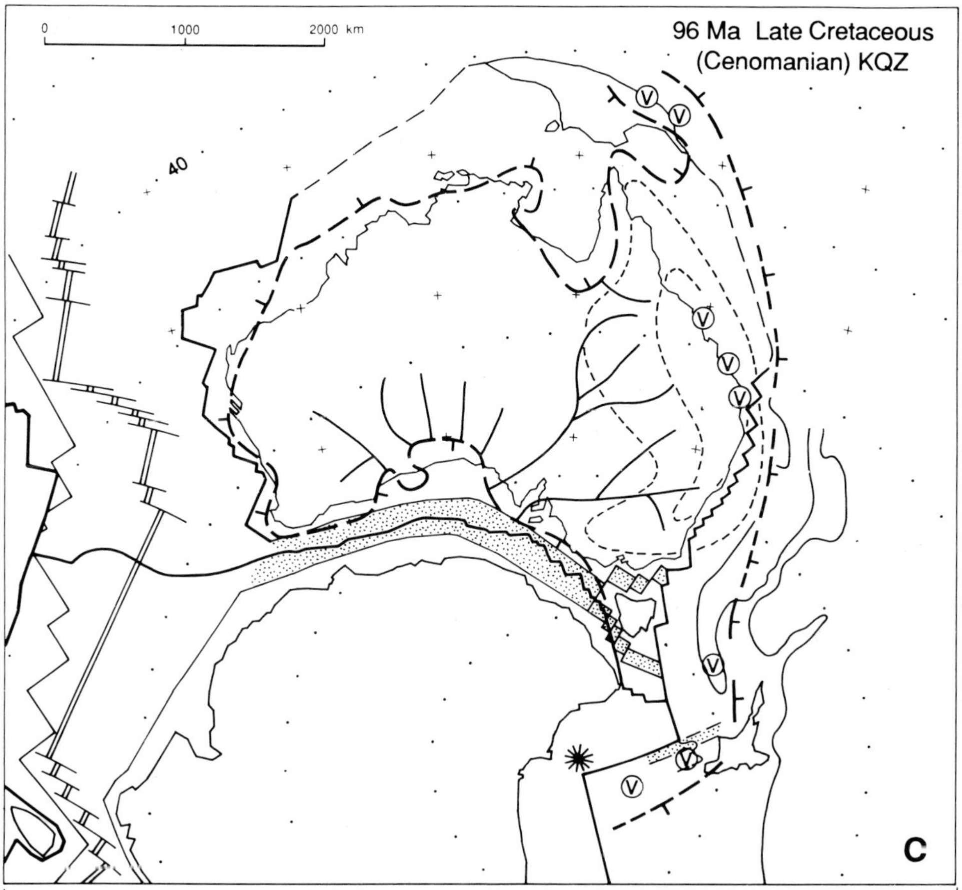
0
1000
2000 km
96 Ma Late Cretaceous
(Cenomanian) KQZ
40
C

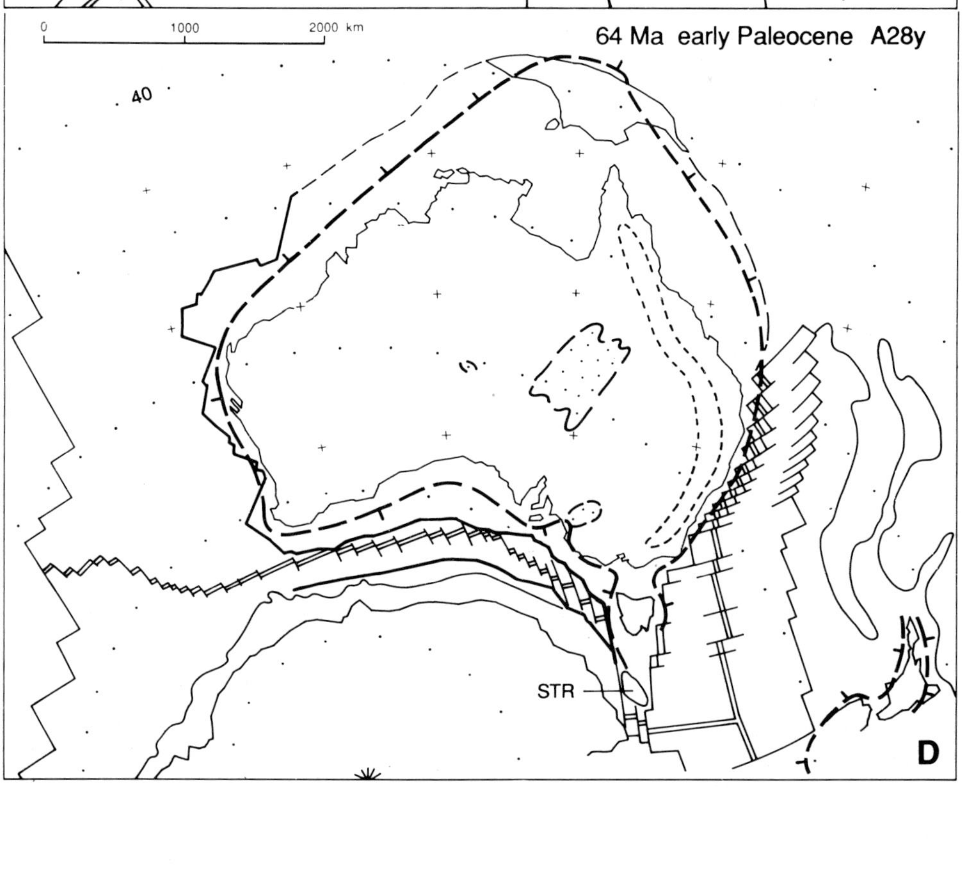
0
1000
2000 km
64 Ma early Paleocene A28y
40
STR
D

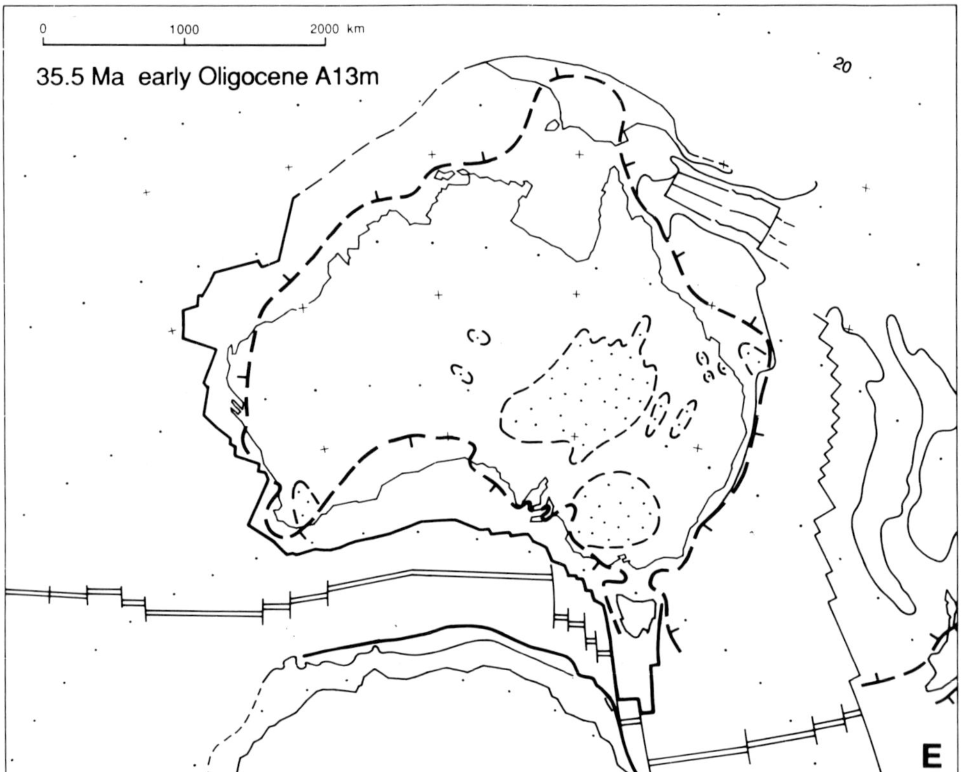
0
1000
2000 km
35.5 Ma early Oligocene A13m
20
E

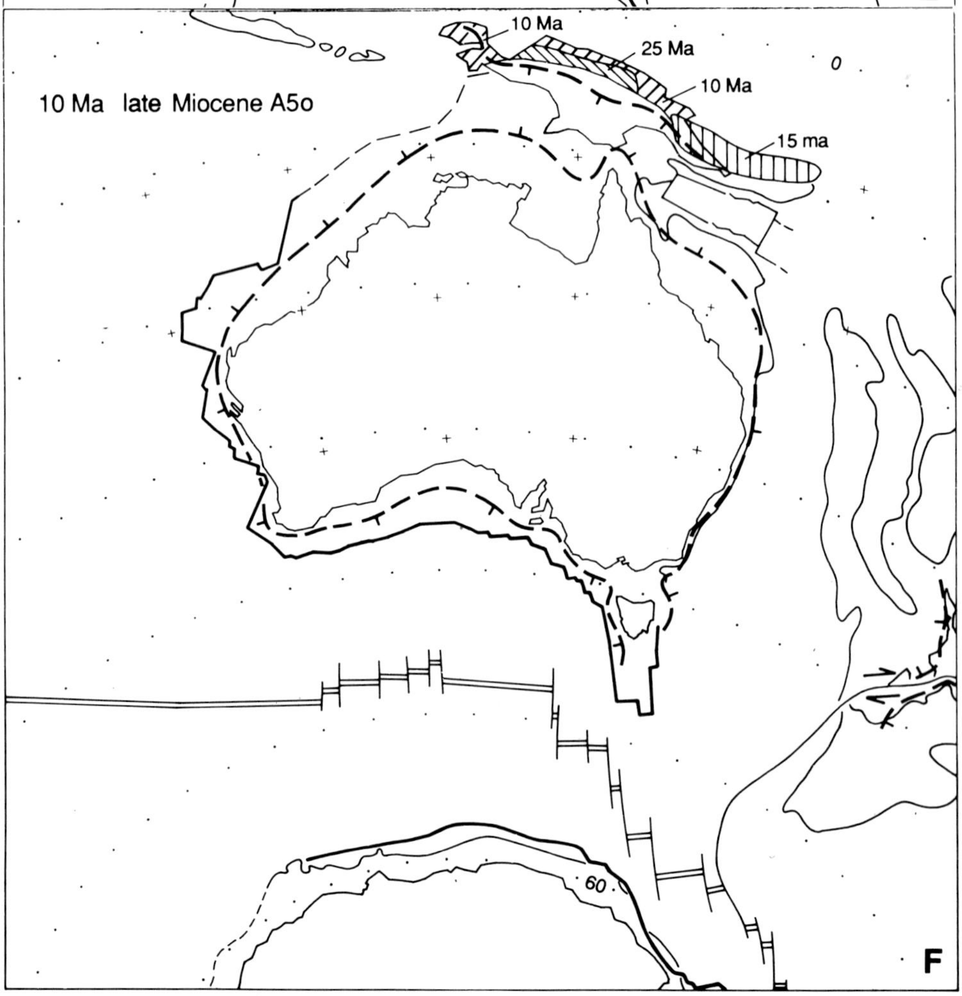
10 Ma late Miocene A5o
10 Ma
25 Ma
10 Ma
15 ma
0
60
F

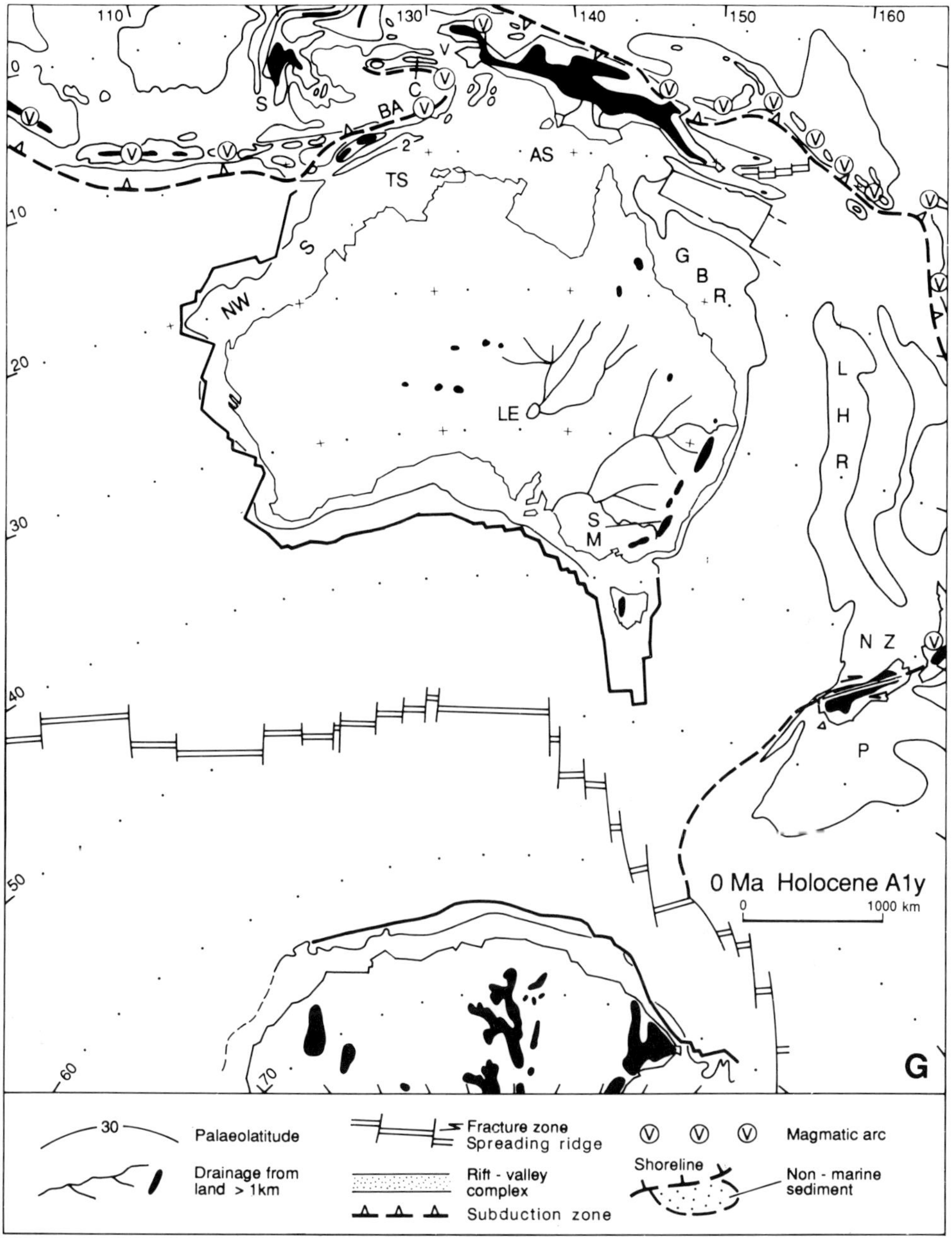

Fig. 3. Regional palaeogeography of Australia and its neighbours in the eastern part of the Gondwanaland province of Pangea from initial breakup in the Late Jurassic (160 Ma) to the present day. Lambert equal-area projection. Continental dispersion and palaeolatitude from Veevers *et al.* (1991) and Veevers and Li (1991); palaeogeography of Australia from Veevers (1984, 1991), of New Zealand from Bradshaw *et al.* (1981), Laird (1981) and Kamp (1986*a*, 1986*b*). Symbols explained in *G*; the land > 1 km in Antarctica is bedrock. In *D*, STR, South Tasman Rise; in *G*, AS, Arafura Sea; BA, Banda Arc; C, Ceram; GBR, Great Barrier Reef; LE, Lake Eyre; LHR, Lord Howe Rise; NW S, Northwest Shelf, NZ P, New Zealand Plateau; SM, Snowy Mountains; TS, Timor Sea; V, Vogelkop.

Outline of Early Phanerozoic Global Palaeogeography from Proto-Pangea to Pangea

The earth alternates in a supercycle 400 Ma long from a single continent (Pangea) and ocean (Panthalassa) with an icehouse climate to many continents and oceans with a greenhouse climate (Fischer 1984). The supercycle is driven by the heat that accumulates beneath the insulator of Pangea, and the greenhouse is made by the excess CO_2 vented from the mantle during the faster rate of plate activity indicated by high sea level (Fig. 1) (Veevers 1990*b*). Fig. 2 shows the stages of (*A*) Proto-Pangea at the point of breakup in the latest Proterozoic/earliest Phanerozoic, (*B*) the dispersed supercontinent Gondwanaland and other continents and (*C*) Pangea. The Austral continents and India remained together in a northern province of Proto-Pangea through Gondwanaland itself to a southern or Austral province of Pangea. Australia, the eastern part of Gondwanaland, occupied low latitudes until its rapid movement (as part of Pangea) southward in the Late Carboniferous–Permian (300 Ma). Its eastern part faced Panthalassa and its northern part Paleo-Tethys, from which Cimmeria disengaged to cross Paleo-Tethys in front of an opening Neo-Tethys (Fig. 2*C*).

The Gondwana sequence of India and its equivalents in Australia, southern Africa, Brazil, and Antarctica constitute the Austral facies of the Pangea supersequence, distinguished by non-marine to marine glacial strata at the base, reflecting mid to high southern latitudes (Veevers and Powell 1987), and succeeded by mainly non-marine strata that contain a provincial biota separated from Boreal Pangea on the east by the equatorial oceanic gulf of Tethys and on the west by the tropical land zone across northern South America/Africa and southern North America/Europe. The Gondwana province glacials and endemic biota were isolated from the rest of Pangea not by wide continental separation but by the tropical continental zone on the west and by the oceanic gulf of Tethys on the east (Anderson 1981). As a free-standing entity, Gondwanaland (Fig. 2*B*) spanned 350 Ma from the fragmentation of a Late Proterozoic supercontinent (Bond *et al.* 1984; Fig. 2*A*) at or about the start of the Palaeozoic (570 Ma) to the coalescence of Gondwanaland and Baltica/Laurentia (Fig. 2*C*) in the mid-Carboniferous (320 Ma). Except for the early loss of Cimmeria (Fig. 2*C*), the Gondwanaland assemblage remained intact within Pangea until incipient dispersal in the Carnian (230 Ma) and definitive dispersal in the Late Jurassic (Veevers 1989). The original Gondwanaland elements of Pangea became independent from the Late Jurassic until the Cenozoic merging of South America in North America and of India and Africa/Arabia (and, in the near future, of Australia) in Eurasia. The south polar Antarctica will apparently remain the only free-standing Gondwanaland element. Note that the Gondwana facies started only after Gondwanaland had merged in Pangea (Veevers 1988*b*).

Austral Palaeogeography since the 160 Ma Breakup of Pangea

Fig. 3 is a set of seven palaeogeographical maps of Australia and its neighbours during the later neotectonic stage ($<$ 160 Ma) of earth history.

Late Jurassic (Fig. 3*A*)

At the initial breakup of Pangea, Australia, Antarctica and Greater India ranged from high (on the east) to low latitudes. On the north-western margin of Australia, a continental sliver ('Argo Land') was on the point of breaking off Australia by the initial seafloor spreading of the Indian Ocean in the Argo Abyssal Plain in a continuation of a pattern established in the Late Palaeozoic by the rifting of Cimmeria (Veevers 1988*a*). Also maintaining a pattern instituted in the Palaeozoic was the activity of a Chilean-type volcanic cordillera on the north-eastern margin behind a zone of Pacific subduction. The cordillera shed detritus westward in tongues of river-borne sand and silt that extended over 1000 km into the continental craton. To the south lay the Lord Howe Rise/New Zealand Plateau.

Aptian (Fig. 3*B*)

By the Aptian, India had separated by seafloor spreading from Australia/Antarctica, which themselves had moved apart by the stretching of continental lithosphere beneath a wide rift-valley complex, the eastern part of which bifurcated around Tasmania (Powell *et al.* 1988). The volcanic arc extended along the east from New Guinea through New Zealand, and much of Australia was covered by a shallow (epeiric) sea.

Cenomanian (Fig. 3*C*)

The terminal climax of the Chilean-type subduction saw the retreat of the epeiric sea by the widening of the eastern cordillera to initiate the asymmetrical drainage of eastern Australia. Soon afterwards, the rift system extended to New Zealand at the same time as seafloor spreading started within the rift system between Australia/Tasmania and Antarctica, and back-arc spreading between Australia and the Lord Howe Rise/New Zealand Plateau. On the other side of Australia, the eastern Indian Ocean was now wide enough to have its own circulation; the increased productivity of the ocean water led to the onset of carbonate deposition on the margin. From this time, Greater India was to drift rapidly northward from a reorganised pattern of seafloor spreading on its way to its collision with Asia about 50 Ma ago.

With radical changes in (*a*) the drainage pattern of eastern Australia, (*b*) the type of subduction along the Pacific margin, (*c*) the inception of spreading between Australia and Antarctica on the south and Lord Howe Rise/New Zealand Plateau on the east and (*d*) the start of deep oceanic circulation in the Indian Ocean the Cenomanian marked a turning point in the palaeogeography of Australia.

Paleocene (Fig. 3*D*)

By the early Paleocene, the eastern cordillera had shrunk in width and height to the size of the present Eastern Highlands and the first sediment since the Cenomanian was deposited in shallow depressions in central–eastern Australia. The back-arc spreading of the Tasman Sea rotated Lord Howe Rise and a centrally emergent New Zealand Plateau away from Australia. Very slow spreading persisted between Australia and Antarctica but the continents remained in contact along the prong of Tasmania and the submarine rise to the south (South Tasman Rise) to impede the flow of seawater around Antarctica. Australia remained in mid to high latitudes until a southward turn in the apparent polar wander path at this time started Australia on a move into lower latitudes that persists to the present day.

Eocene/Oligocene (Fig. 3*E*)

Bottom water from a rapidly cooling Antarctica flowed northward to scour sediment from the western and eastern margins of Australia until the strait between Australia and Antarctica finally opened in the late Oligocene (30 Ma) to allow the unimpeded flow of the Circum-Antarctic Current. The current insulated Antarctica from the rest of the ocean and fostered the growth of an ice sheet. The opening of the Coral Sea in north-eastern Australia rotated the Papuan Peninsula into its present attitude and blocked out the final shape of the north-eastern margin.

Late Miocene (Fig. 3*F*)

By the late Miocene, the oceanic strait between Australia and Antarctica had widened to 1000 km, and the main flow of Antarctic Bottom Water and the main growth of the Antarctic ice sheet started (Kemp 1978). New Zealand had started moving along the Alpine Fault (Kamp 1986*a*). The spotlight now moves to the north: Australia's northward flight brought New Guinea into contact with various Pacific terranes (Pigram and Davies 1987) and north-western Australia approached the Indonesian Arc.

Holocene (Fig. 3*G*)

Continued left-lateral shearing of the Pacific Plate across New Guinea caused rapid uplift of the New Guinea Highlands from 5 Ma and has dragged the Vogelkop, Sulawesi, Ceram and the rest of the Banda Arcs into a megafold in the region of intense biogeographic transfer marked by the Wallace, Weber and Lydekker lines (Powell *et al.* 1980; Talent 1984; Truswell *et al.* 1987). Australia collided with the Inner (volcanic) Banda Arc about 10–5 Ma to form the 3 km high island of Timor and the corresponding 3 km foredeep of the Timor Trough (Audley-Charles 1986).

The Australian continental block now has a mountainous northern rim in the wet equatorial zone backed by the epeiric Timor–Arafura Sea that branches to the south-west and south-east into the broad reef-studded North-west Shelf and Great Barrier Reef. The low-lying mainland of Australia is arid except in the monsoonal north and about the Eastern Highlands, from which the main drainage is directed south-westward into the sump of Lake Eyre and to an outlet in Lake Alexandrina. The mountainous islands of Tasmania and New Zealand lie in the path of the westerly winds from which they receive copious orographic rain.

Very rapid changes have taken place since the start of the Quaternary, 2 Ma ago, epitomised by the difference between glacial and interglacial environments during the most recent climatic cycle. According to Williams (1984), during the last of the 17 glacial maxima 18 thousand years (ka) ago Australia had an even wider arid zone, ice caps in the Snowy Mountains and Tasmania, and a global sea level some 100–150 m lower than that of today, so that New Guinea and Tasmania were joined to the mainland. By 9 ka ago, temperatures were up to 1°C higher than they are today, and by 6 ka the sea attained its present level, again isolating New Guinea and Tasmania.

Other Reviews of Australian Palaeogeography

The palaeoclimatic setting is treated by Frakes (1979) and Frakes and Rich (1982). A *Palaeogeographic Atlas of Australia* is being prepared by the Bureau of Mineral Resources, Canberra; *Volume 1 — Cambrian* (Cook 1988) is the first to be published. *Cenozoic of the Australian region* (edited by M. A. J. Williams and P. DeDeckker) is due to appear in 1991 as a special publication of the Geological Society of Australia.

Acknowledgments

I thank J. A. Talent for comments on the manuscript, and the Australian Research Council for support.

References

Anderson, J. (1981). World Permo-Triassic correlations: their biostratigraphic basis. In 'Gondwana Five'. (Eds M. M. Cresswell and P. Vella.) pp. 3–10. (A. A. Balkema: Rotterdam.)

Audley-Charles, M. G. (1986). Timor-Tanimbar Trough: the foreland basin of the evolving Banda orogen. International Association of Sedimentologists Special Publication **8,** 91–102.

Bond, G. C., Nickerson, P. A., and Kominz, M. A. (1984). Breakup of a supercontinent between 625 Ma and 555 Ma: new evidence and implications for continental histories. *Earth and Planetary Science Letters* **70,** 325–45.

Bradshaw, J. D., Adams, C. J., and Andrews, P. B. (1981). Carboniferous to Cretaceous on the Pacific margin of Gondwana: the Rangitata Phase of New Zealand. In 'Gondwana Five'. (Eds. M. M. Cresswell and P. Vella.) pp. 217–21. (A. A. Balkema: Rotterdam.)

Coney, P. J. (1900). Terranes, tectonics and the Pacific rim. Australasian Institute of Mining and Metallurgy, Pacific Rim 90 Congress. Vol. II, 19–30.

Cook, P. J. (1988). 'Palaeogeographic Atlas of Australia, Volume 1, Cambrian.' (Bureau of Mineral Resources, Geology and Geophysics: Canberra.)

Du Toit, A. (1937). 'Our Wandering Continents.' (Oliver and Boyd: Edinburgh.)

Fischer, A. G. (1984). The two Phanerozoic supercycles. In 'Catastrophes and Earth History'. (Eds W. A. Berggren and J. A. Van Couvering.) pp. 129–50. (Princeton University Press: Princeton.)

Frakes, L. A. (1979). 'Climates Throughout Geologic Time.' (Elsevier: Amsterdam.)

Frakes, L. A., and Rich, P. V. (1982). Paleoclimatic setting and paleogeographic links of Australia in the Phanerozoic. In 'The Fossil Vertebrate Record of Australasia'. (Eds P. V. Rich and E. M. Thompson.) pp. 28–52. (Monash University Offset Printing Unit: Clayton, Victoria.)

Kamp, P. J. J. (1986*a*). Late Cretaceous–Cenozoic tectonic development of the southwest Pacific region. *Tectonophysics* **121,** 225–51.

Kamp, P. J. J. (1986*b*). The mid-Cenozoic Challenger Rift System of western New Zealand and its implications for the age of Alpine fault inception. *Bulletin of the Geological Society of America* **97,** 255–81.

Kemp, E. M. (1978). Tertiary climatic evolution and vegetation history in the Southeast Indian Ocean region. *Palaeogeography, Palaeoclimatology, Palaeoecology* **24,** 169–208.

Laird, M. G. (1981). The Late Mesozoic fragmentation of the New Zealand segment of Gondwana. In 'Gondwana Five'. (Eds M. M. Cresswell and P. Vella.) pp. 311–18. (A. A. Balkema: Rotterdam.)

Morgan, W. J. (1981). Hotspot tracks and the opening of the Atlantic and Indian Oceans. In 'The Oceanic Lithosphere. Volume 7 of The Sea'. (Ed. C. Emiliani.) pp. 443–87. (Wiley-Interscience: New York.)

Pigram, C. J., and Davies, H. L. (1987). Terranes and the accretion history of the New Guinea orogen. *BMR Journal of Australian Geology and Geophysics* **10,** 193–211.

Powell, C. McA., Johnson, B. D., and Veevers, J. J. (1980). Early Cretaceous breakup of eastern Gondwanaland, the separation of Australia and India, and their interaction with southeast Asia. In 'Ecological Biogeography of Australia'. (Ed. A. Keast.) pp. 17–29. (W. Junk: The Hague.)

Powell, C. McA., Roots, P. R., and Veevers, J. J. (1988). Pre-breakup continental extension in East Gondwanaland and the early opening of the Eastern Indian Ocean. *Tectonophysics* **155,** 261–83.

Scotese, C. R., and Sager, W. W. (Eds) (1988). Mesozoic and Cenozoic plate reconstructions. *Tectonophysics* **155,** 1–399.

Scotese, C. R., Bambach, R. K., Barton, C., Van der Voo, R., and Ziegler, A. M. (1979). Paleozoic base maps. *Journal of Geology* **87,** 217–77.

Talent, J. A. (1984). Australian biogeography past and present: determinants and implications. In 'Phanerozoic Earth History of Australia'. (Ed. J. J. Veevers.) pp. 57–93. (Clarendon: Oxford.)

Truswell, E. M., Kershaw, A. P., and Sluiter, I. R. (1987). The Australian–Southeast Asian connection: evidence from the palaeobotanical record. In 'Biogeographical Evolution of the Malay Archipelago'. (Ed. T. C. Whitmore.) pp. 32–49. (Oxford University Press: Oxford.)

Veevers, J. J., Editor (1984). 'Phanerozoic Earth History of Australia.' (Clarendon: Oxford.)

Veevers, J. J. (1988*a*). Morphotectonics of Australia's northwestern margin — a review. In 'The North West Shelf, Australia'. (Eds P. G. Purcell and R. R. Purcell.) pp. 19–27. (Proceedings of Petroleum Exploration Society Australia Symposium: Perth.)

Veevers, J. J. (1988*b*). Gondwana facies started when Gondwanaland merged in Pangea. *Geology* **16,** 732–4.

Veevers, J. J. (1989). Middle/Late Triassic (230±5 Ma) singularity in the stratigraphic and magmatic history of the Pangean heat anomaly. *Geology* **17,** 784–7.

Veevers, J. J. (1990*a*). Pacific rim of Australia: the 600 million year partnership of Gondwanaland and the ocean. Australasian Institute of Mining and Metallurgy, Pacific Rim 90 Congress. Vol. II, 151–8.

Veevers, J. J. (1990*b*). Tectonic-climate supercycle in the billion-year plate-tectonic eon: Permian Pangean icehouse alternates with Cretaceous dispersed-continents greenhouse. *Sedimentary Geology* **68,** 1–16.

Veevers, J. J. (1991). Mid-Cretaceous tectonic climax, Late Cretaceous recovery, and Cenozoic relaxation in the Australian region. Special Publication of the Geological Society of Australia (in press).

Veevers, J. J., and Li, Z. X. (1991). Review of seafloor spreading around Australia II. Marine magnetic anomaly modelling. *Australian Journal of Earth Sciences* **38** (in press).

Veevers, J. J., and Powell, C. McA. (1987). Late Paleozoic glacial episodes in Gondwanaland reflected in transgressive–regressive depositional sequences in Euramerica. *Geological Society of America Bulletin* **98,** 475–87.

Veevers, J. J., Powell, C. McA., and Roots, S. R. (1991). Review of seafloor spreading around Australia I. Synthesis of the patterns of spreading. *Australian Journal of Earth Sciences* **38** (in press).

Wegener, A. (1915). 'Die Entstehung der Kontinente und Ozeane.' (Vieweg: Braunschweig.)

Williams, M. A. J. (1984). Quaternary environments. In 'Phanerozoic Earth History of Australia'. (Ed. J. J. Veevers.) pp. 42–7. (Clarendon: Oxford.)

Zonenshayn, L. P., Kuz'min, M. I., and Kononov, M. V. (1987). Absolute reconstructions of the position of the continents during Paleozoic and Early Mesozoic time. *Geotectonics* **21,** 199–212.

Manuscript received 3 September 1990, accepted 18 January 1991

3 September 1991 *Aust. Syst. Bot.*, **4**, 13–24

Asian and South-western Pacific Continental Terranes Derived from Gondwana, and their Biogeographic Significance

C. Burrett, N. Duhig, R. Berry and R. Varne

Geology Department, University of Tasmania, Box 252C, Hobart, Tas. 7001, Australia.

Abstract

The recent recognition of numerous small geological terranes in the Indo-Pacific region has revolutionised our understanding of geological and biogeographic processes. Most of these terranes rifted from Gondwana. The Shan-Thai terrane rifted from Australia in the Permian and collided with Indo-China in the Triassic. Parts of Sumatra and Kalimantan may have rifted from Australia in the Cretaceous and carried an angiosperm flora north. Other terranes, now dispersed in South-East Asia and in the Pacific were, at various times in the Cenozoic, part of the Australian continent. Faunal and floral mobilism to Fiji via the Solomons and Vanuatu was probably not difficult up to the late Miocene.

Introduction

Plate-tectonic concepts developed 25 years ago have been used with some success to reconstruct in broad terms the palaeogeographical history of parts of the Indo-Pacific region. In particular, the recognition that north-western Australia/New Guinea, in its northward drift, collided during mid-Tertiary times with parts of the Malaysian archipelago provided a plausible tectonic explanation for the presence of faunal transitions, like Wallace's line (Whitmore 1981), where Asian and Australian faunas meet.

However, soon after their general adoption of plate tectonics geologists began to realise that, although present-day continental regions can usually be treated for tectonic purposes as a part of a coherent plate, some orogenic regions are a tectonic collage of crustal fragments, termed terranes, which do not share similar geological histories.

Burrett (1974) early pointed out that many of the small and large continental terranes which now constitute Asia might have been rifted off the Gondwanan supercontinent (e.g. Burton 1970; Ridd 1971; Audley-Charles *et al.* 1972; Crawford 1974*a*, 1974*b*), and in the last few years, detailed geological studies have confirmed a Gondwana provenance for many Asian terranes, and established, in a preliminary way, their rift, drift and collision histories.

Similarly, many continental terranes, both large and small, in the South-western Pacific were derived from Australia during the Cretaceous and Tertiary by the opening of generally eastward-younging basins. The ages of these basins are shown in Fig. 1. Thus, fragments of Gondwana may now be found from Mongolia to Fiji. It has even been suggested that some of the terranes within the western cordilleras of North America, such as the Alexander Terrane of Alaska and Canada, may have been rifted from Gondwana during the Devonian (Gehrels and Saleeby 1987).

The history of the Asian terranes derived from Gondwana during the Palaeozoic has recently been reviewed by Burrett *et al.* (1990) and Nie *et al.* (1990). In this review we will concentrate on those terranes derived from Gondwana (including those derived from Australia after Gondwana breakup) during the Mesozoic and Cenozoic that now lie within the South-eastern Asian and South-western Pacific regions.

1030-1887/91/010013$05.00

Late Palaeozoic–Jurassic

A late Permian reconstruction is shown in Fig. 2. The Shan-Thai terrane (= Sibumasu terrane of Metcalfe 1988) is shown close to north-western Australia having just started to drift away. Recent studies on the widespread, siliciclastic Phuket Group in Thailand by J. Hills and C. Burrett, and on similar sequences in Malaysia (Stauffer and Lee 1986) have shown that these are definitely glacimarine and strongly suggest contiguity of Shan-Thai and a glaciated Gondwana continent in the late Carboniferous–early Permian. Work on the early Palaeozoic faunas of Shan-Thai shows that this continent was to the north-west of Australia (Burrett and Stait 1985). Deep-water deposits are evident in the suture separating Shan-Thai from the Indo-China terrane in the Permian and early and middle Triassic (S. Hada, written communication 1990) and a late Triassic collision is probable.

Fig. 1. Major tectonic units in the South West Pacific–South-East Asian region. Stippled areas are small ocean basins that opened in the Cretaceous–Cenozoic. Spreading axes indicated. The Banda Basin, though, is most likely a slab of early Cretaceous oceanic crust trapped behind the Banda Arc. Age of ocean crust is indicated for each basin. Dashed lines on continental areas are major sutures; solid lines are selected major faults. Compiled from numerous sources: South-East Asia: A, Alor; B1, Bali; B2, Buru; C1, Celebes; C2, Ceram; F, Flores; H, Halmahera; J, Java; K, Kalimantan; L, Lombok; N, Natal terrane; S1, Sikuleh terrane; S2, Sumbawa; S3, Sumba; SU, Sumatra; T, Timor. South-western Pacific: a, New Britain; b, Santa Isabel; c, Guadalcanal; d, Malaita, e, San Cristobal; f, Santa Cruz; g, Espiritu Santo; h, Malekula; i, Efate; j, Viti Levu; k, New Caledonia; l, Loyalty Islands; m, Norfolk Island.

It is difficult to accept the hypothesis of a Cretaceous collision of Shan-Thai and Indo-China (Harbury *et al.* 1990) because the very widespread continental lower Jurassic–Cretaceous sediments of the Khorat Group are virtually undeformed, except where they are in contact with major transcurrent faults. Our structural work with R. H. Findlay in eastern Thailand has shown that the unconformably underlying upper Triassic sediments are strongly deformed and exhibit one phase of steeply plunging folds and the Permian sequences have an extra, probably late Permian deformation phase. Our interpretation is in contrast with that of Audley-

Charles (1987) who would have Shan-Thai rifting from Australia in the Jurassic and colliding with Indo-China in the late Cretaceous.

Veevers (1988) has recognised three main stages of rifting on Australia's north-west shelf: Cambrian, ?late Carboniferous to early Permian, and late Jurassic. Recent results from the ODP Leg 12, however, indicate that the first oceanic crust following the late Jurassic event is early Cretaceous in age (Gradstein and Ludden 1988). This is not conflicting as the breakup and post-breakup stages are usually separated by about 50 Ma (Falvey and Mutter 1981). We would therefore equate the Cambrian event with the rifting of North China (Burrett *et al.* 1990), the Permian event with the rifting of Shan-Thai, and the Jurassic–Cretaceous event as part of a continuous, westward-younging rifting event around the northern and western margins of the Australian craton (Pigram and Panggabean 1984). It is this event that rifted the many Gondwana fragments now found in Indonesia. These terranes are tentatively identified in Fig. 3.

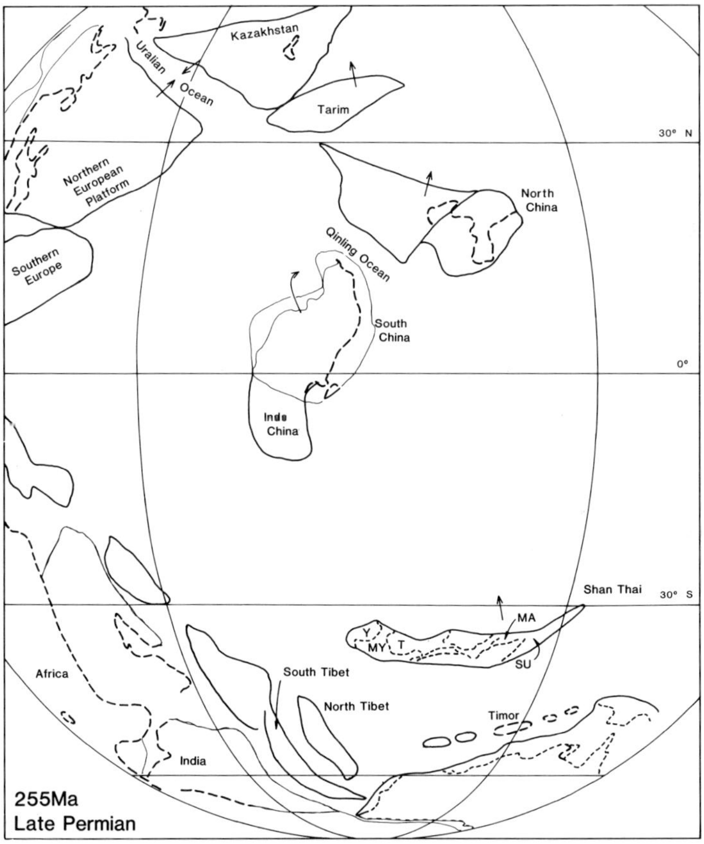

Fig. 2. Reconstruction for the late Permian 255 Ma. MA, Malaysia; MY, Myanmar, Burma; SU, Sumatra; T, Thailand; Y, Western Yunnan. Orthographic projection. Continental blocks shown by solid lines, and modern coasts by dashed lines.

Cretaceous–Cenozoic (Figs 3–7)

Rifting was widespread around Australia in the Cretaceous. Australia rifted from Antarctica and began a northward drift. New Zealand and the associated and extensive continental slivers of the Lord Howe and Norfolk Rises rifted off the eastern margin of Australia and began their eastward drift. Rifting continued during the early Cenozoic and the small ocean basins gradually opened up. The oldest rocks on Fiji are probably late Eocene (Kroenke and Rodda 1984) and Fiji, Vanuatu and the Solomons were in close proximity until the late Miocene (Fig. 7), allowing biotic mobilism between these areas (Michaux 1989).

Pieters and Supriatna (1990) suggest that north and south Kalimantan collided in the Eocene and we have tentatively placed parts of Sumatra and southern Kalimantan close to Australia in the Cretaceous and have them rifting off in the early Cretaceous to collide with the remainder of South-East Asia in the Eocene. Unfortunately, there are few constraints on tectonic scenarios for the Kalimantan–Sumatra area mainly due to a dearth of geological and palaeomagnetic information.

During the Cenozoic the northern margin of the Australian continent also accreted a large number of terranes that now constitute the geological complexity of the island of New Guinea. Pigram and Davies (1987) have identified 32 such terranes and outlined their accretion history.

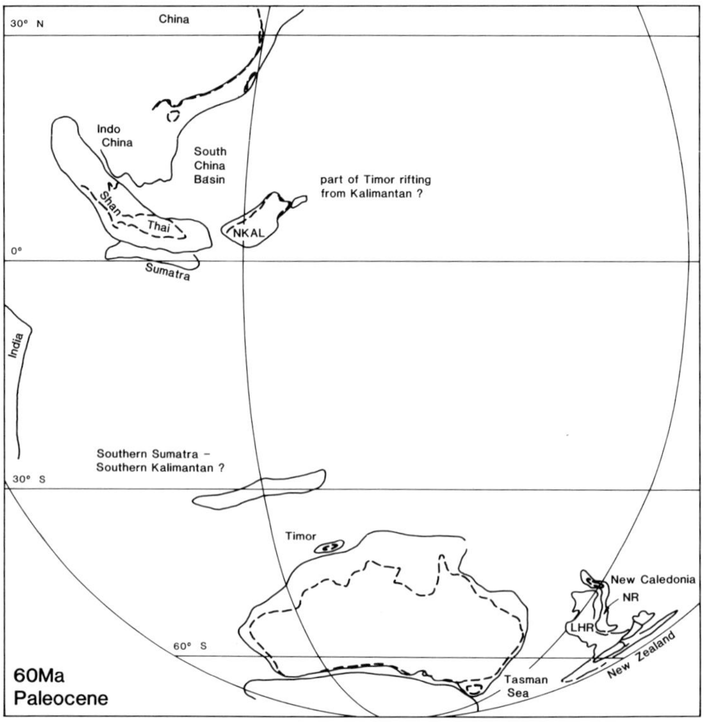

Fig. 3. Reconstruction for Paleocene 60 Ma. LHR, Lord Howe Rise; NKAL, North Kalimantan; NR, Norfolk Rise.

Biogeographic Significance

The series of paleogeographic maps should aid the understanding of biotic distribution in the Indo-Pacific region. The reconstructions show how various terranes were rifted away from the Gondwana continental margin at various times during the Late-Paleozoic–Cretaceous, and then were progressively swept up and accreted to the Asian continental margin or embedded within the Malaysian archipelago. These migrating continental fragments provided 'stepping stones of dry land' for faunal and floral dispersal long before the mid-Miocene collision between South-east Asia and north-western Australia/New Guinea. Whitmore (1987) and Audley-Charles (1987) have pointed out that large emergent

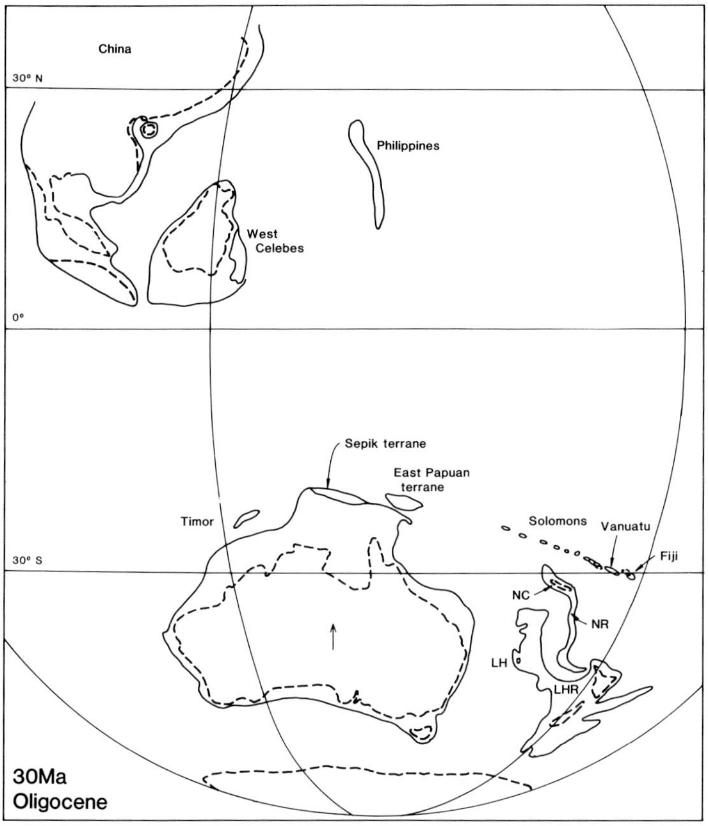

Fig. 4. Reconstruction for Oligocene 30 Ma. LH, Lord Howe; LHR, Lord Howe Rise; NC, New Caledonia; NR, Norfolk Rise.

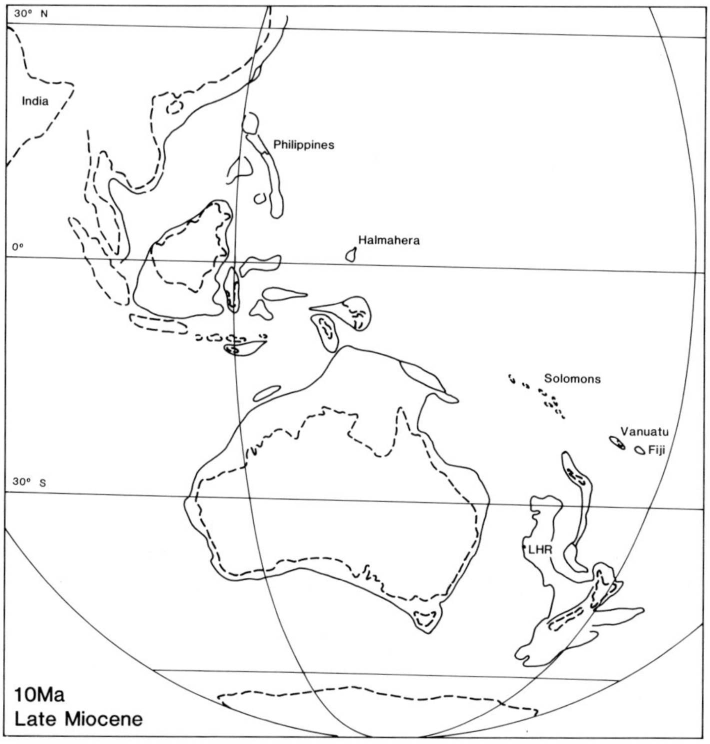

Fig. 5. Reconstruction for the Late Miocene 10 Ma. LHR, Lord Howe Rise.

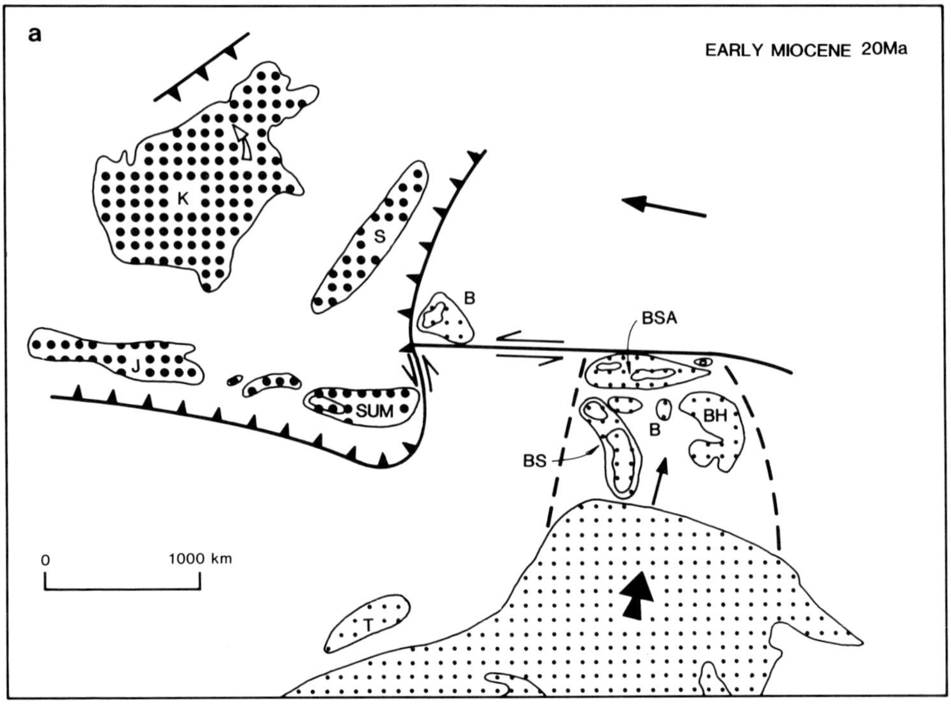
a
EARLY MIOCENE 20Ma
K
S
J
SUM
B
BSA
BH
BS
T
0
1000 km

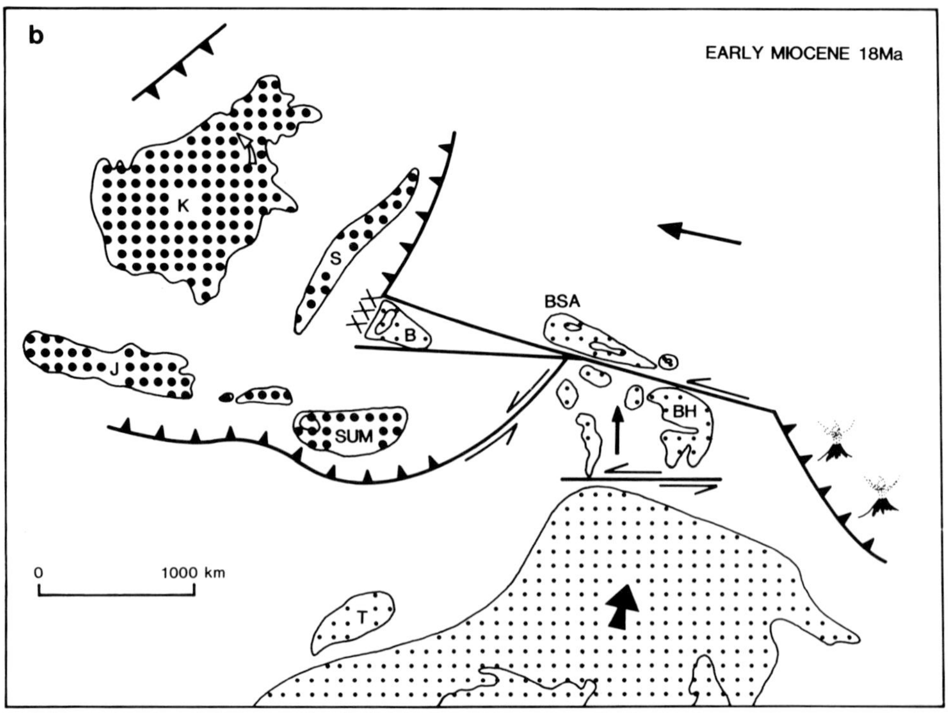
b
EARLY MIOCENE 18Ma
K
S
J
SUM
B
BSA
BH
T
0
1000 km

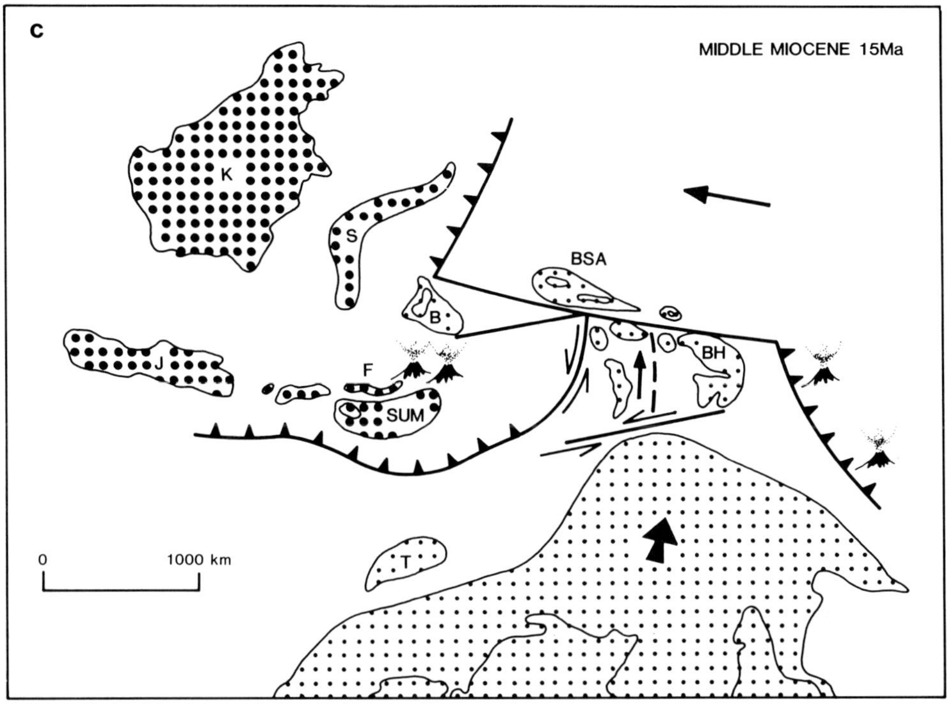
c
MIDDLE MIOCENE 15Ma
K
S
BSA
B
BH
J
F
SUM
T
0
1000 km

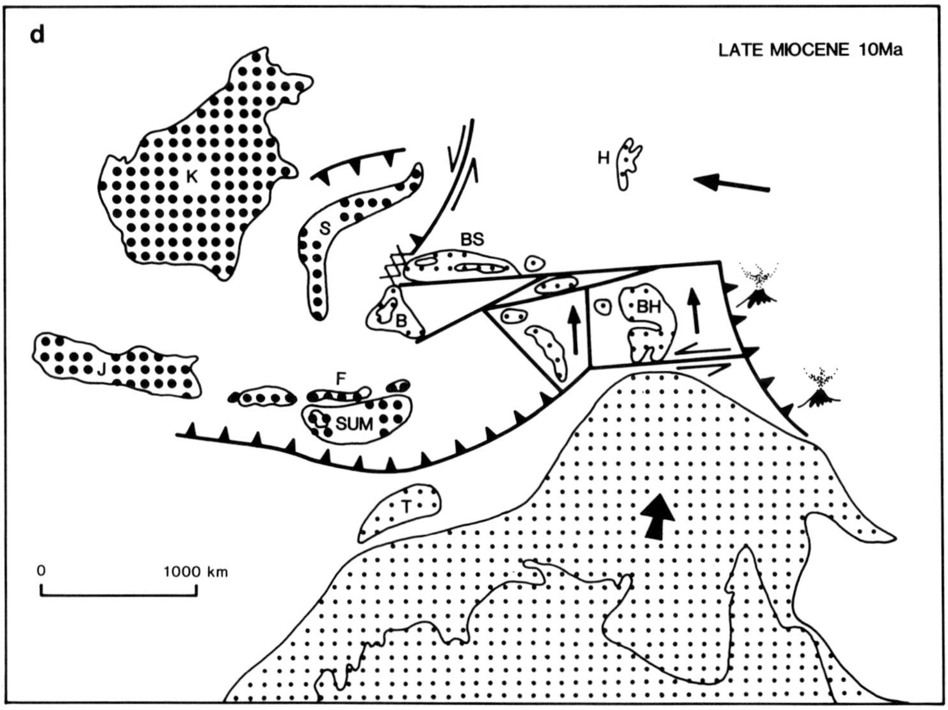
d
LATE MIOCENE 10Ma
K
H
S
BS
B
BH
J
F
SUM
T
0
1000 km

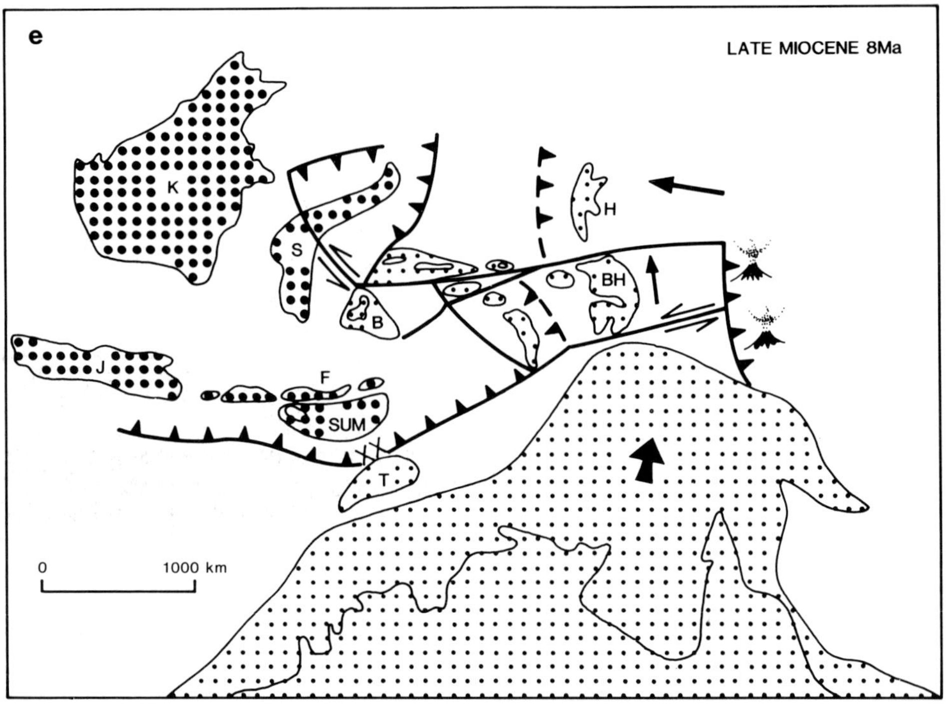
e
LATE MIOCENE 8Ma
K
S
B
H
BH
J
F
SUM
T
0
1000 km

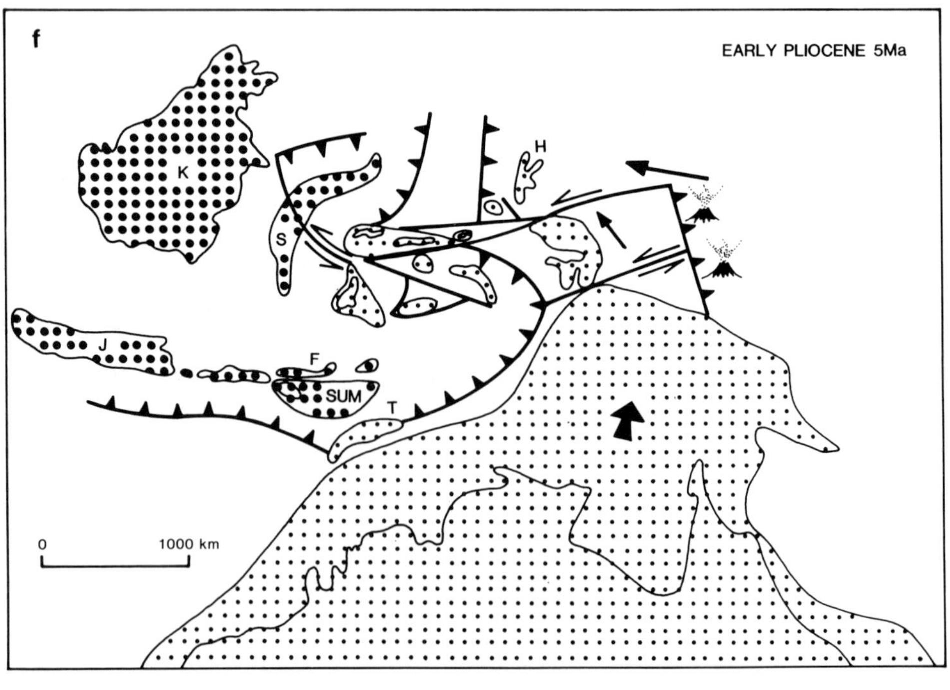
f
EARLY PLIOCENE 5Ma
K
H
S
J
F
SUM
T
0
1000 km

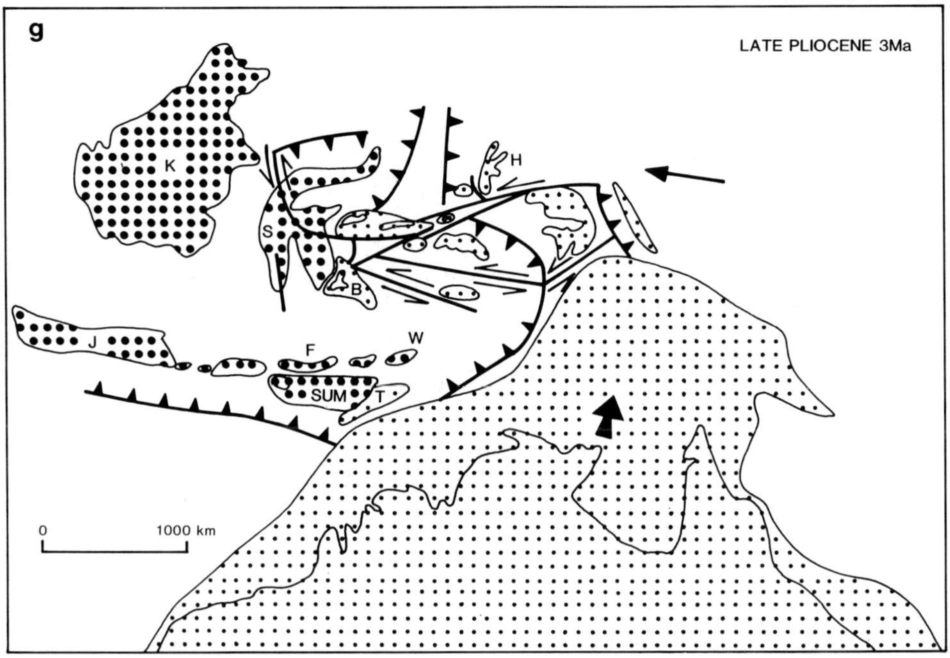

Fig. 6. The present sizes and shapes of the microcontinental blocks are used, although they have undoubtedly changed in the last 20 Ma. The reconstructions have been produced relative to a fixed South-East Asia. The figures are Lambert projections, centred upon 5° S., 125° E., with a 50° field of view. The time frames chosen represent crucial stages in the evolution of eastern Indonesia.
(*a*) 20 Ma. The constraints upon these reconstructions are poor. For the Indonesian region, subduction was active at the Java Trench, the Sulawesi Trench and the North-west Borneo Trench (Hamilton 1979). The Java Trench has been moved south to account for 50 km of back-arc thrusting (Silver *et al.* 1983). The Sumba microcontinent is shown to include the Savu Basin and the Mutis/Paleo thrust sheets now on Timor. Timor is also shown to extend further north than it now does accounting for 80 km of crustal shortening (McCaffrey 1989). The Buton microcontinent has been detached from the microcontinental agglomerate onto the Pacific plate and is steadily moving towards the Sulawesi Trench.
(*b*) 18 Ma. This time-frame shows the collision of Buton into the Sulawesi trench (Smith 1987). The collision initiated the bending of the Sulawesi Arc and the slicing of the Bangaii–Sula and Obi from the microcontinental block, which in turn was removed from the direct influence of the Australian plate by a left-lateral fault. The Banda Arc extension of the Sunda Arc started at about this time, trapping the Banda Sea. The Buton–Sulawesi Trench collision probably played some part in this by altering the triplepoint arrangement shown in Fig. 6*a*.
(*c*) 15 Ma. By this time subduction had almost ceased at the North-west Borneo Trench. Bangaii–Sula continues to approach the now indented Sulawesi arc and the microcontinental block north of Australia continues to move slowly west relative to the Australian craton.
(*d*) 10 Ma. By the middle Miocene the interactions between the major plates and the microplates had become complex. Bangaii–Sula collided with the Sulawesi Trench (Smith 1987), further bending the arc and initiating subduction at the North Sulawesi Trench, the slicing of the Banda Ridges from microcontinental agglomerate and the rotation of the Buru–Seram microcontinent. In the east the inner Melanesian arc began to be obliquely accreted onto the Australian craton (M. Daly *et al.*, personal communication).
(*e*) 8 Ma. The significant event here is the collision of the north coast of Timor with the Banda Trench (Berry and McDougall 1986). This event resulted in the crustal shortening and thickening of Timor and back-arc thrusting and deformation in the Banda Sea (McCaffrey 1989). The relative timing of these events is not clear and is beyond the resolution of these reconstructions. The complex microplate interactions continue in the Banda Sea region, most significantly the 'squeezing' of the Banda Ridges between Bangaii–Sula and the rotating Buru–Seram microcontinent.
(*f*) 5 Ma. Timor, now significantly shortened, overrides the trench, detaching the oceanic slab from the continental lithosphere. The Banda Ridges reach their final positions and Bacan moves along a splay of the Sorong Fault system as east-dipping subduction begins west of Halmahera.
(*g*) 3 Ma. By this time the positions and movements of the microcontinents are much the same as they are at present. Active thrusting beneath the Seram Trough brings the Bird's Head into its present position. The Paleo/Mutis allochthon is thrust over Timor (Harris 1990) and NNW trending, left-lateral faulting in South Sulawesi is generated in response to the continent/arc collision (Berry and Grady 1987).

landmasses now part of Asia may have been rifted from north-western Australia (east Gondwana) in Cretaceous times, carrying with them an angiosperm flora. Our reconstruction differs from that of Audley-Charles (1987): we have already suggested that the Shan-Thai terrane had already collided with Indo-China in the late Triassic, and that the terranes rifted/drifted in Jurassic times are more likely to now form part of either Sumatra and/or southern Kalimantan.

Figs 3–6 show our interpretation of rifting, drifting and collisional episodes in the region in Cretaceous–Cenozoic times. Fig. 6 shows reconstructions of the collision between north-western Australia/New Guinea, and the Indonesian arcs, whereas Fig. 7 shows how various terranes now dispersed in the western Pacific might have been part of the Australian landmass until late Miocene times.

These reconstructions help account, in a general way, for the existence of Gondwana-type (or at least Australian-type) flora on Fiji, and the curious mixed nature of the Sulawesi biota. To go further will probably require detailed geological studies, and it seems to us that some of the critical geological evidence might be overlooked unless geologists are alerted to its significance. For example, the existence of dry land, and the distribution of land and sea in the past, was of intense interest to migrating fauna and flora, yet a thin, scarcely discernible palaeosol (caliche), indicative of former dry land, in a thick marine

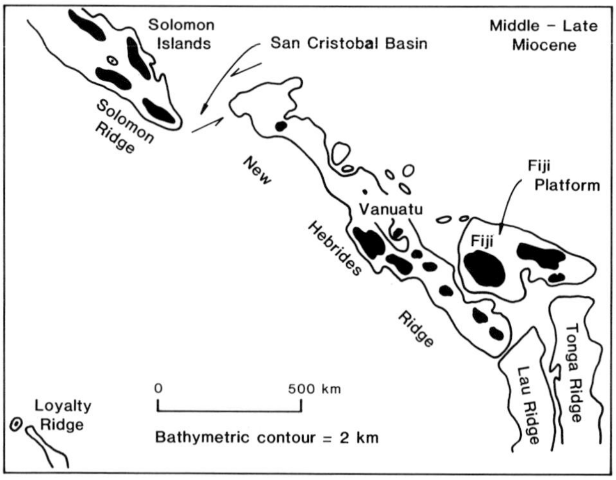

Fig. 7. Reconstruction of the Solomons to Fiji region in the middle to late Miocene based on Falvey and Greene (1988). The San Cristobal basin has been closed up based on the work of R. Musgrave. Clearly, faunal and floral dispersal was possible along this route during the Miocene, but Fiji and Vanuatu would have been isolated by later ocean floor spreading.

limestone sequence might be of little interest to a regional geologist. Similarly, soil horizons in volcanic sequences could easily be missed, yet would be indicative not only of former dry land but also of a hiatus in volcanic activity. The importance of volcanicity as an ecological factor which might influence some faunal distributions seems to have been neglected.

As a general observation, the active volcanoes of Sumatra and Java are built on uplifted continental blocks which expose pre-Miocene marine rocks. The active volcanoes of the Lesser Sunda Islands similarly rest on a bulge of older marine rocks. The volcanicity is a by-product of subduction, and presumably the 'geanticline' is a result of magmatic inflation and thermal uplift of pre-existing marine crust (Hamilton 1988). Thus the subduction-induced magmatism contributes directly and indirectly to the formation of dry land.

Although few dates for the volcanic sequences are available, the dating of the volcanics on Flores (Abbott and Chamalaun 1981) suggests that this island did not exist prior to the Miocene, and that during the Miocene and Pliocene, much of the volcanism was submarine. The oldest known volcanic rocks on Bali are even younger, about 3 Ma, and were submarine (G. Wheller, personal communication), and more recent caldera-forming eruptions from the Balinese volcanoes of Batur and Bratan probably covered most of the island's surface with thick ignimbritic sheets. We therefore suggest that there may have been no easy dispersal route along the line of the Lesser Sunda Islands at and prior to 3 Ma, and that since then volcanic activity would certainly have hindered the spread of some of the fauna and flora, and in particular may help explain the impoverishment of freshwater fish faunas in the region (Cranbrook 1981).

Acknowledgments

We thank numerous colleagues in South-east Asia for their help in the field and Bernard Michaux for a helpful review.

References

Abbott, M. J., and Chamalaun, F. H. (1981). Geochronology of some Banda Arc volcanics. *Indonesia Geological Research and Development Centre Special Publication* **2,** 253–68.

Audley-Charles, M. G. (1987). Dispersal of Gondwanaland: relevance to evolution of the angiosperms. In 'Biogeographical Evolution of the Malay Archipelago'. (Ed. T. C. Whitmore.) 5–25. (Clarendon Press: Oxford.)

Audley-Charles, M. G., Carter, D. J., and Milsom, J. S. (1972). Tectonic development of eastern Indonesia in relation to Gondwanaland dispersal. *Nature, London (Physical Science)* **239,** 35–9.

Berry, R. F., and Grady, A. E. (1987). Mesoscopic structures produced by Plio-Pleistocene wrench faulting in South Sulawesi, Indonesia. *Journal of Structural Geology* **9,** 563–71.

Berry, R. F., and McDougall, I. (1986). Interpretation of ^{40}Ar/^{39}Ar and K/Ar dating evidence from the Aileu Formation, East Timor, Indonesia. *Chemical Geology (Isotope Geoscience Section)* **59,** 43–58.

Burrett, C. (1974). Plate tectonics and the fusion of Asia. *Earth and Planetary Sciences Letters* **21,** 181–9.

Burrett, C., and Stait, B. (1985). South East Asia as a part of an Ordovician Gondwanaland — a palaeobiogeographic test of a tectonic hypothesis. *Earth and Planetary Sciences Letters* **75,** 184–90.

Burrett, C., Long, J., and Stait, B. (1990). Early-Middle Palaeozoic biogeography of Asian terranes derived from Gondwana. In 'Palaeozoic Palaeogeography and Biogeography, Geological Society Memoir No. 12'. (Eds W. S. McKerrow and C. R. Scotese.) 163–74.

Burton, C. K. (1970). The palaeotectonic status of the Malay Peninsula. *Palaeogeography, Palaeoclimatology, Palaeoecology* **7,** 51–60.

Cranbrook, E. O. (1981). The vertebrate faunas. In 'Wallace's Line and Plate Tectonics'. (Ed. T. C. Whitmore.) 57–69. (Clarendon Press: Oxford.)

Crawford, A. R. (1974*a*). The Indus suture line, the Himalaya, Tibet and Gondwanaland. *Geological Magazine* **111,** 369–83.

Crawford, A. R. (1974*b*). A greater Gondwanaland. *Science* **184,** 1179–81.

Falvey, D. A., and Greene, H. G. (1988). Origin and evolution of the sedimentary basins of the New Hebrides arc. In 'Geology and Offshore Resources of Pacific Island Arcs — Vanuatu Region, Circum-Pacific Council for Energy and Mineral Resources Earth Science Series, Volume 8'. (Eds H. G. Greene and F. L. Wong.) 413–42. (Circum-Pacific Council for Energy and Mineral Resources: Houston, Texas.)

Falvey, D. A., and Mutter, J. C. (1981). Regional plate tectonics and the evolution of Australia's passive continental margins. *BMR Journal of Australian Geology and Geophysics* **6,** 1–29.

Gehrels, G. E., and Saleeby, J. B. (1987). Geologic framework, tectonic evolution, and displacement history of the Alexander terrane. *Tectonics* **6,** 151–74.

Gradstein, F., and Ludden, J. (1988). Leg 123, Site 765, Ocean Drilling Program.

Hamilton, W. (1979). Tectonics of the Indonesian region. *United States Geological Survey Professional Paper* **1078,** 1–345.

Hamilton, W. B. (1988). Plate tectonics and island arcs. *Geological Society of America Bulletin* **100,** 1503–27.

Harbury, N. A., Jones, M. E., Audley-Charles, M. G., Metcalfe, I., and Mohamed, K. I. (1990). Structural evolution of Mesozoic Peninsular Malaysia. *Journal of the Geological Society, London* **147**, 11–26.

Harris, R. A. (1990). Time variant distribution of strain in the active Banda orogen: a reconciliation of rival hypotheses. In 'Orogenesis in Action. Symposium at the Geological Society, London, 18–20th April, 1990'. (Eds R. Hall, G. Nichols and C. Rangin.)

Kroenke, L. W., and Rodda, P. (1984). Cenozoic tectonic development of the southwest Pacific. U.N. ESCAP, CCOP/SOPAC Technical Bulletin No. 6.

McCaffrey, R. (1989). Seismological constraints and speculations on Banda Arc tectonics. *Netherlands Journal of Sea Research* **24,** 141–52.

Metcalfe, I. (1988). Origin and assembly of South-East Asian continental terranes. In 'Gondwana and Tethys, Geological Society Special Publication 37'. (Eds M. Audley-Charles and A. Hallam.) 101–18. (Geological Society of London.)

Michaux, B. (1989). Generalized tracks and geology. *Systematic Zoology* **38,** 390–8.

Nie, S., Rowley, D. B., and Ziegler, A. M. (1990). Constraints on the locations of Asian microcontinents in Paleo-Tethys during the Late Paleozoic. In 'Palaeozoic Palaeogeography and Biogeography, Geological Society Memoir No. 12'. (Eds W. S. McKerrow and C. R. Scotese.) 397–409. (Geological Society of London.)

Pieters, P. E., and Supriatna, S. (1990). Late Cretaceous–Early Tertiary continent collision in Borneo. In 'Terrane Analysis of China and the Pacific Rim'. (Eds T. J. Wiley, D. G. Howell and F. L. Wong.) 193–4. (Circum-Pacific Council for Energy and Mineral Resources: Houston, Texas.)

Pigram, C. J., and Davies, H. L. (1987). Terranes and the accretion history of the New Guinea orogen. *BMR Journal of Australian Geology and Geophysics* **10,** 193–211.

Pigram, C.J., and Panggabean, H. (1984). Rifting of the northern margin of the Australian continent and the origin of some microcontinents in eastern Indonesia. *Tectonophysics* **107,** 331–53.

Ridd, M. F. (1971). South-East Asia as a part of Gondwanaland. *Nature* **234,** 531–3.

Silver, E. A., Reed, D., McCaffrey, R., and Joyodiwiryo, Y. (1983). Back arc thrusting in the eastern Sunda Arc, Indonesia: a consequence of arc-continent collision. *Journal of Geophysical Research* **88,** 7429–48.

Smith, R. B. (1987). Diachronous Neogene microcontinent collision in Buton and Eastern Sulawesi, Indonesia. *Geological Society of America Abstracts with Programs* **19,** 850.

Stauffer, P. H., and Lee, C. P. (1986). Late Paleozoic glacial marine facies in Southeast Asia and its implications. *Bulletin of the Geological Society of Malaysia* **20,** 363–97.

Veevers, J. J. (1988). Morphotectonics of Australia's northwestern margin — a review. In 'The North West Shelf, Australia, Proceedings of PESA Symposium, Perth'. 19–27. (Petroleum Exploration Society of Australia: Perth.)

Whitmore, T. C. (1981). 'Wallace's Line and Plate Tectonics.' (Clarendon Press: Oxford.)

Whitmore, T. C. (1987). Introduction. In 'Biogeographical Evolution of the Malay Archipelago'. (Ed. T. C. Whitmore.) 1–2. (Clarendon Press: Oxford.)

Manuscript received 7 December 1990, accepted 15 March 1991

3 September 1991 *Aust. Syst. Bot.*, **4**, 25–36

Distributional Patterns and Tectonic Development in Indonesia: Wallace Reinterpreted

B. Michaux

Private Bag, Kaukapakapa, New Zealand.

Abstract

A. R. Wallace's name will always be linked with what was known as the Malay Archipelago, and thus it seems appropriate to reassess his contribution to Indonesian biogeography. Wallace is often regarded as the founder of 'dispersalist' biogeography. However, this simplified view seriously misrepresents Wallace's position, which was that animal and plant distributions in Indonesia could be understood in terms of the region's geological history. This view is contrasted with that of dispersalist biogeography, which is that present distributions are the result of unique events, that the present is a consequence of past accidents. The results of Wallace's work are outlined and are reinterpreted in the light of modern evidence about the tectonic development of this region. It is concluded that Wallace's observations are broadly intelligible in the light of what is known about Indonesia's geological history. There is much to gain by biogeographers and geologists working together more closely in Indonesia, for there are many outstanding problems that require resolution. This then is one future direction that may be taken as we stand at the crossroads.

Introduction

There are many examples in science where one person's name has dominated a particular sphere of study. A. R. Wallace occupies that central position with respect to Indonesian biogeography through his pioneering work in the region between 1854 and 1862.

A reading of Wallace's (1880) book 'The Malay Archipelago', in which he gives an account of his travels, clearly demonstrates the best and worst sides of Victorian natural history. Descriptions of his collecting expeditions are particularly offensive to modern sensibilities. Wallace's account of collecting orangutan, in what was then Borneo (Wallace 1880, pp. 39–64), must rank as one of the worst cases of wanton destruction ever carried out in the name of science. Yet to dismiss him as a gun-happy chauvinist is to ignore the fact that he was a remarkably astute and thorough observer of nature. It is this latter characteristic that gives his work a timeless quality that makes it as relevant today as it was when published. Indeed, as a record of a world long since transformed through colonialism and the pressures of burgeoning human population, it is priceless.

Today Wallace is regarded as the founder of 'dispersalist biogeography' (e.g. Croizat 1982). However, this simplified view seriously misrepresents his position. Dispersalist biogeographers support the theory that history is, by and large, a chapter of accidents, a series of unique events determined by chance coincidences. For example, it can be observed that birds and insects are regularly blown across the Tasman Sea from Australia to New Zealand. These events are

1030–1887/91/010025$05.00

unique happenings brought about by a fortuitous interplay of events. From this observation it is concluded that these birds or insects may one day establish themselves in New Zealand and, by becoming part of the New Zealand biota, evolve independently of their parental stock. The general conclusion that follows is that *all* distributional patterns are the result of unique events.

The fallacy inherent in this type of argument is demonstrated by the example of 'Cleopatra's nose' discussed in Carr (1970) and outlined in Table 1. Carr (1970, p. 108) considers that 'the dual and reciprocal function of history . . . [is] to promote our understanding of the past in the light of the present and of the present in the light of the past. Anything which, like Anthony's infatuation with Cleopatra's nose, fails to contribute to this dual purpose is from the point of view of the historian dead and barren'. This is so because no general conclusion can be drawn, which may promote our understanding either of the past or of the present, from unique events. To do so is simply to fall into the trap of arguing that all generals lose battles because they are infatuated with beautiful women.

Table 1. The fallacy of 'Cleopatra's Nose' or the dangers of arguing from the unique to the general

Example	Unique event	Specific result	General conclusion
History	Mark Antony becomes infatuated with Cleopatra	Loss of the battle of Actium	Generals lose battles because of their infatuation with beautiful women
Biogeography	Accidental dispersal from A to B	Establishment at B and evolution into a new species	Distribution patterns are the result of accidental dispersal

Wallace did not try to explain distribution patterns by invoking the occurrence of unique events but rather by recourse to general principles. For Wallace that general principle was geological change. There are numerous passages in Wallace (1880) that confirm his appreciation of the importance of geological change in understanding distribution patterns in Indonesia. His discussion of the faunal relationship between the Malay Peninsula and the islands of Borneo, Sumatra, and Java is detailed and provides a clear statement of his position. Having noted the overall similarity of this area's flora and fauna to that of India, he continued by describing his views of the biotic relatedness of these large islands. He commented on the greater similarity of the mammals and birds of Borneo and Sumatra than those of Borneo and Java, and on the high degree of endemism of the Javan fauna. Noting that the pattern of distribution is similar to other groups with poorer dispersal powers, he continued:

> 'We learn from this fact, that the seas which separate the islands from each other are wide enough to prevent the passage of even flying animals, and that we must look to the same causes as having led to the present distribution of both groups. The only sufficient cause we can imagine is the former connexion of all the islands with the continent, and such a change is in perfect harmony with what we know of the earth's past history, and is rendered probable by the remarkable fact that a rise of only 300 feet would convert the wide seas that separate them into an immense winding valley or plain about 300 miles wide and 1200 long.' (Wallace 1880, p. 143)

He even used the idea of geological change to understand problematical distributions on a quite detailed scale. The 'Tin Isle' of Bangka, which lies just off the coast of Sumatra, had a number of endemic species not found on its larger neighbour but relatives of which are found on the Malay Peninsula. He speculated that Bangka is a geological extension of the Malay Peninsula, noting that granites are found on both but not on Sumatra, and that the east coast area of this latter island had been formed only in recent history. He continued:

> 'As the southern part of Sumatra extended eastward and formed the narrow straits of Banca, many birds and insects and some Mammalia would cross from one to the other, and thus produce a general similarity of productions, while a few of the older inhabitants remained, to reveal by their distinct forms their different origins'.

Wallace believed that evolution was a slow and gradual process and that a kind of 'morphological clock' existed. Thus he related genealogical relatedness of organisms to geological relatedness of areas, a remarkably 'modern' idea. It is precisely this connection that led him, on the basis of observable distributions, to reject the idea of long-distance dispersal, even for animals that could fly.

'Even the bats furnish an additional argument, if one were needed, to show that the islands could not have been peopled from each other and from the continent without some former connexion. For if such had been the mode of stocking them with animals, it is quite certain that creatures which can fly a long distance would be the first to spread from island to island, and thus produce an almost perfect uniformity of species over the whole region. But no such uniformity exists, and the bats of each island are almost, if not quite, as distinct as other mammals' (Wallace 1880, pp. 142–3).

These are hardly the sentiments of one who subscribes to the idea that present distributions can be understood in terms of a series of accidental events that occurred in the past. Indeed, rather than being regarded as the founder of 'dispersalist biogeography', a case could be made that he was a forerunner of modern biogeographic theories that use the idea of geological change as an explanatory principle (Nelson and Platnick 1981, p. 400).

Wallace's Indonesian Biogeography

A summary of Wallace's biogeographic conclusions is presented in Fig. 1. On the basis of his observations he divided the archipelago into five regions — the Indo-Malay Islands (Fig. 1 : 1), Celebes (Fig. 1 : 2), the Timor group (Fig. 1 : 3), the Moluccas (Fig. 1 : 4), and the Papuan group (Fig. 1 : 5). On the basis of the avian and mammalian faunas he also recognised a fundamental biotic discontinuity between Indo-Malay and Australian forms across the narrow Straits of Lombok. This discontinuity was later traced to the north along the Macassar Strait and became known as Wallace's Line (Fig. 1).

The general biogeographic characteristics of the Indo-Malay islands (Fig. 1 : 1) have already been noted. His explanations of distribution patterns among the islands of this group, based largely on sea-level changes, are more plausible here than anywhere else in the archipelago. There is, however, one interesting anomaly that he discusses at some length, concerning the complete absence of orangutan from Sarawak where apparently suitable habitat widely existed. He argued that because orangutan require swamp forest the inland mountains surrounding Sarawak acts as an effective ecological barrier to westward expansion.

Wallace regarded the fauna of Sulawesi (Fig. 1 : 2) as unusual, not only because it was rather impoverished, despite the island's central position in the archipelago, but also because of its uniqueness. The Sulawesian biota seemed to have more in common with the Philippines and Africa than its closest neighbours. The Sula Islands and Buton, which are geographically close to Sulawesi, were regarded by Wallace as essentially Moluccan.

The biota of the Timor region (Fig. 1 : 3) was characterised by a mixture of Indo-Malay and Australian elements. At the western end of this chain of islands there was a greater percentage of Indo-Malay forms, while Australian forms dominated in Timor. Several Moluccan bird species were also recorded. Wallace recorded no endemic bird genera from this region and thus viewed the avian fauna as derivative.

Wallace recognised two subregions in the Moluccas (Fig. 1 : 4 and 4′). The biota of the northern islands, centred around Halmahera, had affinities with the Papua region (Fig. 1 : 5), while the southern group, centred around Seram, had a biota related more to Australia.

Although this is a rather brief summary of the distribution patterns that Wallace recognised in Indonesia, it does, I believe, accurately reflect the major patterns he observed. His interpretations of these patterns, based as they were on an incomplete understanding of the dynamic nature of the earth's surface, are only really of historical interest. Wallace's attempt to understand distributional patterns in terms of geological change was doomed from the start because neither he nor anyone else at the time realised that land could move laterally as well as vertically. It wasn't until the publication of Wegener's 'Die Entstehung der Kontinentder Ozeane' in 1915 that anyone suggested otherwise (Wegener 1929). In any event the concept was largely ignored until the 1960s when Wegener's central thesis was confirmed and extensively developed.

An Overview of the Tectonic Development of Indonesia

Indonesia, like the Caribbean and Mediterranean, has developed in response to continental–continental collision. Such collisions are characterised by complex structural developments as plate boundaries adjust to changing forces. Although there are many difficulties, both practical and theoretical, in trying to unravel the tectonic development of an area such as Indonesia, the impetus to do so has been provided by the desire to develop the mineral potential of the region. As a result of this effort there exists an extensive literature dealing with the geology

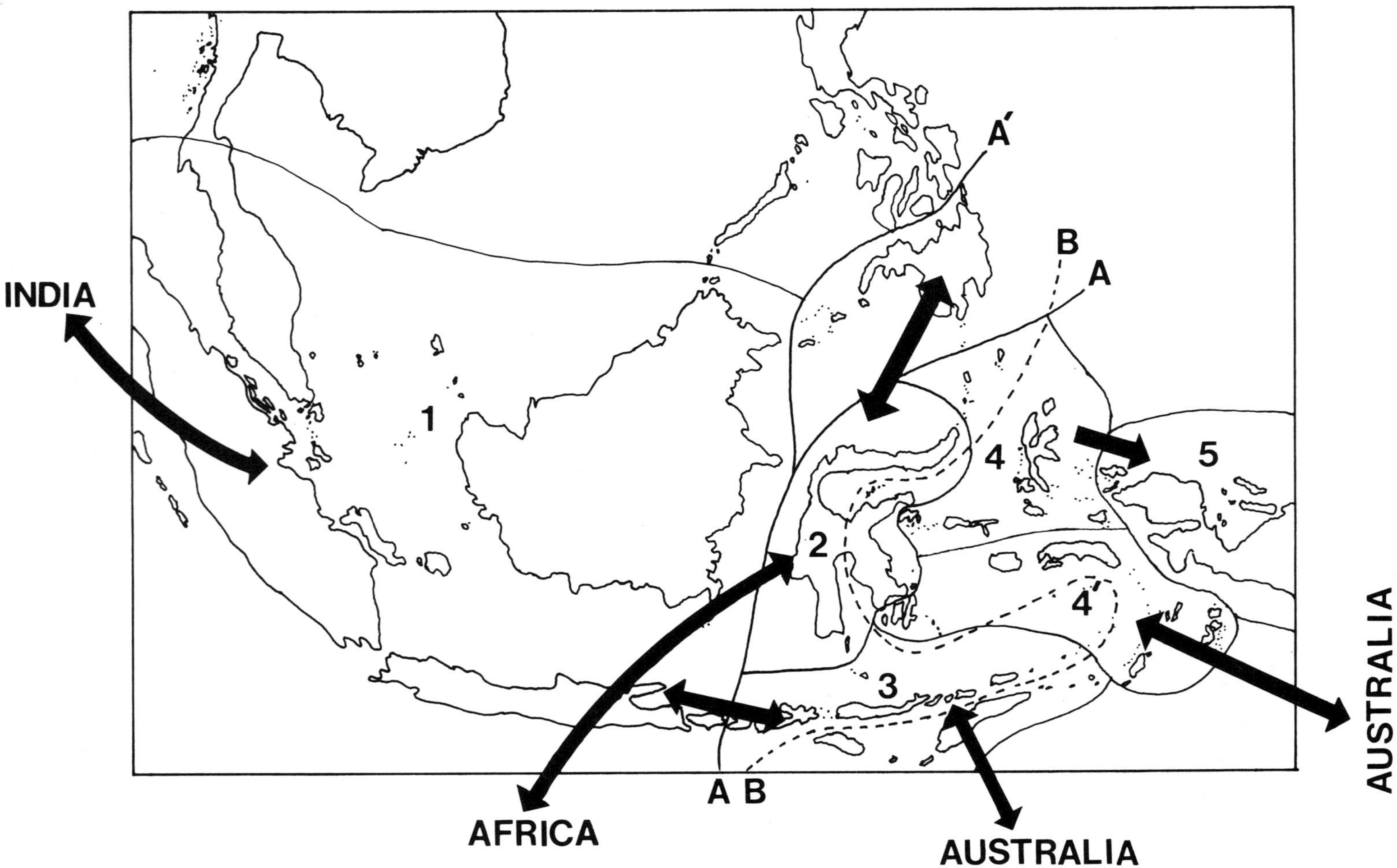

Fig. 1. A summary of the biogeographic regions recognised by Wallace in Indonesia. 1 = Indo-Malay Islands; 2 = Celebes (Sulawesi); 3 = Timor; 4 = northern Moluccas, 4′ = southern Moluccas; 5 = Papuan Region. A—A = Wallace's Line, A—A′ = modified Wallace's Line; B—B = geological partition of Asian and Australian land masses proposed by Carey in 1938 (Carey 1975).

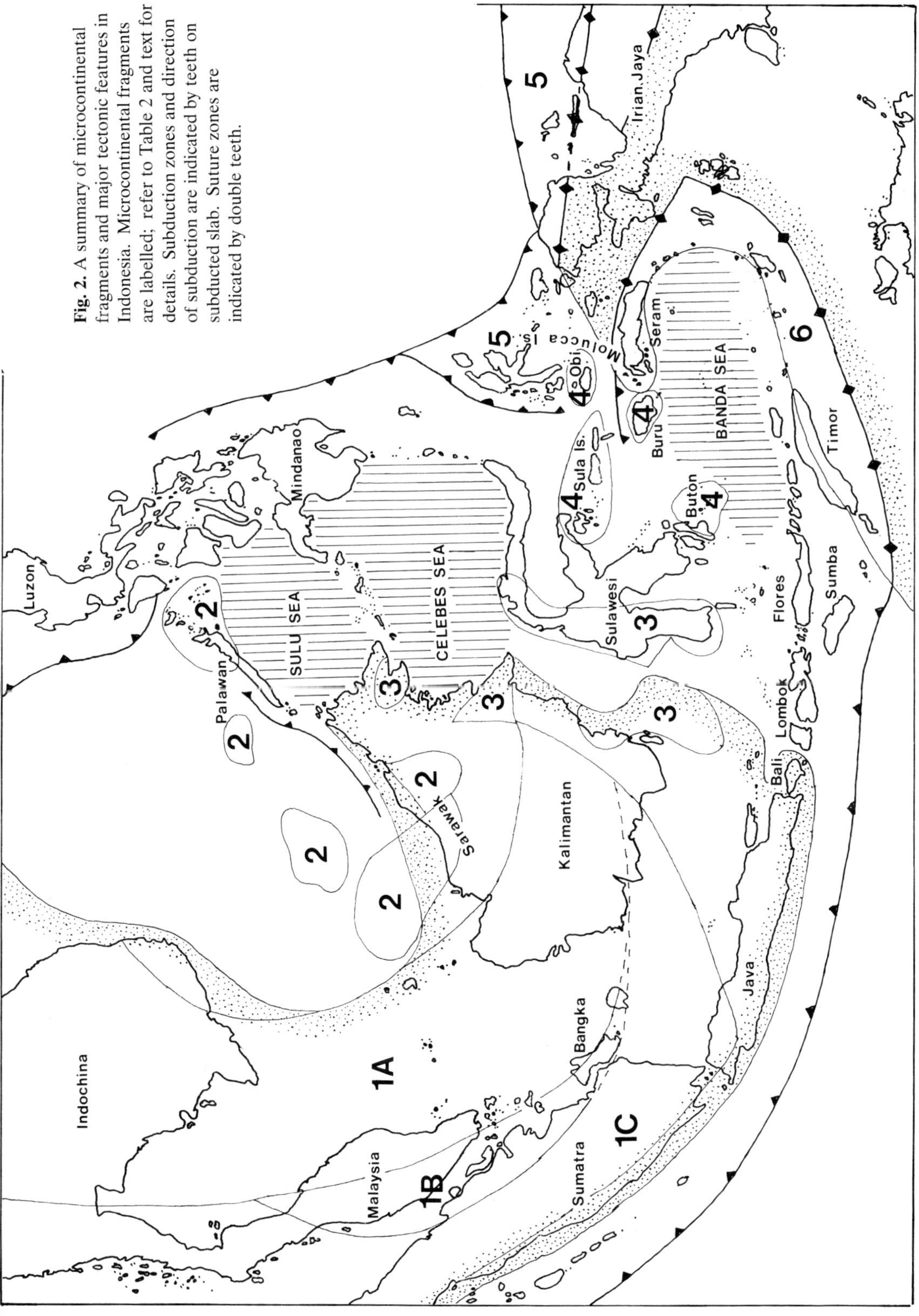

Fig. 2. A summary of microcontinental fragments and major tectonic features in Indonesia. Microcontinental fragments are labelled; refer to Table 2 and text for details. Subduction zones and direction of subduction are indicated by teeth on subducted slab. Suture zones are indicated by double teeth.

of the region and, more importantly for those of us interested in biogeography, with its tectonic interpretation.

Tectonic plates, despite being 100 km thick, do not act as rigid structures, particularly so at their margins. In structurally complex situations, such as exist in continental–continental collisions or during the breakup of continents, large plates become fragmented giving rise not so much to plate tectonics as 'flake tectonics'. Hutchinson (1989) has discussed the role that thermal heating has in this process of fragmentation. Convection currents within the upper mantle can cause strong thermal heating in overlying continental crust, which results in thermal relaxation. Rocks under these heat flow conditions act as a fluid and respond to extension or shear pressures by stretching. This can lead to rifting and the formation of marginal seas. During this breakup phase microcontinental fragments can be formed and transported elsewhere.

Indonesia is bounded by three major plates. The Pacific plate forms an eastern boundary, the Indo-Australian plate occurs to the west and south, and the Indo-China plate to the north. Rifting occurred along the edge of the Indo-China plate shortly after subduction of oceanic crust beneath it ceased at the end of the Cretaceous. As new oceanic crust formed and the rifts widened, continental fragments migrated southwards. To the south, on the Gondwanic margin, continental blocks had been rifted away and drifting to the north since the early Palaeozoic. Indonesia formed where these Asian and Gondwanic fragments collided. The eastern margin, formed by the Pacific plate, acted as a barrier containing these fragments within the Indonesian region. Australia and its associated rift systems separated from Antarctica during the early Tertiary. Australia's subsequent northward drift into the Indonesian region, a process still occurring today, resulted in further modifications and tectonic displacements within the region. A model for this three-plate system is provided by Charlton (1986). As Hutchinson (1989) remarked:

> 'on-going processes are now at work in Southeast Asia, and we see that as a major Australian Craton approaches China–Indochina, continental scraps rift off both cratons and drift away to meet in the middle ground characterized by island arcs. The complexity resulting from rifting and displacement of the terranes and their final compression together in a future orogenic belt will result in a jigsaw puzzle of insolvable complexity'.

Table 2. A summary of the source areas and ages of derivation and collision of microcontinental fragments in Indonesia

Microcontinent	Source area	Derivation age	Collision age
1A	Gondwana	Lower Palaeozoic	Early Triassic
1B, 1C	Gondwana[a]	Middle Carboniferous	Late Triassic
		Jurassic	Late Cretaceous[f]
2	Indochina	Late Cretaceous	Middle Miocene
3	Gondwana[b]	Jurassic/Cretaceous?	Late Cretaceous?
4	Gondwana[c]	Early Tertiary	Late Miocene
5	Gondwana[d]	Late Cretaceous	Late Miocene
6	Gondwana[e]	Jurassic	Pliocene

[a] east coast of India (Burton 1970); [b] between India and Western Australia?; [c] western part of Irian Jaya (Audley-Charles *et al.* 1972; Hamilton 1979; Norvick 1979) or north-eastern New Guinea–Australia (Pigram and Panggabean 1984); [d] northern New Guinea–Australian margin; [e] north-western Australia; [f] Harbury *et al.* (1990).

Fig. 2 shows the approximate extent of the south-east Asian and Australian cratons (stippled) and the *melée* of rifted microcontinents, island arcs, and marginal seas in the middle. Microcontinental fragments are numbered. A summary of the source areas, and ages of derivation and collision of these microcontinental blocks is provided in Table 2. Inspection of Fig. 2 reveals that the 'regions' Wallace identified in Indonesia (Fig. 1) are geological composites, composed of either large continental blocks or smaller microcontinental fragments. These continental fragments were derived — at different times — from a number of distinct areas of Gondwana, or from Indo-China. Wallace's lack of any conception of lateral movement of land made informed interpretation of biogeographic patterns in Indonesia impossible. In the following section I would like to interpret what he observed in the light of the information summarised in Fig. 2, and thus to indulge in some speculation about what Wallace himself may have made of the data now available.

Wallace Reinterpreted

Indo-Malay Islands

The Indo-Malay Islands are part of continental south-east Asia. Three schools of thought exist as to how this region became assembled (Metcalfe 1988). All agree that the blocks from which south-east Asia have been assembled are Gondwanic. The evidence for Gondwanic origin includes extensive Permian glacial deposits thought to have formed in a rift system (Gatinsky and Hutchinson 1986), palaeontological evidence (Buffetaut 1982; Burrett and Stait 1985), and palaeomagnetic data (Achache *et al.* 1983). The source area was probably in that region of Gondwana between India and Australia ('Sinus Australis').

However, the different schools of thought differ markedly from one another on when these blocks assumed their present relative positions. Helmke (1985) considered that south-east Asia had amalgamated by the late Palaeozoic, while others consider that the blocks were derived during the early Triassic and assembled at their present latitudes at the end of the Triassic (e.g. Burton 1970; Acharyya 1979; Parker and Gealey 1983; Liew and McCulloch 1985; Metcalfe 1988; Hutchinson 1989). Thus, in the opinion of these latter authors, south-east Asia was assembled in the early Mesozoic. Yet a third view is favoured by Harbury *et al.* (1990), who suggest that the Sibumasu block (Fig. 2: terranes 1B and 1C) rifted from Gondwana much later than other south-east Asian blocks in the Jurassic, and docked with the rest of south-east Asia in the mid- to late Cretaceous. It is possible, as suggested by Burton (1970), that the Sibumasu block was adjacent to India prior to the Jurassic breakup of this part of Gondwana.

Wallace's biogeographic work indicated a close biotic relationship between the Indo-Malay islands and India. It is possible that this relationship dates from a period of mobilism initiated by mountain building during the collision of India with Eurasia. Under this hypothesis, animals and plants would have moved down newly created dispersal routes along these rising mountain chains.

A more intriguing possibility is that the Sibumasu block was, as Harbury *et al.* (1990) suggested, derived from Gondwana during the Jurassic. Biogeographic evidence would favour a source area close to the Indian sector of Gondwana. The east African–Madagascar–Indian–Malayan 'track' identified by Croizat (1964, fig. 24) is intelligible in the light of what is known of the breakup of Gondwana and the Cretaceous arrival of the Sibumasu block, with its associated biota, in south-east Asia.

Borneo

The island of Borneo is, as Croizat suggested, a geological composite (Croizat 1958, fig. 200, reproduced as Fig. 3). South-west Kalimantan is interpreted by geologists as either an extension of the Malay Peninsula (Hartono and Tjokrosapoetro 1986), or as a continental fragment derived from Indo-China either during the middle Cretaceous (Hutchinson 1989), or late Cretaceous (Gatinsky and Hutchinson 1986 ; Metcalfe 1988).

Hutchinson (1989) regarded south-east Kalimantan as a microcontinental fragment which, prior to the opening of the Macassar Strait during the mid-Tertiary (Oligocene according to McCabe and Cole 1989), formed a single unit with south-west Sulawesi. As indicated in Fig. 3, Croizat (1958) linked these latter two areas to Sumatra.

The third microcontinental block forms the state of Sarawak. For much of the Mesozoic the continental edge of south-east China and the east coast of Indochina was an Andean-type volcanic margin under which oceanic crust was being subducted (Hutchinson 1989 and references therein). By the close of the Mesozoic this subduction had ceased and portions of thinned continental crust had started to rift away from the margin. These rifted microcontinents (Fig. 2 : 2) were transported south during the early Miocene as a spreading centre developed to the north, forming what is now the South China Sea.

Older Mesozoic crust on the southern boundary of these rifted blocks was subducted under a trench system in the Palawan–northwest Borneo region. This trench system ceased to operate in the mid-Miocene when these microcontinents derived from the north collided with other continental fragments in the Borneo region (Hutchinson 1989). In Fig. 3 Croizat (1958) identified this event with a track connecting China with Sarawak. The occurrence of orangutan in Kalimantan and Sumatra but not in Sarawak, which Wallace (1880) interpreted in terms of ecological factors, may thus have its origin (as opposed to its perpetuation) in the different source areas from which the various terranes of Borneo were derived. Orangutan appear still to be restricted to areas derived from Gondwana.

Sulawesi

The island of Sulawesi is composite, with mid- to late Tertiary island arcs in the north and east sutured to a microcontinental fragment in the south. In addition, the Sula Islands and Buton are microcontinental fragments derived from the east (Hartono and Tjokrosapoetro 1986). The Sula Islands terrane is reported to have collided with eastern Sulawesi as it was translated westwards in the late Miocene (McCabe and Cole 1989). The source area for the Sula Islands has variously been identified as north of Misool (Audley-Charles *et al.* 1972), the Bird's Head of Irian Jaya (Norvick 1979), the Bird's Neck of Irian Jaya (Hamilton 1979), or from further east from central New Guinea (Pigram *et al.* 1985). Hartono and Tjokrosapoetro (1986) link Buton with the Buru–Seram region. Wallace's observation that the fauna of the Sula Islands and Buton were related to the 'Moluccas', rather than to Sulawesi, is therefore correct.

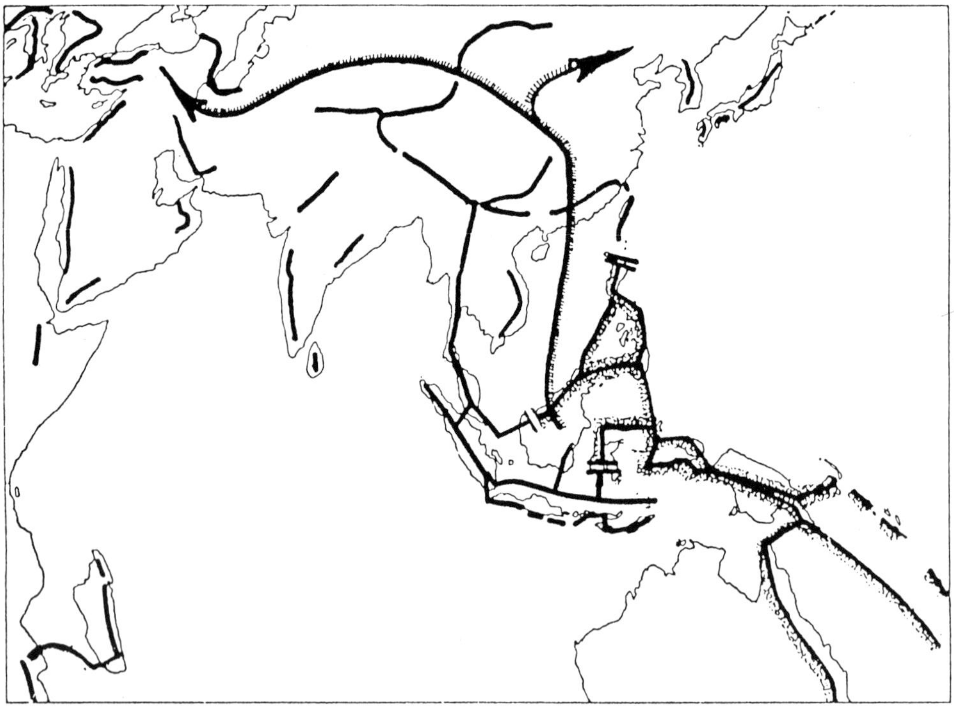

Fig. 3. Croizat's (1958, fig. 200) summary of Indonesian biogeography.

The 'anomalous' biota of Sulawesi, with its apparent African connection, does not lend support to Hutchinson's (1989) hypothesis that the microcontinental fragments labelled 3 in Fig. 2 were derived from the north. Inspection of Fig. 3 shows that Croizat (1958) linked south-west Sulawesi and south-east Kalimantan with Sumatra. If this biogeographic hypothesis is correct then these areas shared a common geological history. Harbury *et al.*'s (1990) hypothesis regarding the relatively late addition of the Sibumasu block to south-east Asia would account for this shared history.

One would have to hypothesise that an already differentiated biota occurred over that region of Gondwana composed of east Africa–Madagascar–India–Sibumasu prior to breakup. As Sibumasu rifted away from Gondwana the composition of its biota would have reflected this original pre-rift differentiation which, it seems reasonable to suggest, reflected the original geographic distances separating the component blocks. Because the Sibumasu block was originally nearer to India its biota was, in a general sense, more closely related to Indian forms. However, for more specific cases the genealogical similarities may have been with African forms, either reflecting the original pre-rift differentiation, or resulting from late phase mobilism along newly created rift systems formed as India moved away from Gondwana.

Wallace (1880) also suggested that the biotas of Sulawesi and the Philippines showed direct similarities. With the exception of the North Palawan microcontinent, the remainder of the Philippines archipelago is thought to be composed of Cretaceous to Recent oceanic sediments and volcanics (Karig 1983). This, together with the lack of sediments with a continental character, implies an island-related history since the Cretaceous (McCabe and Cole 1989). However, the diverse biota of the Philippines strongly suggests that the geological history of these islands may have been more complex than this. In particular, either the Philippines were adjacent to continental areas, or microcontinental fragments are present.

Croizat (1958) suggested that Sulawesi and the Philippines were linked by a generalised track via Halmehera (Fig. 3). Cladistic analyses of groups occurring in south-west Sulawesi–Halmahera–Mindanao may clarify the nature of the biotic relationship between Sulawesi and the southern Philippines. If a direct relationship is demonstrated to exist, as suggested by Wallace (1880), then one could speculate that parts of Mindanao may be continental in origin and derived from elsewhere. The Banda, Sulu, and Celebes Seas — the later of which is adjacent to Mindanao — are composed of old oceanic crust surrounded by younger, actively spreading oceanic crust (McCabe and Cole 1989). This anomalous situation indicates that these oceanic basins, which according to McCabe and Cole (1989) were originally a single unit, have become trapped in their present position. The implication is, of course, that they have also been derived from elsewhere. According to this hypothesis the three areas of old oceanic crust, the microcontinental fragments labelled 3 in Fig. 2, and parts of Mindanao formed a single tectonic unit.

Timor and the Banda Arc

The Banda arc is a composite unit, the inner arc of islands being a volcanic extension of western Sumatra and Java, the outer arc being a fragment rifted from north-western Australia. On the island of Timor, representative rocks of both units are present (Hartono and Tjokrosapoetro 1986; McCabe and Cole 1989). Wallace observed that the biota of the Banda arc islands was a mixture of Australian and Malayan elements, and that the proportion of Australian elements increased on going eastwards while the proportion of Malayan elements decreased.

Once again the general biogeographic patterns of this area are quite intelligible in the light of what is known about the area's geological history. There is a strong possibility that the Australian element of the biota is temporarily composite (Michaux 1989), an original component linked to the separation of the Timor–Seram rift system during the Jurassic, and a younger element derived when this rift system collided with continental Australia during the Pliocene. As Timor was compressed between the inner Banda arc to the north and the Australian craton to the south, a period of mobilism was initiated as once disjunct regions became physically united. Audley-Charles and Hooijers' (1973) discussion of the migration of pigmy elephants between Flores and Timor during the Pleistocene is an example of this mobilism.

The islands of the inner Banda arc are a structural continuation of Java. Their biota shows a strong relationship with Java and this indicates that, although these islands are now physically isolated both from Java and each other, connections existed in the past. During the Pliocene–Pleistocene phase of mobilism initiated by the collision of the outer and inner Banda arcs, intermixing appears to have been restricted to the area of collision. This resulted in a 'purer' Malay biota in the west and a 'purer' Australian biota in the east. The faunal connection with the 'Moluccas' is a consequence of the geological relationship between Timor and Seram which occupy extreme ends of the same tectonic unit rifted from Australia in the Jurassic.

The Molucca Islands

Considering the complex geological history of this 'area', which is composed of numerous terranes, it is not surprising that Wallace had difficulty in interpreting the biogeography. His recognition of two distinctive faunal provinces within this region is a consequence of the differences in geological relatedness of the 'northern' and 'southern' islands of the Moluccas. The northern group of islands, centred around Halmahera, is geologically related to the northern New Guinea/Australian margin, whilst the southern group, centred around Seram, is part of the rift system connected to north-western Australia.

In the light of what is now known, further refinements can be made. The Moluccas are composed of terranes from three distinct source areas. The island of Halmahera is part of Parker and Gealey's (1983) outer Pacific arc, and as such is related biotically to that part of Irian Jaya north of the Sorong fault (Dow and Sukamto 1984), or its structural equivalents, and thence eastwards to the Solomon Islands and Vanuatu. Many of the other islands of Wallace's northern province, such as Bacan (Hartono and Tjokrosapoetro 1986) and Obi, as well as Buru from Wallace's southern area, are related to what was once the foreland of the New Guinea/Australian craton. As already detailed, the island of Seram is part of a Jurassic rift system derived from north-western Australia.

Conclusions

There are a number of specific conclusions that can be drawn from this study. The major conclusion is that plant and animal distributions in Indonesia result from the transportation of ancestral taxa on continental fragments derived from both Gondwana and Indo-China during the Mesozoic, and Australia/New Guinea's mid-Tertiary collision with the region.

In many instances the distributional evidence and reconstructions of tectonic development are perfectly congruent. Thus Wallace's (1880) discussion of the biota of the Sula Islands or of Timor, and Croizat's (1958) identification of a Sarawak/Indo-China track and composite geological nature of Borneo, are in perfect agreement with independently derived tectonic data. In other instances the two sets of data are either contradictory or one or other is equivocal. These cases are of interest because they can help focus further research effort.

Examples from Indonesia where biogeographic and geological evidence contradict each other are few — the source area for the microcontinental fragments of south-east Kalimantan and south-west Sulawesi is one. Incongruencies appear more commonly to result from data that are inadequate to resolve complex questions. For example, detailed distributional data combined with cladistic analyses may help geologists resolve the source areas for those microcontinental fragments translated westwards from Australia/New Guinea during the mid-Tertiary (Fig. 2 : 4). A further example, again from eastern Indonesia, concerns the relationship between Mindanao, Sulawesi and the northern Moluccas. Resolution of the conflicting biogeographic evidence detailed previously may well provide important information to geologists in their attempts to understand, in detail, the tectonic development of eastern Indonesia.

The last specific conclusion concerns Wallace. As detailed in the introduction, it is an oversimplification to regard him as a dispersalist. The fundamental modern opposition in biogeography is not between dispersalism and vicariance *per se*, because for vicariant form making to occur an ancestral taxon must have become widely distributed during a period of mobilism. The distinction between mobilism and dispersal is clearly only one of degree. Rather, the fundamental opposition — as in all historical studies — is between those who wish to interpret history as a sequence of unique happenings and those who interpret history in terms of some underlying principle. Wallace attempted the latter.

Brundin (1966) regarded Wallace as a champion of the view that animal distributions were due to origins in northern centres with subsequent dispersal toward the southern antipodes. This analysis is correct, Wallace certainly championed this view in later writings. However, as I hope I have demonstrated, he originally attempted — unsuccessfully as it turned out — to interpret distributions in terms of geological history. Given the paucity of geological knowledge and a cultural context of imperialism and colonial expansion at the time, this change is not altogether surprising.

Finally, this study demonstrates that it is not possible to understand the distribution of plants and animals without a knowledge of how that distribution came about. Empirical studies concerned with describing distributions and theories concerned with reconstructing genealogical relationships between taxa are necessary, but not sufficient, to allow us to understand how the present has come to be. Area cladograms describe a sequence of splitting events and tracks link areas with common geological histories, but neither describes what has happened. This is so because of their 'one-to-many' relationship to geological development, i.e. one area cladogram or track is consistent with many geological histories. If biogeographers believe that geographic distributions are the result of tectonic events, then their studies should contain an in-depth analysis and discussion of pertinent geological evidence.

Acknowledgments

My thanks to Jack Grant-Mackie, Peter Ballance, David Lambert, Hans Duffels, Ronald Vonk and Garry Nelson for their help and encouragement, and for providing much appreciated feedback. The generous financial assistance provided by the Australian Systematic Botany Society and the Evolutionary Genetics Laboratory, which allowed me to attend the ASBS conference, is gratefully acknowledged. This is publication number 43 in conjunction with the EGL.

References

Achache, J., Courtillot, C., and Besse, J. (1983). Paleomagnetic constraints on the late Cretaceous and Cenozoic tectonics of southeastern Asia. *Earth and Planetary Science Letters* **63,** 123–36.

Acharyya, S. K. (1979). India and Southeast Asia in Gondwanaland fit. *Tectonophysics* **56,** 261–76.

Audley-Charles, M. G., Carter, D. J., and Milson, J. S. (1972). Tectonic development of eastern Indonesia in relation to Gondwana dispersal. *Nature* **239,** 35–9.

Audley-Charles, M. G., and Hooijer, D. A. (1973). Relation of Pleistocene migrations of pigmy stegodonts to island arc tectonics in eastern Indonesia. *Nature* **241,** 197–8.

Buffetaut, E. (1982). Mesozoic vertebrates from Thailand and their palaeobiological significance. *Terra Cognita* **2,** 27–34.

Burrett, C., and Stait, B. (1985). South East Asia as part of an Ordovician Gondwanaland — a palaeobiogeographic test of a tectonic hypothesis. *Earth and Planetary Science Letters* **75,** 184–90.

Burton, C. K. (1970). The palaeotectonic status of the Malay Peninsula. *Palaeogeography, Palaeoclimatology, Palaeoecology* **7,** 51–60.

Brundin, L. (1966). Transantarctic relationships and their significance as evidenced by chironomid midges. *Kunglia Svenska Vetenkapsakademiens Hanlingar, 4 Serien* **11,** 1–472.

Carey, S. W. (1975). Tectonic evolution of Southeast Asia. Fourth Indonesian Petroleum Association Congress, Jakarta, pp. 1–31.

Carr, E. H. (1970). 'What is History?' (Penguin: London.)

Charlton, T. R. (1986). A plate tectonic model of the eastern Indonesian collision zone. *Nature* **319,** 394–6.

Craw, R. C., and Weston, P. (1984). Panbiogeography: a progressive research program? *Systematic Zoology* **33,** 1–13.

Croizat, L. (1958) 'Panbiogeography.' Vols. 1, 2a, 2b. (Published by author: Caracas.)

Croizat, L. (1964). 'Space, Time, Form.' (Published by author: Caracas.)

Croizat, L. (1982). Vicariance/vicariism, panbiogeography, 'vicariance biogeography', etc.: a clarification. *Systematic Zoology* **31,** 291–304.

Dow, D. B., and Sukamto, R. (1984). Western Irian Jaya: the end product of oblique plate convergence in the Late Tertiary. *Tectonophysics* **106,** 109–39.

Gatinsky, Y. G., and Hutchinson, C. S. (1986). Cathaysia, Gondwanaland, and the paleotethys in the evolution of continental Southeast Asia. GEOSEA Proceedings Vol. 2, *Geological Society of Malaysia Bulletin* **20,** 179–99.

Hamilton, W. (1979). Tectonics of the Indonesian region. U.S. Geological Survey Professional Paper No. 1078, 345 pp.

Harbury, N. A., Jones, M. E., Audley-Charles, M. G., Metcalfe, I., and Mohamed, K. R. (1990). Structural evolution of Mesozoic Peninsula Malaysia. *Journal of the Geological Society of London* **147,** 11–26.

Hartono, H. M. S., and Tjokrosapoetro, S. (1986). Geological evolution of the Indonesian Archipelago. GEOSEA Proceedings Vol. 2, *Geological Society of Malaysia Bulletin* **20,** 97–136.

Helmke, D. (1985). The orogenic evolution (Permian–Triassic) of central Thailand. Implications on paleogeographic models for mainland SE-Asia. *Memoire du Société Géologique de France* **147,** 83–91.

Hutchinson, C. S. (1989). Displaced terranes of the southwest Pacific. In 'The Evolution of the Pacific Ocean Margins'. (Ed. Z. Ben-Avraham.) pp. 161–75. Oxford Monographs in Geology and Geophysics Volume 8. (Oxford University Press: New York.)

Karig, D. E. (1983). Accreted terranes in the northern part of the Philippine Archipelago. *Tectonics* **2,** 211–36.

Liew, T. C., and McCulloch, M. T. (1985). Genesis of granitoid batholiths of Peninsular Malaya and implications for models of crustal evolution: evidence from Nd–Sr isotopes and U–Pb zircon study. *Geochimica et Cosmochimica Acta* **49,** 587–600.

McCabe, R., and Cole, J. (1989). Speculations on the late Mesozoic and Cenozoic evolution of the Southeast Asian margin. In 'The Evolution of the Pacific Ocean Margins.' (Ed. Z. Ben-Avraham.) pp. 143–60. Oxford Monographs in Geology and Geophysics Volume 8. (Oxford University Press: New York.)

Metcalfe, I. (1988). Origin and assembly of south-east Asian continental terranes. In 'Gondwana and Tethys'. (Eds M. G. Audley-Charles and A. Hallam.) pp. 101–18. Geological Society Special Publication No. 37. (Oxford University Press: New York.)

Michaux, B. (1989). Generalized tracks and geology. *Systematic Zoology* **38,** 390–8.

Nelson, G., and Platnick, N. (1981). 'Systematics and Biogeography. Cladistics and Vicariance.' (Columbia University Press: New York.)

Norvick, M. S. (1979). The tectonic history of the Banda Arcs, eastern Indonesia, a review. *Journal of the Geological Society of London* **136,** 519–27.

Parker, E. S., and Gealey, W. K. (1983). Plate tectonic evolution of the western Pacific–Indian Ocean region. Proceedings of the EAPI/ASCOPE/CCOP/IOC workshop on the geology and hydrocarbon potential of the South China Sea and possibilities of joint development, Honolulu.
Pigram, C. J., and Panggabean, H. (1984). Rifting of the northeastern margin of the Australian continent and the origin of some microcontinents in eastern Indonesia. *Tectonophysics* **107,** 331–53.
Pigram, C. J., Surono, V., and Supandjono, J. B. (1985). Origin of the Sula platform, eastern Indonesia. *Geology* **13,** 246–8.
Wallace, A. R. (1880). 'The Malay Archipelago.' Seventh edition. (Macmillan: London.)
Wegener, A. (1929). 'The Origin of Continents and Oceans.' [Reprint of English translation, 1966.] (Dover Publications: New York.)

Manuscript received 3 September 1990, accepted 31 January 1991

3 September 1991 *Aust. Syst. Bot.*, **4**, 37–40

Implications of Shallow Tethys and the Origin of Modern Oceans

K. G. McKenzie

Geology Department, University of Melbourne, Parkville, Vic. 3052, Australia.

Cladograms resemble directed Prim networks or Wagner trees, and are usually programmed by parsimony, i.e. the program searches for the least number of steps to achieve a cladistic synthesis. Large matrices can yield an almost infinite number of possible trees; and even when differences of only a few steps are involved there are numerous plausible solutions. Parsimony and other factors, such as Nelson and Platnick's assumptions, act as Occam's razor, reducing these to a manageable few.

In the cladistic analysis of fossil taxa, a number of other techniques are available also to minimise the plethora of choice. When the fossil record is good to adequate, then first appearance in geological time (the chorological principle) can be used to refine ancestor/descendant relationships. Secondly, when the taxa concerned have a well known palaeobiogeography, their spatial nearness/distance may be helpful. And thirdly, when the ecological associations of the taxa being analysed are known, then environmental factors (facies relationships) can be combined with stratigraphic and distributional data to arrive at an overall best-fit result. Cladists cannot afford to ignore such considerations.

Well worked-out examples include the cladistic analyses of evolution in some Middle Devonian species of the ostracode genera *Ctenoloculina* (Beyrichicopida) and *Euglyphella* (Podocopida) for which the stratigraphy, distribution and facies of occurrence are all well- to reasonably well- understood (Collier 1971; Al-Abdul-Razzaq 1973). Taxa not included initially in such analyses, because of poor quality specimens or a paucity of material or incomplete literature search or the description of new species, can be used to regulate the final cladogram (Kesling 1973).

Shallow Tethys can be defined in geological time; and its palaeobiogeography and environmental associations are likewise well known. Tethys proper initiated during the later part of the Hercynian tectonic epoch and its initial phase (Palaeo-Tethys) lasted until the close of the Permian. During this period, Tethys was uniformly shallow to epibathyal and had the character of an epicontinental sea rather than an ocean in the modern sense. Its greatest depths were exemplified by basinal facies and, in the Devonian–Carboniferous, by Thuringia-type assemblages which were probably deposited at no more than 300–500 m (McKenzie 1987, plus references cited therein).

During the Mesozoic, Tethys s.s. was more truly oceanic and locally deeper. Suess (1893) noted the juxtaposition of relatively deeper water radiolarian cherts and contemporary massive white shallow water limestones in the Triassic of the eastern Alps that epitomises this phase. Kristan-Tollmann (1987, 1988) has demonstrated the remarkable uniformity of Triassic Tethyan faunas from the Americas to Europe, Turkey, Iran, Burma and the Far East (Fig. 1). By the Jurassic, there is evidence for a wider, ocean-like sea in the region of western China, with different faunas on its northern and southern shores (Wen Shi-xuan 1987). But the prevailing picture, especially in the central part of Tethys, is complex; it comprises emergent blocks surrounded by broad epicontinental shelves with interconnecting sinuous deeper zones, and some small confined basins (McKenzie 1987, plus references cited therein). The well documented Cretaceous marine extinctions (Kaufmann 1985) correlate

1030-1887/91/010037$05.00

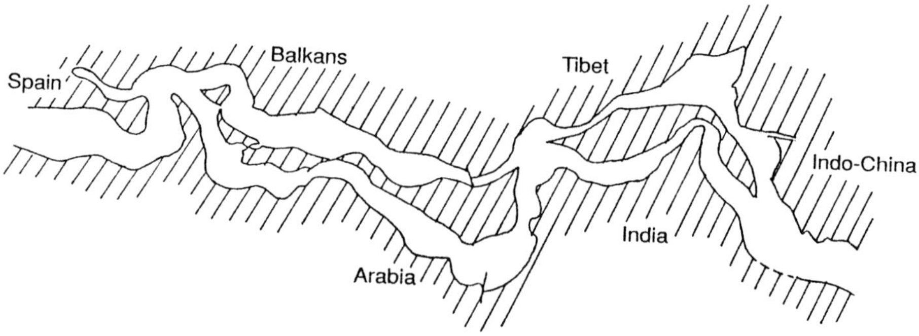

Fig. 1. Tethys s.s. during the Triassic, modified after Kristan-Tollmann (1987). This pattern of broadly latitudinal shores and large central islands typified Tethys s.s. throughout the Mesozoic.

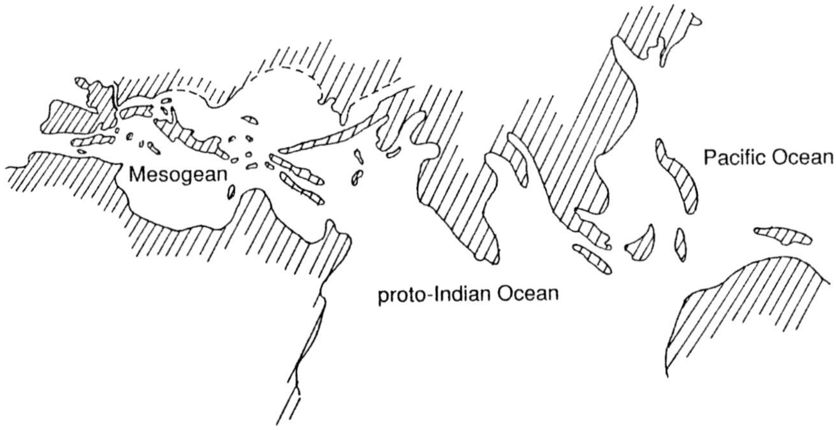

Fig. 2. Neo-Tethys during the Early Oligocene, modified after Piccoli *et al.* (1987).

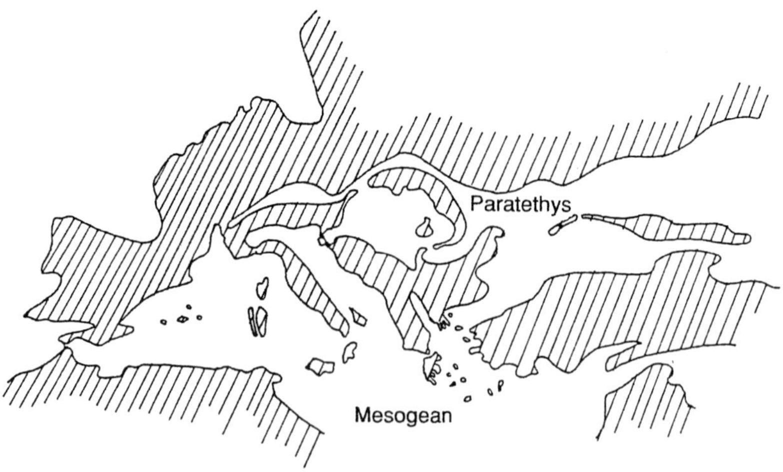

Fig. 3. Paratethys and the Mesogean (proto-Mediterranean) during the later part of the Miocene, modified after Roegl and Steininger (1984).

significantly with the end of this phase. There was no uniformly broad and deep Tethyan ocean as simplistically mapped by some palaeocartographers based in plate tectonics.

During the Cainozoic (Fig. 2), as the Atlantic and Indian Oceans increased in extent, Neo-Tethys achieved psychrospheric oceanic depths throughout (Benson 1975; Pokorny 1979), and functioned as a corridor for the dispersal of deep- as well as shallow-adapted taxa, from the Caribbean, through West Africa and the Mesogean (proto-Mediterranean) to the Indopacific and Australasia (McKenzie 1967; Piccoli *et al.* 1987; Whatley and Ayress 1988). Tethys disrupted finally in the Neogene (Fig. 3), forming Paratethys and, subsequent to the Messinian Event and a Lago Mare phase, the modern Mediterranean.

The history outlined so briefly above has the following interpretative implications for cladists.

(1) Prior to about the mid-Cretaceous, Tethyan faunas generally were relatively shallow. Tethyan stratigraphy, facies and palaeobiogeography embraced punctuations in the geological record due to tectonism, eustasy and changes in climate (Fischer 1985). Any adaptive radiations that followed such environmental changes can be interpreted as effects of local biogeographic and facies differences. When the inferred evolutionary rate was more sedate, speciation was probably controlled by the grosser environmental disequilibria. This view of evolution as a response to ecosystem change is inherent to the Red Queen and stasis hypotheses as analysed by Stenseth and Maynard Smith (1984).

(2) Psychrospheric deepwater faunas do not antedate the Eocene (Benson 1975; Steineck *et al.* 1984). Therefore, Cainozoic deepwater species of long-ranging genera evolved from relatively shallow water Cretaceous–Palaeocene or earlier forebears, e.g. the Recent deepwater ostracode *Nemoceratina* sp. Nohara 1987 found off Okinawa—the genus originated in the Triassic. See also Newman (1985).

(3) The modern Mediterranean fauna, including deepwater species (Bonaduce *et al.* 1983), due to a combination of the effects of the Messinian Event and the presence of the Gibraltar Sill, has evolved from shallow water Neogene and later Atlantic taxa, not from Indopacific species (Benson 1985). Exceptions include Lessepsian migrants since the late eighteenth century via the Suez Canal (Por 1971); and other species dispersed incidentally by humans via discharged ballast sludges, especially since about 500 years ago, following the invention of bulkheads and shipboard pumps for ocean-going vessels (Needham 1971).

(4) Paratethyan faunas are characterised by rapid radiation and turnover in a number of more or less interconnected basins during relatively brief epochs (Karpatian, Badenian, Sarmatian, Pannonian, Pontian, Dacian, Romanian) in which tectonics-induced changes in basinal salinities were the most critical environmental factor (Roegl and Steininger 1984). Because of the rapid turnover, cladistic analysis *within* these assemblages, with their numerous endemic genera, owes little to stratigraphy or palaeobiogeography.

(5) Following the disruption of Tethys and termination of the Tethyan corridor, interpretation of the cladistic analysis of widely distributed Neogene to Recent marine taxa is likely to be more influenced by provincialism than by stratigraphy (Titterton and Whatley 1988).

References

Al-Abdul-Razzaq, S. (1973). Evolution of Middle Devonian species of *Euglyphella* as indicated by cladistic analysis. *Contributions from the Museum of Paleontology University of Michigan* **24**(6), 47–64, 12 text-figs.

Benson, R. H. (1975). The origin of the psychrosphere as recorded in changes of deep-sea ostracode assemblages. *Lethaia* **8**, 69–83, 10 figs.

Benson, R. H. (1985). The Phanerozoic 'crisis' as viewed from the Miocene, In 'Catastrophes in Earth History: the New Uniformitarianism'. (Eds W. A. Berggren and J. A. Van Couvering.) pp. 436–46. (Princeton University Press: Princeton.)

Bonaduce, G., Ciliberto, B., Masoli, M., Minichelli, G., and Pugliese, N. (1983). The deep-water benthic ostracodes of the Mediterranean, In 'Applications of Ostracoda'. (Ed. R. F. Maddocks.) pp. 459–71. (Houston Geosciences: Houston.)

Collier, W. W. (1971). Phylogeny of the Devonian ostracod genus *Ctenoloculina* Bassler. *Contributions from the Museum of Paleontology University of Michigan* **23**(23), 357–76, 6 text-figs, 3 pls.

Fischer, A. G. (1985). The two Phanerozoic suypercycles. In 'Catastrophes and Earth History: the New Uniformitarianism'. (Eds W. A. Berggren and J. A. Van Couvering.) pp. 129–50. (Princeton University Press: Princeton.)

Kaufmann, E. G. (1985). The fabric of Cretaceous marine extinctions. In 'Catastrophes and Earth History: the New Uniformitarianism'. (Eds W. A. Berggren and J. A. Van Couvering.) pp. 151–246. (Princeton University Press: Princeton.)

Kesling, R. V. (1973). *Euglyphella bellensis*, a new Middle Devonian ostracod from Michigan. *Contributions from the Museum of Paleontology University of Michigan* **24**(7), 65–7, 1 pl.

Kristan-Tollman, E. (1987). Triassic of the Tethys and its relations with the Triassic of the Pacific Realm. In 'Shallow Tethys 2'. (Ed. K. G. McKenzie.) pp. 169–86. (Balkema: Rotterdam.)

Kristan-Tollmann, E. (1988). Unexpected microfaunal communities within the Triassic Tethys. In 'Gondwana and Tethys'. (Eds M. G. Audley Charles and A. Hallam.) Geological Society Special Publication No. 37, pp. 213–23, 9 figs.

McKenzie, K. G. (1967). The distribution of Caenozoic marine Ostracoda from the Gulf of Mexico to Australasia. In 'Aspects of Tethyan Biogeography'. (Eds C. G. Adams and D. V. Ager.) pp. 219–38. (Systematics Association: London.)

McKenzie, K. G. (1987). Tethys and her progeny. In 'Shallow Tethys 2'. (Ed. K. G. McKenzie.) pp. 501–23. (Balkema: Rotterdam.)

Needham, J. (1971). Nautical technology. In 'Science and Civilisation in China, Volume 4, Part 3'. (Ed. J. Needham.) pp. 475–535. (Cambridge University Press: Cambridge.)

Newman, W. A. (1985). The abyssal hydrothermal vent invertebrate fauna: a glimpse of antiquity? Biological Society of Washington Bulletin No. 6, pp. 231–42, 4 tables.

Piccoli, G., Sartori, S., and Franchino, A. (1987). Benthic molluscs of Shallow Tethys and their destiny. In 'Shallow Tethys 2'. (Ed. K. G. McKenzie.) pp. 333–73. (Balkema: Rotterdam.)

Pokorny, V. (1979). *Abyssocypris (Proabyssocypris) palavensis* subg. n., sp. n. (Ostracoda, Crust.) from the Paleogene of Czechoslovakia. *Casopis pro mineralogii a geologii* **24**(2), 195–200, 3 figs, 1 pl.

Por, F. D. (1971). One hundred years of Suez Canal—a century of Lessepsian migration; retrospect and viewpoints. *Systematic Zoology* **20**, 138–59.

Roegl, F., and Steininger, F. F. (1984). Neogene Paratethys, Mediterranean and Indo-pacific seaways. In 'Fossils and Climate'. (Ed. P. Brenchley.) pp. 171–200. (John Wiley: New York). p. xv, 352.

Steineck, P. L., Breen, M., Nevins, N., and O'Hara, P. (1984). Middle Eocene and Oligocene deep-sea Ostracoda from the Oceanic Formation, Barbados. *Journal of Paleontology* **58**, 1463–96, 19 figs.

Stenseth, N. C., and Maynard Smith, J. (1984). Coevolution in ecosystems: Red Queen evolution or stasis? *Evolution* **18**(4), 422–6, 2 figs.

Suess, E. (1893). Are great ocean depths permanent? *Natural Science* **2**, 180–7.

Titterton, R., and Whatley, R. C. (1988). The provincial distribution of shallow water Indo-Pacific marine Ostracoda: origins, antiquity, dispersal routes and mechanisms. In 'Evolutionary Biology of Ostracoda'. (Eds T. Hanai, N. Ikeya and K. Ishizaki.) pp. 759–86. (Elsevier, Kodansha: Amsterdam, Tokyo).

Wen Shi-xuan (1987). Jurassic bivalve faunas of the Qinghai–Xizang (Tibet) Plateau in west China and their palaeo-biogeography. In 'Shallow Tethys 2'. (Ed. K. G. McKenzie.) pp. 247–52. (Balkema: Rotterdam).

Whatley, R. C., and Ayress, M. (1988). Pandemic and endemic distribution patterns in Quaternary deep-sea Ostracoda. In 'Evolutionary Biology of Ostracoda'. (Eds T. Hanai, N. Ikeya and K. Ishizaki.) pp. 739–55. (Elsevier, Kodansha: Amsterdam, Tokyo).

Manuscript received 10 September 1990, accepted 3 June 1991

3 September 1991 *Aust. Syst. Bot.*, **4**, 41–58

Standard Assumptions for Biogeographic Analysis

Gareth Nelson[A] *and Pauline Ladiges*[B]

[A]American Museum of Natural History, New York, N.Y. 10024, U.S.A.
[B]School of Botany, The University of Melbourne, Parkville, Vic. 3052, Australia.

Abstract

Reanalysis of Mayden's data on distribution and relationships of North American freshwater fishes reveals weaknesses of the analytical protocol termed Brooks' Parsimony Analysis (BPA). Standard assumptions 2, 1, and 0 are explored with reference to suites of 3-area statements contained in cladograms of species. Component analysis proved effective for all assumptions for the freshwater fish data whereas BPA yielded results not optimal for any assumption. For Rosen's data on *Heterandria* and *Xiphophorus*, component analysis proved effective for assumption 2 whereas results from other methods proved effective for assumptions 1 and 0. Analysis of widespread species indicates that, when they are used to build area cladograms, they cause spurious results. Assumptions 1 and 0 are sensitive to these spurious effects.

Introduction

Recent discussion of methods in biogeography promotes parsimony programs (PAUP, Hennig86) to resolve area relationship. A protocol termed Brooks' Parsimony Analysis (BPA) uses a data matrix coding geographic distributions of species as well as cladograms relating the species (Wiley 1988; Brooks 1990). Mayden (1988) used BPA to resolve relationships of 34 areas inhabited by 29 freshwater fishes in central North America. We reanalyse Mayden's data and reveal weaknesses of BPA, in comparison with results for current standards: assumptions 2 (A2), 1 (A1), and 0 (A0). For the assumptions themselves we explore formalisation as suites of 3-area statements derived from species-cladograms and geographic distributions.

Assumptions 2, 1, and 0 offer different assessments of area relationship derived from species-cladograms which can be ambiguous because of incongruence as well as widespread, redundant, and missing taxa and areas. An example of a simple case (Fig. 1) includes three taxa (1–3) occupying four areas (A–D), with taxon 1 widespread in areas AB, and taxa 2 (endemic in C) and 3 (endemic in D) as closest relatives. For assumption 2 there are three possibilities of area relationship. The first is that areas C and D are related more closely to each other than to area A: A2:1 = A(CD), with no information about B (which could be placed in any of five positions). The second is that areas C and D are related more closely to each other than to area B: A2:2 = B(CD), with no information about A (which could be placed in any of five positions). The third is the sum of the first two possibilities: A2:3 = AB(CD), where areas C and D are related more closely to each other than to areas A and B. For A1 there is only one possibility: A1 = A2:3 = AB(CD). For A0 there is only one possibility: A0 = (AB)(CD).

The three assumptions are sometimes described in terms of the status of the areas (AB) occupied by the widespread taxon: A2 sees AB as possibly polyphyletic; A1, as possibly paraphyletic; A0, as monophyletic. The three assumptions are also sometimes described in terms of the status of the widespread taxon (1); A2 sees taxon 1 as possibly polyphyletic; A1, as possibly paraphyletic; A0, as monophyletic.

1030-1887/91/010041$10.00

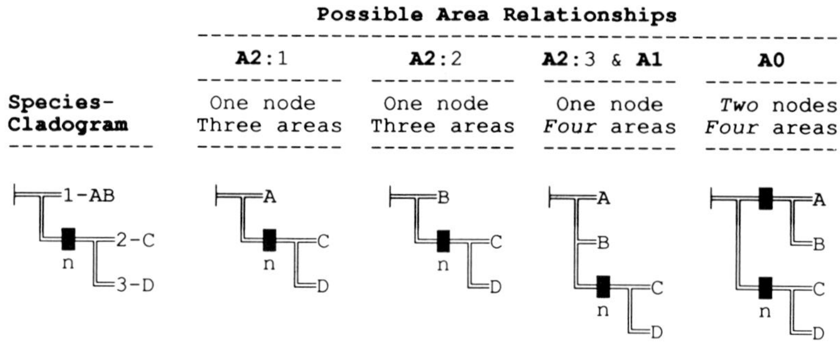

Fig. 1. Hypothetical example of three species (1–3) in four areas (A–D), showing area relationships according to assumptions 2 (A2), 1 (A1), and 0 (A0). For A2 there are three possibilities, A2:1–3. For A1 (=A2:3) and A0 there is only one possibility. A0 treats species 1, widespread in areas A and B, as equivalent to a node.

However described, the three assumptions are logically interrelated in the sense that two possibilities of A2 (A2:1–2) are encompassed by the third (A2:3), which equals that of A1 which encompasses that of A0. A1 in effect is the upper limit of informativeness of A2. A0, however, treats a widespread taxon as if it were an additional node. Interpretations based on A0 sometimes differ from those based on A2 and A1 because A0 treats widespread taxa as additional nodes but A2 and A1 do not.

Interrelationships of the three assumptions may also be seen in terms of their 3-area statements. For A2 there is only one such statement: x(CD), where area x = A+B. For A1 there are two statements: A(CD) and B(CD). For A0 there are four statements: (AB)C, (AB)D, A(CD), B(CD). From this standpoint the assumptions may be understood as specifying the minimum number of 3-area statements that embodies all nodes of a species cladogram (A2) or the maximum number (A1); A0 adds to the maximum by treating widespread taxa as nodes. We explore the 3-area approach as a possible alternative to the formalisation of Nelson and Platnick (1981), which used set theory to differentiate between assumptions 2 and 1.

Methods

Data for analysis are phylogenetic relationships (R) and geographic distributions (D) of 29 species of seven species-groups of North American freshwater fishes (Fig. 2*A–G*, based on Mayden 1988, table 1). The BPA protocol specifies a matrix of pooled data, including both classes (R+D) of information. In analyses below we treat the data of each class separately, and produce an R area-cladogram based on relationships of species and a D area-cladogram based on distributions of species.

Computation of machine-generated area-cladograms based on parsimony analysis was performed with the MSDos program Hennig86 (version 1.5) having a special provision of 320KB for tree files, for which we are indebted to J. S. Farris. Hennig86 was run on Toshiba T1200FB and NorthGate 386 personal computers. All runs were duplicated on both machines. Times of runs are those for Hennig86 option bb* (following option mh*) on the 9.54MHz Toshiba, with times for the 33MHz NorthGate shorter by a factor of ten. Abbreviations are CI, consistency index; RI, retention index; T, time in seconds.

Data matrices of Appendices 1 and 2 (based on Mayden 1988, table 1) contain an outgroup (OTG) and areas (A01–34) ordered numerically following Mayden (1988, fig. 3), but not his table 1 (which deviates in the order of areas 1–2, 7–8, 10, 12). The seven groups of columns correspond to the seven species-groups. The first one or two columns in each group are multistate characters coding species-cladograms. A question mark denotes absence of the group from an area. The other columns are single characters coding a species as (1) present in or (0) absent from an area.

Suites of 3-area statements appropriate for assumptions 2, 1, and 0 were derived from the seven species-cladograms (Fig. 2). Area cladograms based on different classes of data or on different assumptions, or generated by the BPA protocol or by hand were evaluated by counting the number of 3-area statements from species-cladograms contained in each area-cladogram (Tables 5–8). Fit of Preglacial and modern drainage patterns to the species cladograms was evaluated in the same way (Table 6). Counts were made by hand and checked at least once; however, because of the large number of statements it would be remarkable if all counts proved exact.

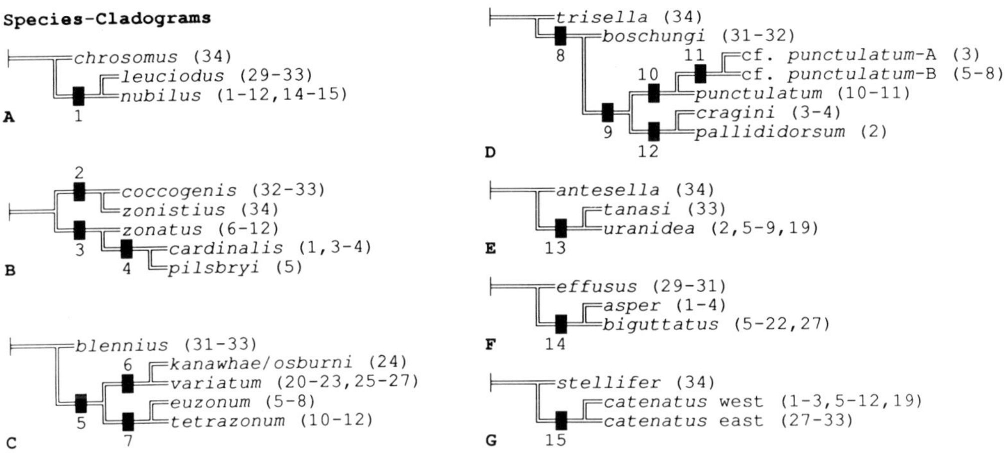

Fig. 2. Cladograms and distributions of seven species-groups of North American freshwater fishes. *A*, *Notropis leuciodus* group. *B*, *Luxilus zonatus* group. *C*, *Etheostoma variatum* group. *D*, subgenus *Ozarkia* of *Etheostoma*. *E*, subgenus *Imostoma* of *Percina*. *F*, *Nocomis biguttatus* group. *G*, *Fundulus catenatus* group (after Mayden 1988). Drainages numbered as in Fig. 3 (also Mayden 1988, fig. 3).

Results

RD Cladogram

When analysed by the BPA protocol the pooled RD data, based both on phylogenetic relationships and on distributions of species (Mayden 1988, table 1), yield an area-cladogram of 16 informative nodes (Fig. 3*A*:2–17). Mayden argued that this RD cladogram, because of the fit of the relationships of some of its areas with the Pre-Pleistocene drainage pattern of North American rivers, is evidence that distribution of fishes in central North America was determined significantly by Pre-Pleistocene factors.

A2 Cladogram

An alternative area-cladogram (Fig. 3*B*), resolved by hand, is based on assumption 2 (A2). The A2 cladogram contains eight informative nodes (a–h). Two nodes occur in both A2 and RD cladograms (nodes c=3, d=7). Six nodes (a–b, f–h), unique to the A2 cladogram, contradict nodes of the RD cladogram. The following analyses help to explain incongruity of RD and A2 cladograms.

Analysis of RD Cladogram: R and D Cladograms

The R cladogram (Fig. 4*A*) has three nodes, two of which are found also in the RD cladogram (Fig. 3*A*); the D cladogram (Fig. 4*B*) has six nodes, five found also in the RD cladogram. The successive weighting feature of Hennig86 (xsw) improves resolution in the latter but not the former case: the Dxsw cladogram (Fig. 4*C*) has nine nodes, six found also in the RD cladogram. Of the 16 nodes of the RD cladogram, it would seem from this analysis that two nodes (2, 7) are derived from phylogenetic relationships of species (R), six nodes (5, 13–17) are derived from geographic distributions of species (D) and eight nodes (3–4, 6, 8–12) are the result of interactions between these two classes of data.

Analysis of A2 Cladogram: Stepwise Combination

Resolution of the A2 cladogram may be illustrated in a stepwise combination of all seven species-cladograms (Fig. 5*A*–*G*). The final resolution (Fig. 5*G*, right) includes eight nodes, three of which (B,C,E) represent more than one node of species-cladograms, five of which represent single nodes (2,4,6,11,12) of species cladograms. The final resolution includes all seven species cladograms (Fig. 2*A*–*G*) in all of their detail. Conflicting elements, areas 3–4 and 32–33 (Fig. 6*A*), involve widespread versus endemic species. A2 resolves such conflict on

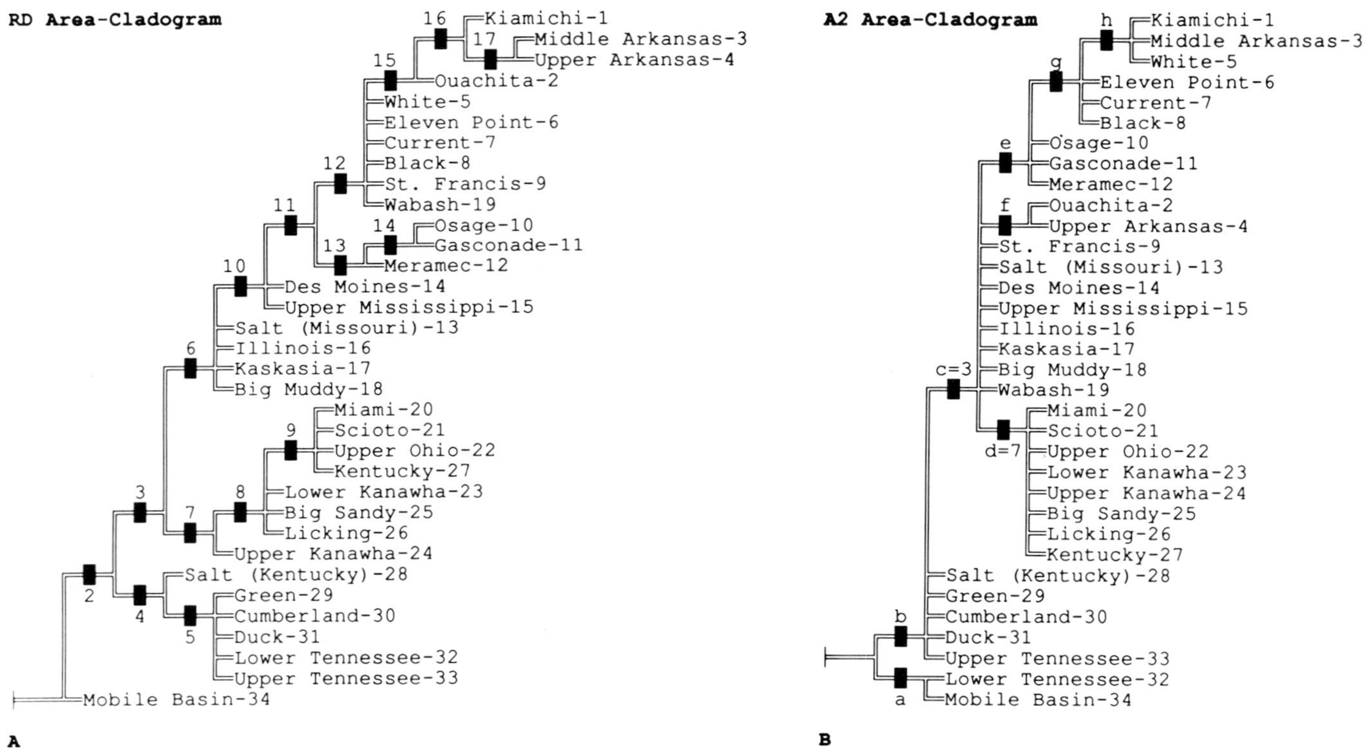

Fig. 3. *A*, RD cladogram of 34 North American drainages from Brooks Parsimony Analysis (BPA), based both on species-cladograms (weighted ×2 by Mayden) and species distributions (characters 1–37 in Appendix 1*A*–*C*; also Mayden 1988, table 1). For computer analysis Mayden consolidated drainages with identical character strings: 6=7=8, 10=11, 13=16=17=18, 14=15, 20=21=22, 23=25=26, 29=30. For the consolidated matrix Mayden's PAUP (version 2.4.1) resulted in this strict consensus of 33 most parsimonious trees (length=88, CI=0·84). For the consolidated matrix (Appendix 1, right, above), Hennig86 finds 33 most parsimonious trees (length=90, CI=0·82, RI=0·87, T=44). With correction of the published data matrix (Appendix 1, right, below: character 12: 31-Duck=1, 32-Lower Tennessee=1, 33-Upper Tennessee=1, 34-Mobile Basin=?), Hennig86 finds 33 trees (length=88, CI=0·84, RI=0·89, T=43). The full uncorrected matrix (Appendix 1, left) gives the same consensus of 1455 trees (length=90, CI=0·82, RI=0·91, T=3738, 1560 trees if areas are ordered as in Mayden 1988, table 1), except that Eleven Point, Black and Current Rivers compose an additional node, and that Green and Cumberland Rivers compose an additional node. The full corrected matrix (Appendix 1, left and right, below) gives the same consensus of 2703 trees (length=88, CI=0·84, RI=0·92, T=6808), except that Green and Cumberland Rivers compose an additional node. *B*, A2 cladogram from component analysis (assumption 2). Two of eight nodes are the same in the RD area-cladogram (Fig. 3*A*, node d=7, c=3). Six nodes conflict with the RD area-cladogram. Of 16 nodes in the RD cladogram, six are compatible but not resolved in this A2 cladogram (8–11, 13–14).

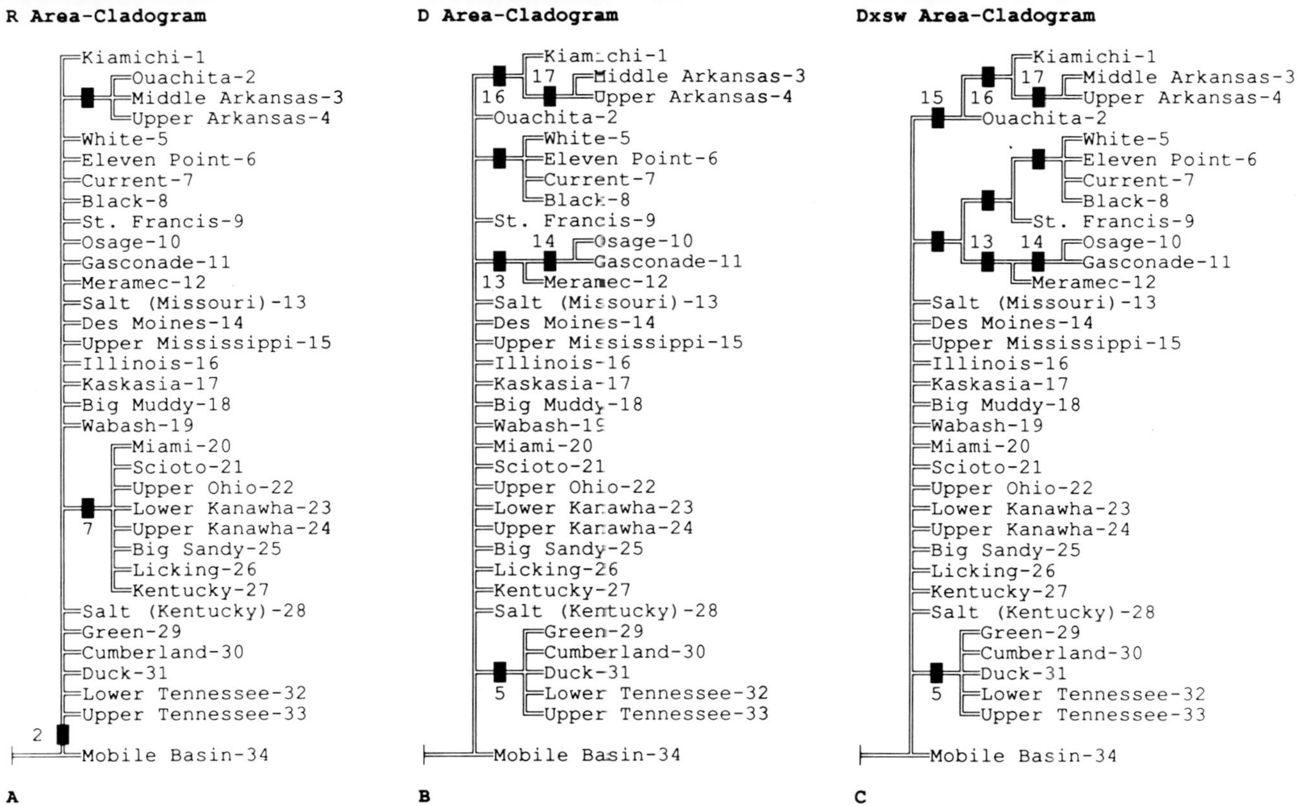

Fig. 4. *A*, R cladogram from parismony analysis, based only on species-cladograms (columns 1, 5, 11–12, 18–19, 27, 31, 35 of Appendix 1, also Mayden 1988, table 1). For the consolidated and uncorrected matrix, Hennig86 finds 7850+ (overflow) most parsimonious trees (length=25, CI=0·92, RI=0·95, T=2858); for the full uncorrected matrix, 5300+ (overflow) trees (length=25, CI=0·92, RI=0·96, T=3306). Both matrices give this strict consensus tree. For the consolidated and corrected matrix, Hennig86 finds 7850+ (overflow) trees (length=24, CI=0·95, RI=0·97, T=2857); for the full corrected matrix, 5300+ (overflow) trees (length=24, CI=0·95, RI=0·98, T=3744). Both matrices give this strict consensus tree but do not resolve node 7. For all four matrices successive weighting results in no further resolution. *B*, D cladogram from parsimony analysis, based only on species distributions (characters 2–4, 6–10, 13–17, 20–26, 28–30, 32–34, 36–37 of Appendix 1A; also Mayden, 1988, table 1). For the consolidated matrix, Hennig86 finds 7674 most parsimonious trees (length=39, CI=0·71, RI=0·78, T=7654); for the full matrix, 5300+ (overflow) trees (length=39, CI=0·71, RI=0·87, T=6668). Both matrices give this strict consensus tree. *C*, Dxsw cladogram from parsimony analysis after successive weighting (xsw), based only on species distributions (see Fig. 6). For the consolidated matrix, Hennig86 finds 90 most parsimonious trees (length=299, CI=0·84, RI=0·88, T=1308, 122); for the full matrix, 1345 trees (length=309, CI=0·83, RI=0·91, T=3800, 4044). Both matrices give this strict consensus tree.

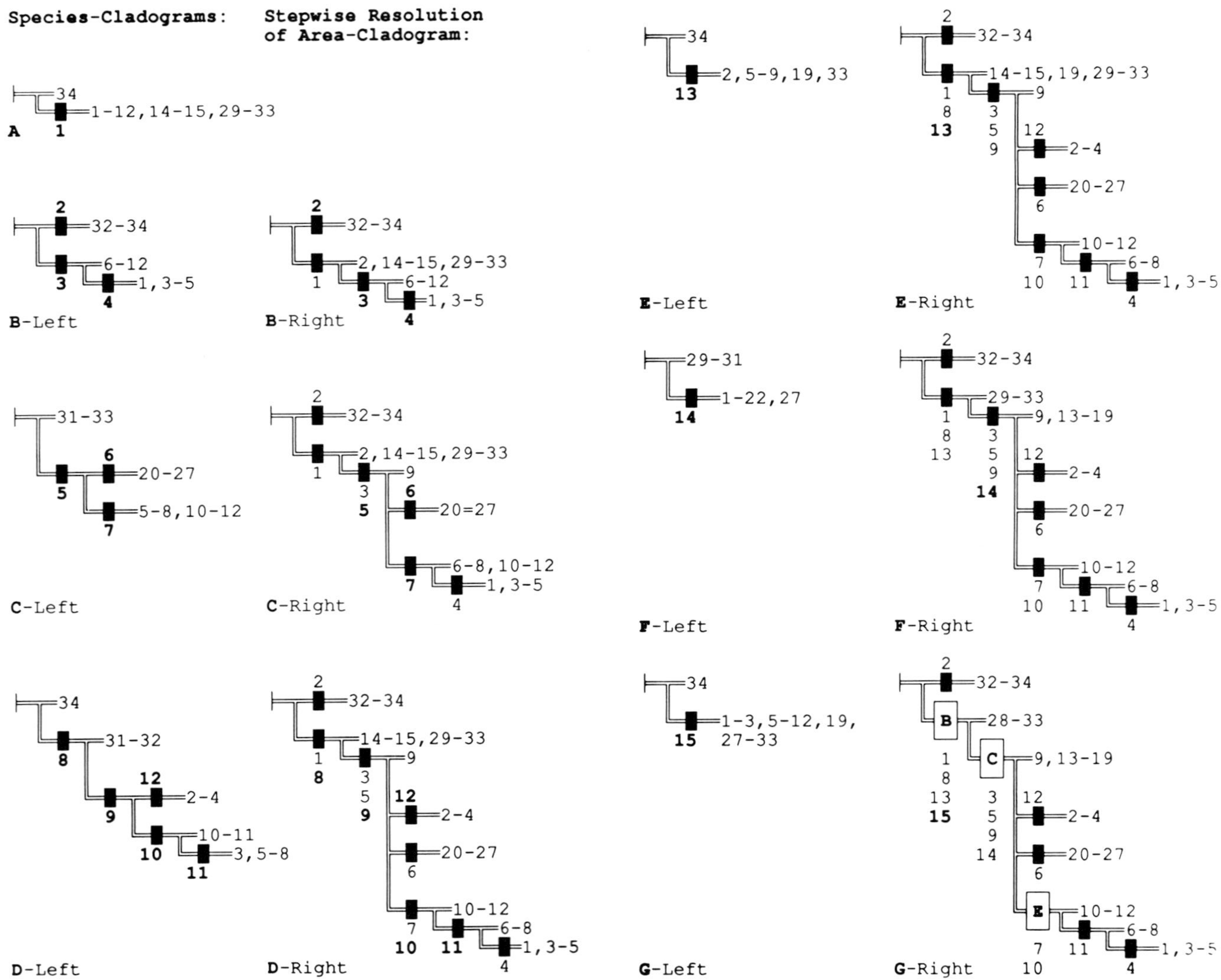
Species-Cladograms:
Stepwise Resolution of Area-Cladogram:
A
34
1-12,14-15,29-33
B-Left
32-34
6-12
1,3-5
B-Right
32-34
2,14-15,29-33
6-12
1,3-5
C-Left
31-33
20-27
5-8,10-12
C-Right
32-34
2,14-15,29-33
9
20=27
6-8,10-12
1,3-5
D-Left
34
31-32
2-4
10-11
3,5-8
D-Right
32-34
14-15,29-33
9
2-4
20-27
10-12
6-8
1,3-5
E-Left
34
2,5-9,19,33
E-Right
32-34
14-15,19,29-33
9
2-4
20-27
10-12
6-8
1,3-5
F-Left
29-31
1-22,27
F-Right
32-34
29-33
9,13-19
2-4
20-27
10-12
6-8
1,3-5
G-Left
34
1-3,5-12,19,
27-33
G-Right
32-34
B
28-33
C
9,13-19
2-4
20-27
E
10-12
6-8
1,3-5

behalf of endemic species (Fig. 6*B*). Resolution is equivalent to believing that certain species have enlarged their range by dispersal and, accordingly, to eliminating areas into which species are believed to have dispersed. Eliminating such areas from the matrix of species-cladograms analysed by parsimony yields an R cladogram of 100% consistency. This R cladogram has two nodes (Fig. 6*C*). Both nodes occur in the A2 cladogram (Fig. 3*B*).

Nodes B, C, and E are corroborated by different species-cladograms. The degree of corroboration is analysed in Table 1. Ideally, each area might be represented in each species-cladogram by an endemic species. From each of seven species-cladograms (Fig. 2), however, some areas are missing; and in each species-cladogram some species are widespread. On average only 49% of the areas that might have occurred actually do so, represented by only 12% of the species that might have been there. Overall, corroboration is only 31% of what it might have been in an ideal case.

Analysis of D46 Cladogram

Mayden considered 17 species for which he had no cladograms. Adding their geographic distributions to the matrix of 29 species and analysing all 46 distributions by parsimony yields a more completely resolved area-cladogram (Fig. 7*A*). This D46 cladogram includes 21 informative components, of which seven (5, 8–9, 13–15, 17) occur also in the RD cladogram (Fig. 3*A*), and of which none occurs in the A2 cladogram (Fig. 3*B*).

Analysis of A2+D46 Cladogram

The A2 cladogram may be further resolved by adding to it those nodes from the D46 cladogram that are wholly (Fig. 7*A*: nodes 8, 13, 14, q) or partly combinable (Fig. 7*A*: nodes 5, p, r, s, t). The resulting A2+D46 cladogram includes 18 nodes (Fig. 7*B*).

Discussion

Species Distributions as Predictors of Nodes in Area-cladograms

Distributions of the 46 species may be compared with area-cladograms by scoring each species as monophyletic, possibly monophyletic, or non-monophyletic according to its fit to the A2, A2+D46, RD, D46 cladograms (Table 2). Apparently, species are poor predictors of nodes (0 nodes for the A2 cladogram) unless species distributions are used as data to build area-cladograms, either in part (A2+D46, RD cladograms) or completely (D46 cladogram).

If species distributions are poor predictors of nodes of area-cladograms, should species distributions be used as data to build area-cladograms? Using species distributions makes it easy to build area-cladograms, since no species-cladograms are necessary. If species-cladograms are unnecessary for biogeographical analysis, the paradox is created of 'cladistic' biogeography without species-cladograms, as if using a parsimony procedure is in itself enough to render any result cladistic, i.e. generally informative of historical pattern.

Comparison of Cladograms and Drainage Patterns: Monophyletic Nodes and Species

For central North America, Page (1990, fig. 14) rendered 'Preglacial' and 'Today's' drainage patterns as cladograms (Fig. 8*A*, *B*). Drainage patterns may be compared with the four area-cladograms (A2, A2+D46, RD, and D46) by scoring each area-cladogram node as monophyletic or non-monophyletic according to its fit with the pattern (Table 3). The A2 cladogram fits both drainage patterns equally well as one node is monophyletic in each. The A2–D46, RD and D46 cladograms fit Today's pattern better than the Preglacial: more nodes are monophyletic

---

Fig. 5. Stepwise resolution (including conflict) of area-cladogram. *A–G* left, species-cladogram as in Fig. 2. *B–G* right, combination of preceding species-cladograms. Each species-cladogram (*A–G*, left) shows distributions distal to nodes, and distributions of related species are combined. For the *Notropis leuciodus* group in *A*, for example, the distributions of *N. leuciodus* (29–33) and *N. nubilus* (1–12, 14–15) are combined (1–12, 14–15, 29–33). The *leuciodus* (*A*) and *zonatus* (*B*, left) groups combine in the area-cladogram *B*, right. Adding the *Etheostoma variatum* group (*C*, left) gives *C*, right; adding the subgenus *Ozarkia* (*D*, left) gives *D*, right; adding the subgenus *Imostoma* (*E*, left) gives *E*, right; adding the *Nocomis biguttatus* group (*F*, left) gives *F*, right; adding the *Fundulus catenatus* group (*G*, left) gives the final resolution (*G*, right). Each of nodes *B*, *C*, *E* of the final resolution represents nodes from different species-cladograms: node *B* represents nodes 1 (from cladogram *A*), 8 (*D*), 13 (*E*) and 15 (*G*); node *C* represents nodes 3 (*B*), 5 (*C*), 9 (*D*) and 14 (*F*); node *E* represents nodes 7 (*C*) and 10 (*D*). Each of five nodes of the final resolution (2, 4, 6, 11, 12 of *G*, right) represents a single node from one species-cladogram.

in Today's pattern (3, 4, 9 respectively) than in the Preglacial (2, 2, 1). These results conflict with Mayden's interpretation that the RD cladogram fits the Preglacial pattern better than it does Today's pattern.

Drainage patterns may alternatively be evaluated through their comparison with species distributions independent of the cladograms interrelating the species. Of 46 species of Mayden's study, 31 are widespread and 15 are endemic. Distributions of widespread species may be scored as monophyletic, possibly monophyletic, or non-monophyletic according to their fit with the drainage pattern (Table 4). Species distributions fit Today's pattern better than the

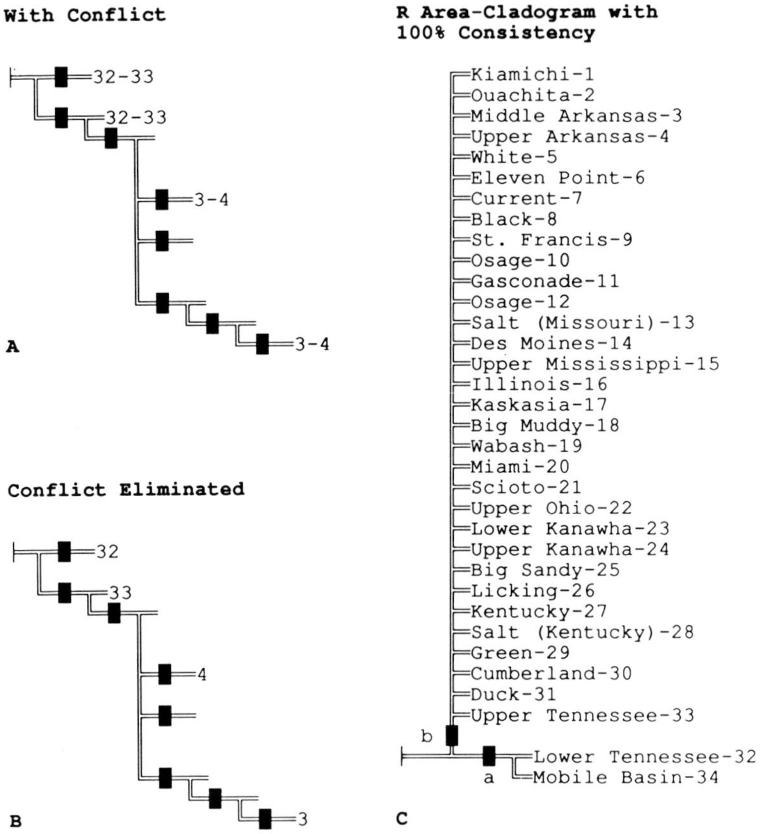

Fig. 6. *A*, *B*, topology of final resolution showing conflicting areas (*A*) and their resolution under assumption 2 (*B*). In all cases conflict is between widespread and endemic taxa, such that the widespread taxon might be supposed to have expanded its range into the area of the endemic. Widespread taxa with distributions conflicting with endemics include *Etheostoma cragini* (in area 3), *Luxilus cardinalis* (4), *Notropis leuciodus* (32), *Fundulus catenatus* eastern form (32), *Etheostoma boschungi* (32) and *Luxilus coccogenis* (33). *C*, R cladogram from parsimony analysis of species cladograms with conflicting distributions eliminated (characters 1, 5, 11, 12, 18, 19, 27, 31, 35 of Appendix 2, right). For the corrected and consolidated matrix, Hennig86 finds 7850+ (overflow) most parsimonious trees (length=23, CI=100, RI=100, T=2809); for the corrected full matrix, 5300+ (overflow) trees (length=23, CI=100, RI=100, T=3107). Both matrices give this consensus tree.

Preglacial: more distributions are monophyletic in Today's pattern (10) than in the Preglacial (2). These results also conflict with Mayden's interpretation that the RD cladogram fits the Preglacial pattern better than it does Today's pattern.

There is reason to believe, nevertheless, that relationships of fishes fit the Preglacial pattern better than they fit Today's, not as revealed by analysis of nodes of area-cladograms, or of distributions of species, but rather by analysis of 3-area information contained in species-cladograms. The following analysis allots 3-area information into suites according to standard assumptions (A2, A1, A0). The suites permit evaluation of area-cladograms according to a

Table 1. Analysis of degree of confirmation or replicated nodes (B, C, E) in A2 cladogram (assumption 2)

Ideally each replicate (1, 3, 7 etc.) might include all possible areas, each represented by an endemic taxon. Overall only 49% of possible areas and 12% of possible taxa occur, amounting to 31% of possible confirmation

Nodes	Areas		Taxa		Total
	No.	%	No.	%	%
b	32:	100	32:	100	100
1	18:	56	2:	6	31
8	10:	31	6:	19	25
13	8:	25	2:	6	16
15	18:	56	2:	6	31
1+8+13+15	54:	42	12:	9	26
c	27:	100	27:	100	100
3	11:	41	3:	11	26
5	14:	52	4:	4	33
9	8:	30	5:	19	24
14	23:	85	2:	7	46
3+5+9+14	56:	52	14:	13	32
e	9:	100	9:	100	100
7	7:	78	2:	22	50
10	8:	89	3:	33	61
7+10	15:	83	5:	28	56
b+c+e	254:	100	254:	100	100
1+3+5+7–10+13–15	125:	49	31:	12	31

Table 2. Comparison of distributions of 31 widespread species with four area-cladograms

A2, from component analysis (assumption 2) (Fig. 3*B*); A2+D46, derived from A2 cladogram by addition of nodes from D46 cladogram (Fig. 7*B*); RD, from Brooks Parsimony Analysis (BPA) of species-cladograms and species distributions combined (Fig. 3*A*); D46, from parsimony analysis only of species distributions (Fig. 7*A*). Each widespread species is scored as monophyletic (Mono), possibly monophyletic (?), or non-monophyletic (Non-Mono) in accordance with its fit with each area-cladogram. A species is scored ? if its distribution occupies some but not all of the areas distal to a node of an area-cladogram, provided that the missing areas do not occur in the distributions of other species of that species group

Cladogram	Mono	?	Non-Mono
A2	0	10	21
A2+D46	4	4	23
RD	8	15	8
D46	18	4	9

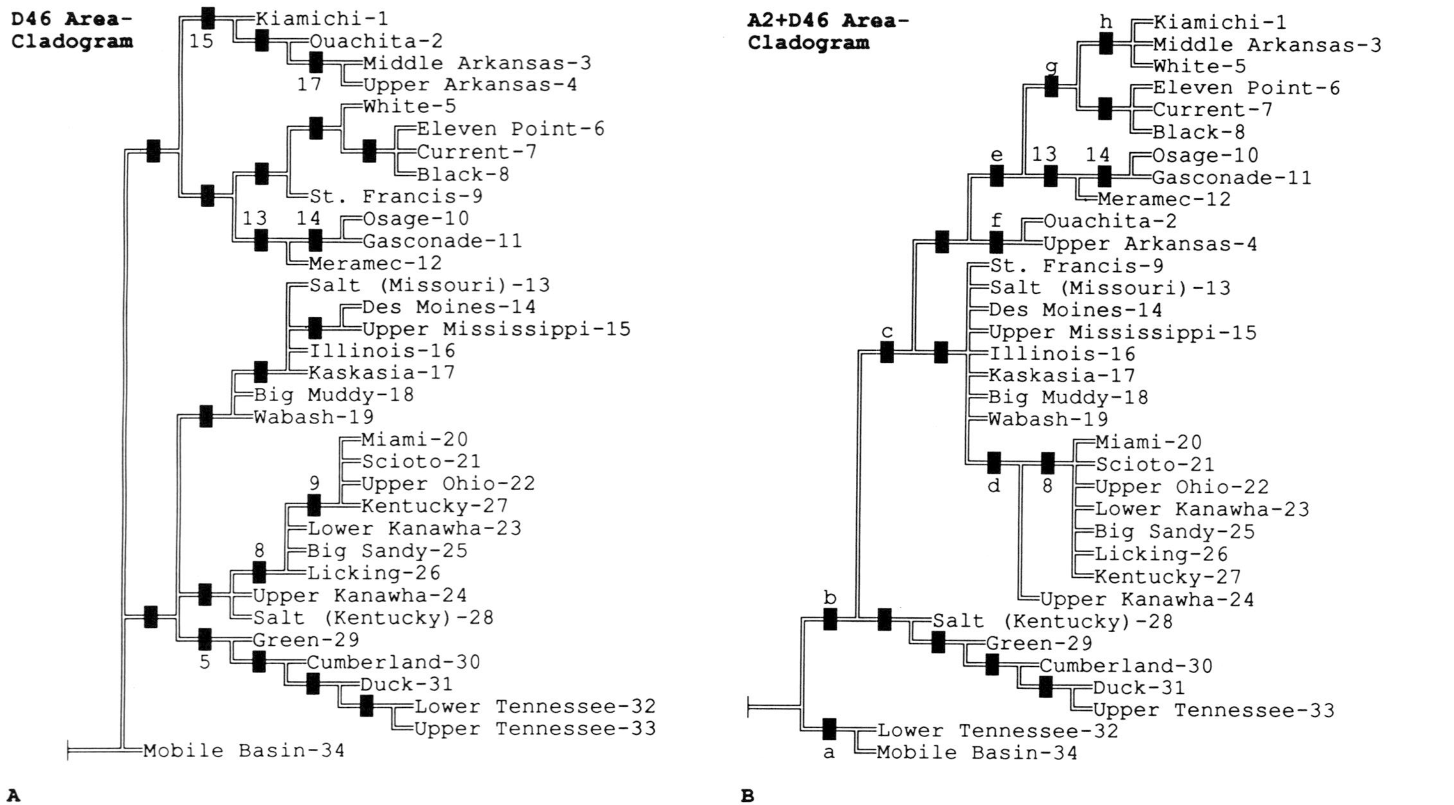

Fig. 7. *A*, D46 cladogram from parsimony analysis, based only on species distributions (including characters 38–40, Mayden 1988, table 1). Eastern and western forms of *Fundulus catenatus* were scored as separate species. Each of 17 species of characters 38–40 was scored as a binary character, with ? replaced by 0. Hennig86 finds 10 most parsimonious trees (length=63, CI=0·73, RI=0·88, T=24). After successive weighting Hennig86 finds two trees (length=404, CI=0·89, RI=0·95, T=6, 6). The two trees give this consensus tree. *B*, A2+D46 cladogram, concocted by transferring compatible or nearly compatible nodes from the D46 cladogram (Fig. 7*A*) to the A2 cladogram (Fig. 2*B*), without disrupting any node of the A2 cladogram.

criterion seemingly more accurate than scoring nodes of area-cladograms and distributions of widespread species.

Alternative Formalisation of Assumptions 2, 1, and 0

A cladistic approach to biogeography focuses on information about area relationship contained in one or more cladogram of taxa. For seven species-cladograms of Mayden's study (Fig. 2), information may be reckoned as 3-area statements according to assumptions 2, 1, and 0 (Table 5).

Assumption 2. The number of 3-area statements for A2 is calculable by treating all taxa as if they were endemic. In cladogram A, for example, there are two widespread and one endemic species: *leuciodus* (in areas 29–33), *nibilus* (1–12, 14–15), and *chrosomus* (in 34) (Fig. 2*A*; Table 5: cladogram A, node 1). If for A2, x=areas 29–33 and y=areas 1–12, 14–15, then information on area relationship reduces to one general statement 34(x,y), i.e. areas x and y are related more closely to each other than either is to area 34. Consider three hypothetical taxa: taxon 1 in area 34, taxon 2 in area 29 and taxon 3 in area 1; their information on area relationship is 34(29,1), which is an instance of 34(x,y). In other words, 34(29,1) satisfies A2. Other possible information would also satisfy A2, e.g. 34(30,2). Although there are at least 70 different instances of 34(x,y), for A2, instances are equivalent and no single instance is specified.

In cladogram B, there are three widespread and two endemic species: *coccogenis* (in areas 32–33), *zonatus* (6–12), *cardinalis* (1,3–4), *zonistius* (34) and *pilsbryi* (5) (Fig. 2*B*, Table 5: cladogram B, nodes 2–4). If for A2, y=32–33, y=6–12 and z=1,3–4, then information on area relationship reduces to (x,34)(y(z,5)), which yields ten 3-area statements: y(z,5), y(x,34), z(x,34), 5(x,34), x(z,5), x(y,z), x(y,5), 34(z,5), 34(y,z), 34(y,5). The ten statements form three groups (Table 5), each associated with a node of the cladogram: y,z,5(x,34) with node 2; x,34(y(z,5)) with node 3; and x,34,y(z,5) with node 4. The information (32,34) (6(1,5)) is an instance of (x,34) (y(z,5)). There are at least 136 different instances, all equivalent under A2.

Over the seven species-cladograms (Fig. 2) there are 59 3-area statements for A2 (Table 5: Total).

Table 3. Comparison of nodes of four area-cladograms with river cladograms (Fig. 8)

Cladograms as in Table 2

Cladogram	No. of nodes	Preglacial		Today	
		Mono	Non-Mono	Mono	Non-Mono
A2	8	1	7	1	7
A2+D46	18	2	16	3	15
RD	16	2	14	4	12
D46	21	1	20	9	12

Assumption 1. A species-cladogram with widespread species is seen as inherently more informative by A1 than by A2. The number of 3-area statements for A1 is the same as the number of instances of A2; but for A1, instances are not equivalent, and each is specified. Cladogram A, where information of area relationships is 34(29–33) (1–12,14–15), contains seventy 3-area statements: 34(29,1), 34(30,1), 34(30,2), 34(31,3) etc.: five areas of 29–33 times 14 areas of 1–12,14–15 times one area of 34, 5×14×1=70 (Fig. 2*A*, Table 5: cladogram A, node 1).

In cladogram B, information on area relationship is ((32–33)(34)) ((6–12) ((1,3–4) (5))), which yields 136 statements. As for A2, the statements form three groups, each associated with one node of the cladogram: (32–33) (34)6–12:1,3–4:5 with node 2; 32–33:34(6–12) (1,3–4:5) with node 3; 32–33:34:6–12(1,3–4) (5) with node 4 (Fig. 2*B*; Table 5: cladogram B, nodes 2–4).

Over the seven species-cladograms (Fig. 2) there are 1080 3-area statements for A1 (Table 5: Total).

Assumption 0. A species-cladogram with widespread species is seen as inherently more informative by A0 than by A1. To the 3-area statements of A1, A0 adds statements associated

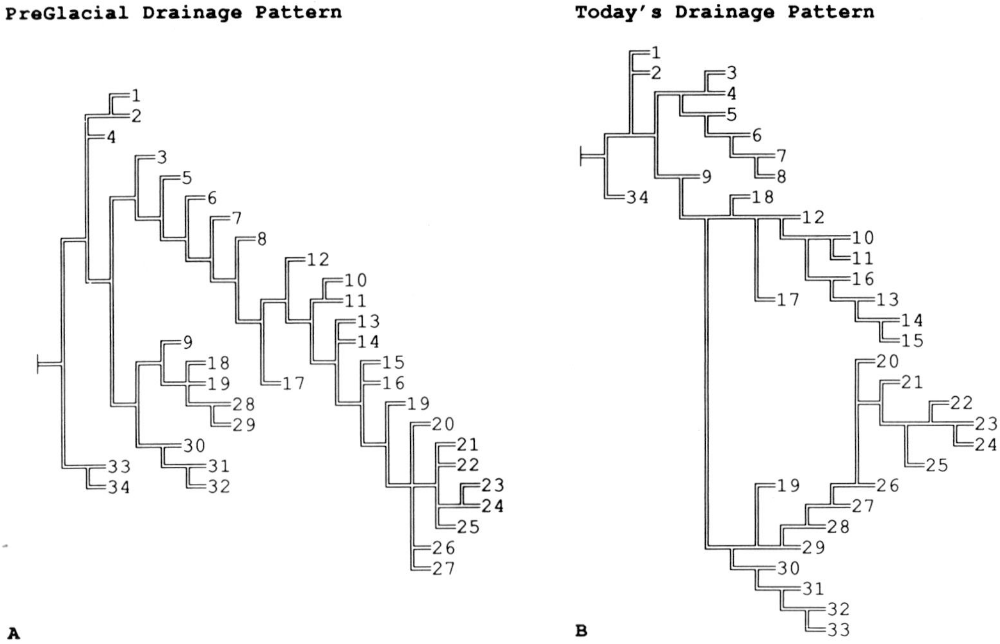

Fig. 8. Area-cladograms based on interconnection of rivers in (*A*) Preglacial and (*B*) Today's times (after Page 1990, fig. 14, with consolidation of areas 1=Little River and 2=Kiamichi River). Area 19 (20 in Page) in Fig. 8*A* occurs twice.

Table 4. Comparison of species distributions with river cladograms (Fig. 8)

Species scored as in Table 2

Species		Preglacial			Today		
Endemic	Widespread	Mono	?	Non-Mono	Mono	?	Non-Mono
15	31	2	3	26	10	4	17

Table 5. Three-area information contained in 15 nodes of seven species-cladograms (Fig. 2*A–G*)

Cladogram	Node	Structure of node	3-area statements under assumptions 2	1	0
A	1	34(29–33) (1–12,14–15)	1	70	766
B	2	1,3–4:5:6–12(32–33) (34)	3	22	36
	3	32–33:34(6–12) (1,3–4:5)	4	84	231
	4	6–12:32–33:34(1,3–4) (5)	3	30	63
C	5	31–33(5–8:10–12) (20–23,25–27:24)	4	168	213
	6	5–8:10–12:31–33(20–23,25–27) (24)	3	70	301
	7	20–23,25–27:24:31–33(5–8) (10–12)	3	132	261
D	8	34(2:3:3–4:5–8:10–11) (31–32)	5	20	31
	9	31–32:34(2:3–4) (3:5–8:10–11)	12	60	60
	10	2:3–4:31–32:34(3:5–8) (10–11)	8	58	69
	11	2:3–4:10–11:31–32:34(3) (5–8)	5	28	82
	12	3:5–8:10–11:31–32:34(2) (3–4)	5	19	30
E	13	34(2,5–9,19) (33)	1	7	49
F	14	29–31(1–4) (5–22,27)	1	228	1626
G	15	34(1–3,5–12,19) (27–33)	1	84	885
Total	15		59	1080	4701

with each widespread species. In cladogram A, for example, information of area relationship is 34(29–33) (1–12,14–15). For the widespread *leuciodus*, A0 adds information 1–12,14–15,34(29,30), 1–12,14–15,34(29,31) etc., amounting to 150 3-area statements (10 possible pairs of areas within 29–33 times 15 possible roots within 1–12,14–15,34: 10×15=150). For the widespread *nubilus*, A0 adds information 29–34(1,2), 29–34(1,3) etc., amounting to 546 3-area statements (91 possible pairs of areas within 1–12,14–15 times 6 posible roots within 29–34: 91×6=546). Adding 150+546=690 statements to the original 70 statements from A1 gives 696+70=766 total (Fig. 2*A*; Table 5: cladogram A, node 1).

Table 6. Summary of information content of two drainage patterns (Fig. 8*A*,*B*) and four area-cladograms, showing number and percentage of 3-area statements () contained in seven species-cladograms (Fig. 2*A*–*G*) under assumptions 2, 1 and 0

Cladograms as in Table 2

	Assumption					
	2 (59)		1 (1080)		0 (4701)	
	No.	%	No.	%	No.	%
Drainage pattern						
Preglacial	46	78	672	62	3249	69
Today	36	61	420	39	2600	55
Area-cladogram						
A2	59	100	1012	94	3224	69
A2+D46	59	100	1012	94	3886	83
RD	49	83	1009	93	3814	81
D46	38	64	543	50	2755	59

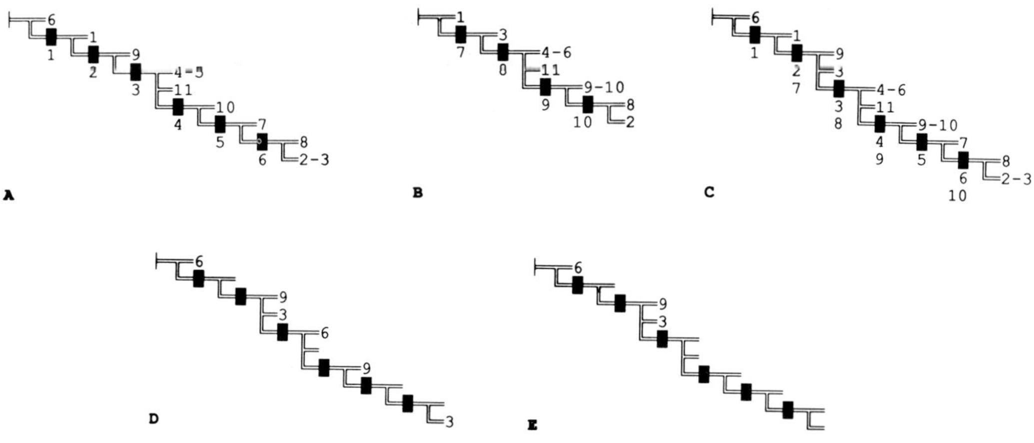

Fig. 9. *A*,*B*, species-cladograms of (*A*) *Heterandria* and (*B*) *Xiphophorus*. *C*, Combination of species-cladograms *A* and *B*, above. *D*,*E*, topology of resolution showing (*D*) conflicting areas and (*E*) their resolution under assumption 2. Widespread taxa with distributions conflicting with endemics include (as listed in Zandee and Roos 1987:319) *Heterandria bimaculata* (in area 3), *Xiphophorus alverezi* (6), and *Xiphophorus* 'PMH' (9).

Over the seven species-cladograms (Fig. 2) there are 4701 3-area statements for A0 (Table 5: Total).

Comparison of Drainage Patterns and Area Cladograms: 3-area Statements for A2, A1, and A0

The drainage patterns, Preglacial and Today, may be compared by determining how many 3-area statements from species cladograms (Table 5) are contained in each pattern. The Preglacial pattern contains more statements under each assumption (Table 6). If significantly different from those of Today's pattern, the greater numbers of the Preglacial pattern are evidence that distribution of fishes in central North America was determined significantly by pre-Pleistocene factors.

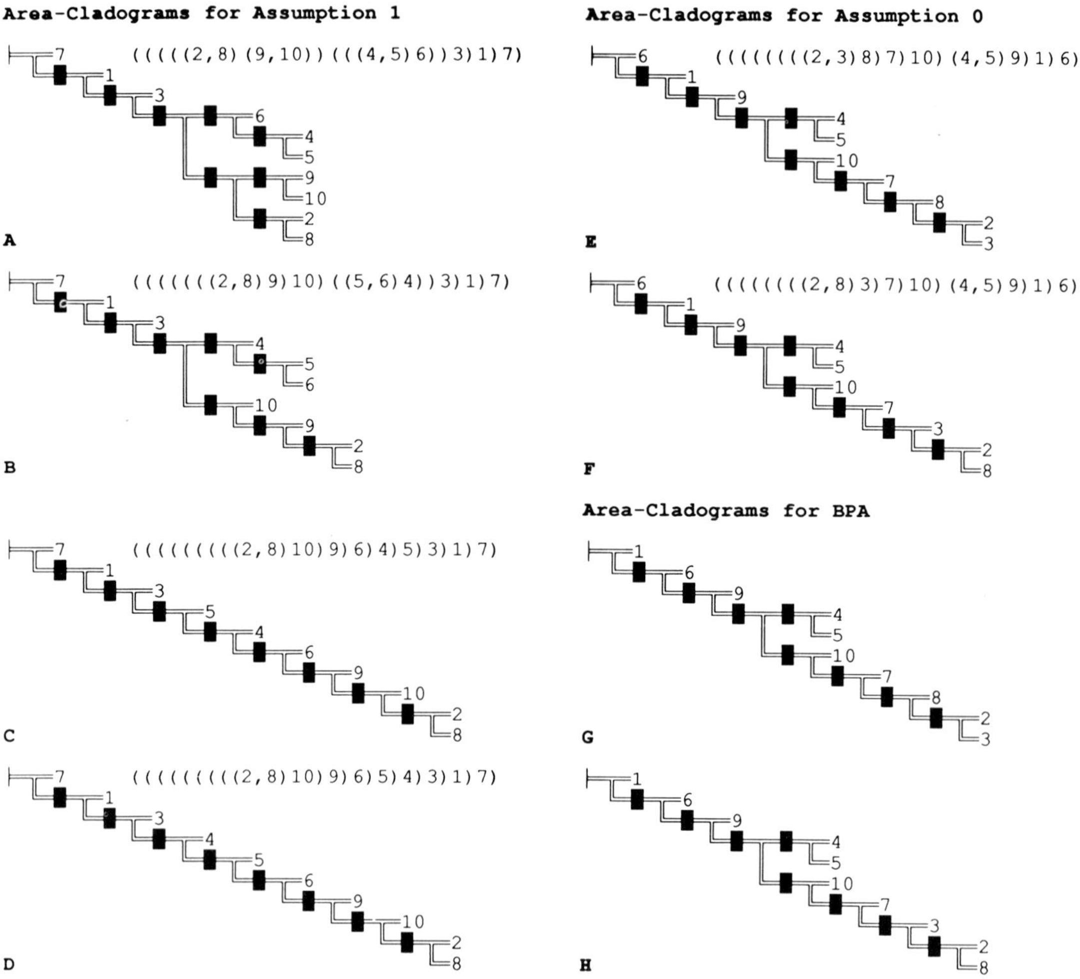

Fig. 10. *A–D*, area-cladograms suggested as optimal for *Heterandria* + *Xiphophorus* for assumption 1 (Zandee and Roos 1987:328, fig. 11a–d; parenthetical notation as published by Zandee and Roos). *A,E,F*, area-cladograms suggested as optimal for *Heterandria* + *Xiphophorus* for assumption 0 (Zandee and Roos 1987:321, fig. 8a–c; parenthetical notation as published by Zandee and Roos). *G,H*, area-cladograms from BPA of data matrix for *Heterandria* and *Xiphophorus* (columns 4, 8–14 in (*A*) and columns 6–7, 12–15 in (*B*) of Zandee and Roos 1987, table 16). Hennig86 finds 10 most parsimonious trees (length=22, CI=68, RI=74), which with successive weighting reduce to these two (length=163, CI=76, RI=82).

Area-cladograms may be compared by noting how many 3-area statements from species-cladograms (Table 5) are contained in each area cladogram. Comparisons among area-cladograms A2, A2+D46, RD, and D46 reveal that the A2+D46 cladogram performs as well or better than any other under all assumptions, that the A2 cladogram performs as well under A2 and A1 but not A0, and that the RD and D46 cladograms are not optimal under any assumption (Table 6).

Heterandria and Xiphophorus Revisited

Cladograms for *Heterandria* and *Xiphophorus*, first analysed by Rosen (1979), combine in a similar fashion (Fig. 9*A–C*). Conflicting elements, areas 3, 6 and 9, involve widespread *versus* endemic species (Fig. 9*D*). A2 resolves such conflict on behalf of endemic species (Fig. 9*E*), yielding the area-cladogram for A2 first published by Platnick (1981, fig. 6). Other workers have offered alternative area-cladograms. Zandee and Roos (1987) suggested four cladograms as optimal under A1 (Fig. 10*A–D*, their 11a–d) and three cladograms as optimal under A0 (Fig. 10*A,E,F*, their 8a–c). We found two cladograms to result from BPA (Fig. 10*G–H*), in agreement with preliminary findings of Platnick (1988).

Table 7. Three-area information contained in 10 nodes of two species-cladograms (Fig. 9*A,B*)

Following Zandee and Roos (1987:318), area 11 is omitted from analysis

Cladogram	Node	Structure of node	3-area statements under assumptions 2	1	0
A	1	6(1) (2–3:4–5:7:8:9:10)	6	8	8
	2	1:6(9) (2–3:4–5:7:8:10)	10	14	14
	3	1:6:9(4–5) (2–3:7:8:10)	12	30	38
	4	1:4–5:6:9(10) (2–3,7:8)	12	20	20
	5	1:4–5:6:9:10(7) (2–3:8)	10	18	18
	6	1:4–5:6:7:9:10(8) (2–3)	6	14	22
B	7	1(3) (2:4–6:8:9–10)	4	7	7
	8	1:3(4–6) (2:8:9–10)	6	24	42
	9	1:3:4–6(9–10) (2:8)	6	20	27
	10	1:3:4–6:9–10(2) (8)	4	7	7
Total	10		76	162	203

Table 8. Summary of information content of 10 area-cladograms, showing number and percentage of 3-area statements () contained in two species-cladograms (Fig. 9*A,B*) under assumptions 2, 1 and 0

The Platnick (1981, fig. 6) area-cladogram is Fig. 9*E*. Z&R8a–c area-cladograms are Fig. 10*E,A,F*. Z&R11a–d area cladograms are Fig. 10*A–D*. BPA1–2 area cladograms are Fig. 10*G,H*

Area-cladogram	Assumption 2 (76) No.	%	1 (162) No.	%	0 (203) No.	%
Platnick	76	100	131	81	149	73
Z&R 8a	70	92	136	84	160	79
Z&R 8b	42	55	93	57	128	63
Z&R 8c	71	93	137	85	160	79
Z&R 11a	42	55	93	57	128	63
Z&R 11b	42	55	93	57	125	62
Z&R 11c	44	58	95	59	111	55
Z&R 11d	44	58	95	59	111	55
BPA1	64	84	133	82	159	78
BPA2	65	86	134	83	159	78

The species-cladograms of *Heterandria* and *Xiphophorus* (Fig. 9*A,B*) comprise 10 nodes, which contain 76, 162 and 203 3-area statements under assumptions 2, 1 and 0 respectively (Table 7: Total).

Area-cladograms suggested for *Heterandria* and *Xiphophorus* may be compared by noting how many 3-area statements from species-cladograms (Table 7) are contained in each area cladogram (Table 8). Comparisons reveal that Platnick's A2 cladogram is best for A2, that one of Zandee and Roo's A0 cladograms (Fig. 10*F*, their 8c) is best for A1 and that two of their A0 cladograms (Fig. 10*E,F*, their 8a,c) are best for A0. Cladograms suggested by Zandee and Roos as optimal for A1 perform the worst of any evaluated for A1. Cladograms resolved by BPA are not optimal for any assumption, and for all assumptions perform less well than two cladograms (Fig. 10*E,F*, their 8a,c) suggested by Zandee and Roos as optimal for A0.

Conclusions

We do not enthuse about possibilities offered by assumptions 0 and 1, which depend upon widespread terminal taxa and their ability to predict historical nodes of area-cladograms. Their predictive ability may be studied empirically. Our results indicate poor performance.

Standard assumptions (2,1,0) have not been well understood by recent commentators. We offer as possibly helpful clarifications, the formal identity of these assumptions with suites of 3-area statements, noting that those of assumption 2 are encompassed by those of assumption 1, which are encompassed by those of assumption 0.

There is a tendency to believe that using 'all of the data' is best, with a bias in favour of assumption 0. This bias, evidently, has been uncritically misplaced. All of the data, simply put, do not apply to the problem. Widespread species might become relevant to future geographic differentiation if in fact the species were to differentiate geographically. But does a widespread species reveal past differentiation except in comparison with related taxa? What, after all, does an undifferentiated species tell us, by itself, about past differentiation?

To see evidence of geographic differentiation — relationship — where none is known transforms symplesiomorphy into 'evidence'. We are familiar with the transformation, which is the heart of phenetics and its enthusiasm to embrace all of the data whether they be relevant or not.

Acknowledgments

We are grateful to Richard Mayden for information and comment, to Peter Weston for critical review, to the University of Melbourne for a Visiting Scholar Grant (to G.N.) and to the School of Botany, University of Melbourne, for facilities.

References

Brooks, D. R. (1990). Parsimony analysis in historical biogeography and coevolution: methodological and theoretical update. *Systematic Zoology* **39,** 14–30.

Mayden, R. L. (1988). Vicariance biogeography, parsimony, and evolution in North American freshwater fishes. *Systematic Zoology* **37,** 329–55.

Nelson, G., and N. Platnick. (1981). 'Systematics and Biogeography: Cladistics and Vicariance.' (Columbia University Press: New York.)

Page, R. D. M. (1990). New Zealand and the new biogeography. *New Zealand Journal of Zoology* **16,** 471–83.

Platnick, N. I. (1981). Widespread taxa and biogeographic congruence. In 'Advances in Cladistics: Proceedings of the First Meeting of the Willi Hennig Society'. (Eds V. A. Funk and D. R. Brooks.) pp. 223–7. (New York Botanical Garden: Bronx.)

Platnick, N. I. (1988). Systematics, evolution, and biogeography: A Dutch treat. *Cladistics* **4,** 308–13.

Rosen, D. E. (1979). Fishes from the uplands and intermontane basins of Guatemala: revisionary studies and comparative geography. *Bulletin of the American Museum of Natural History* **162,** 267–376.

Wiley, E. O. (1988). Vicariance biogeography. *Annual Review of Ecology and Systematics* **19,** 513–42.

Zandee, M., and M. C. Roos. (1987). Component-compatibility in historical biogeography. *Cladistics* **3,** 305–32.

Manuscript received 2 October 1990, accepted 8 March 1991

Appendix 1A

Area	Cladograms and Distributions of Species						
OTG	0000	100000	0100000	000000000	0000	0000	000
A01	2010	310000	??00000	??0000000	?000	2100	301
A02	2010	?00000	??00000	4?1000000	2100	2100	301
A03	2010	310000	??00000	430110000	?000	2100	301
A04	2010	310000	??00000	4?0100000	?000	2100	?00
A05	2010	301000	1010000	320001000	2100	2010	301
A06	2010	200100	1010000	320001000	2100	2010	301
A07	2010	200100	1010000	320001000	2100	2010	301
A08	2010	200100	1010000	320001000	2100	2010	301
A09	2010	200100	??00000	??0000000	2100	2010	301
A10	2010	200100	1001000	310000100	?000	2010	301
A11	2010	200100	1001000	310000100	?000	2010	301
A12	2010	200100	1001000	??0000000	?000	2010	301
A13	?000	?00000	??00000	??0000000	?000	2010	?00
A14	2010	?00000	??00000	??0000000	?000	2010	?00
A15	2010	?00000	??00000	??0000000	?000	2010	?00
A16	?000	?00000	??00000	??0000000	?000	2010	?00
A17	?000	?00000	??00000	??0000000	?000	2010	?00
A18	?000	?00000	??00000	??0000000	?000	2010	?00
A19	?000	?00000	??00000	??0000000	2100	2010	301
A20	?000	?00000	1200100	??0000000	?000	2010	?00
A21	?000	?00000	1200100	??0000000	?000	2010	?00
A22	?000	?00000	1200100	??0000000	?000	2010	?00
A23	?000	?00000	1200100	??0000000	?000	?000	?00
A24	?000	?00000	1300010	??0000000	?000	?000	?00
A25	?000	?00000	1200100	??0000000	?000	?000	?00
A26	?000	?00000	1200100	??0000000	?000	?000	?00
A27	?000	?00000	1200100	??0000000	?000	2010	201
A28	?000	?00000	??00000	??0000000	?000	?000	201
A29	2100	?00000	??00000	??0000000	?000	1001	201
A30	2100	?00000	??00000	??0000000	?000	1001	201
A31	2100	?00000	1000001	200000010	?000	1001	201
A32	2100	000001	1000001	200000010	?000	?000	201
A33	2100	000001	1000001	??0000000	2010	?000	201
A34	1001	000010	?000000	100000001	1001	0000	110

Appendix 1B

Area	Cladograms and Distributions of Species						
OTG	0000	100000	0100000	000000000	0000	0000	000
A01	2010	310000	??00000	??0000000	?000	2100	301
A02	2010	?00000	??00000	4?1000000	2100	2100	301
A03	2010	310000	??00000	430110000	?000	2100	301
A04	2010	310000	??00000	4?0100000	?000	2100	?00
A05	2010	301000	1010000	320001000	2100	2010	301
A06	2010	200100	1010000	320001000	2100	2010	301
A09	2010	200100	??00000	??0000000	2100	2010	301
A10	2010	200100	1001000	310000100	?000	2010	301
A12	2010	200100	1001000	??0000000	?000	2010	301
A13	?000	?00000	??00000	??0000000	?000	2010	?00
A14	2010	?00000	??00000	??0000000	?000	2010	?00
A19	?000	?00000	??00000	??0000000	2100	2010	301
A20	?000	?00000	1200100	??0000000	?000	2010	?00
A23	?000	?00000	1200100	??0000000	?000	?000	?00
A24	?000	?00000	1300010	??0000000	?000	?000	?00
A27	?000	?00000	1200100	??0000000	?000	2010	201
A28	?000	?00000	??00000	??0000000	?000	?000	201
A29	2100	?00000	??00000	??0000000	?000	1001	201
A31	2100	?00000	1000001	200000010	?000	1001	201
A32	2100	000001	1000001	200000010	?000	?000	201
A33	2100	000001	1000001	??0000000	2010	?000	201
A34	1001	000010	?000000	100000001	1001	0000	110

Appendix 1C

Area	Cladograms and Distributions of Species						
A31	2100	?00000	1**1**00001	200000010	?000	1001	201
A32	2100	000001	1**1**00001	200000010	?000	?000	201
A33	2100	000001	1**1**00001	??0000000	2010	?000	201
A34	1001	000010	?**?**00000	100000001	1001	0000	110

Appendix 1. *A*, unconsolidated marix. *B*, consolidated matrix. *C*, corrections (areas ?1–34) for both unconsolidated and consolidated matrices. See Methods for explanation.

Appendix 2A

Area	Distributions of Species
OTG	000 00000 00000 0000000 000 000 000 000 000000 00000000
A01	010 10000 00000 0000000 000 100 010 000 000000 00000000
A02	010 00000 00000 1000000 100 100 010 000 000001 00000001
A03	010 10000 00000 0110000 000 100 010 000 000001 00000001
A04	010 10000 00000 0100000 000 100 000 000 000001 00000001
A05	010 01000 10000 0001000 100 010 010 001 000001 00000010
A06	010 00100 10000 0001000 100 010 010 001 000010 00000010
A07	010 00100 10000 0001000 100 010 010 001 000010 00000010
A08	010 00100 10000 0001000 100 010 010 001 000010 00000010
A09	010 00100 00000 0000000 100 010 010 001 000010 00000010
A10	010 00100 01000 0000100 000 010 010 001 000100 00000100
A11	010 00100 01000 0000100 000 010 010 001 000100 00000100
A12	010 00100 01000 0000000 000 010 010 001 000100 00000100
A13	000 00000 00000 0000000 000 010 000 010 000100 00000000
A14	010 00000 00000 0000000 000 010 000 010 000100 00000000
A15	010 00000 00000 0000000 000 010 000 010 000100 00000000
A16	000 00000 00000 0000000 000 010 000 010 000100 00000000
A17	000 00000 00000 0000000 000 010 000 010 000100 00000000
A18	000 00000 00000 0000000 000 010 000 010 000000 00000000
A19	000 00000 00000 0000000 100 010 010 010 000000 00000100
A20	000 00000 00100 0000000 000 010 000 010 001000 00001000
A21	000 00000 00100 0000000 000 010 000 010 001000 00001000
A22	000 00000 00100 0000000 000 010 000 010 001000 00001000
A23	000 00000 00100 0000000 000 000 000 010 001000 00001000
A24	000 00000 00010 0000000 000 000 000 010 000000 00001000
A25	000 00000 00100 0000000 000 000 000 010 001000 00001000
A26	000 00000 00100 0000000 000 000 000 010 001000 00001000
A27	000 00000 00100 0000000 000 010 001 010 001000 00001000
A28	000 00000 00000 0000000 000 000 001 010 000000 00001000
A29	100 00000 00000 0000000 000 001 001 010 010000 00010000
A30	100 00000 00000 0000000 000 001 001 010 100000 00100000
A31	100 00000 00001 0000010 000 001 001 010 100000 01000000
A32	100 00001 00001 0000010 000 000 001 010 100000 01000000
A33	100 00001 00001 0000000 010 000 001 010 100000 01000000
A34	001 00010 00000 0000001 001 000 100 100 000000 10000000

Appendix 2B

Area	Cladograms
OTG	010100000
A01	23?????23
A02	2???4?223
A03	23???3?23
A04	2???4???2?
A05	231032223
A06	221032223
A07	221032223
A08	221032223
A09	22?????223
A10	221031?23
A11	221031?23
A12	2210???23
A13	????????2?
A14	2???????2?
A15	2???????2?
A16	????????2?
A17	????????2?
A18	????????2?
A19	??????223
A20	??12????2?
A21	??12????2?
A22	??12????2?
A23	??12??????
A24	??13??????
A25	??12??????
A26	??12??????
A27	??12????22
A28	????????2
A29	2??????12
A30	2??????12
A31	2?1120?12
A32	?011?0???
A33	2?11???2?2
A34	10??10101

Appendix 2C

Area	Cladograms
OTG	010100000
A01	23?????23
A02	2???4?223
A03	23???3?23
A04	2???4???2?
A05	231032223
A06	221032223
A09	22?????223
A10	221031?23
A12	2210???23
A13	????????2?
A14	2???????2?
A19	??????223
A20	??12????2?
A23	??12??????
A24	??13??????
A27	??12????22
A28	????????2
A29	2??????12
A31	2?1120?12
A32	?011?0???
A33	2?11???2?2
A34	10??10101

Appendix 2. *A*, matrix of distributions of 46 species. *B*, unconsolidated matrix of species-cladograms (corrected). *C*, consolidated matrix of species-cladograms (corrected). See Methods for explanation.

3 September 1991 *Aust. Syst. Bot.*, **4**, 59–71

Biogeography without Area?

Ian M. Henderson

Department of Botany and Zoology, Massey University, Palmerston North, New Zealand.

Abstract

Recent methodological developments in historical biogeography generally treat biogeographic distribution as synonymous with occupancy of 'areas'. The aim of biogeographic analysis has been to determine the historical relationships of these areas using information from the distributions and phylogenetic relationships of animals and plants. While this may be of interest to geologists, it is of little interest to most biologists since it offers no direct insight into the historical processes that generate biogeographic patterns. Attempts to use relationships of areas (obtained from biogeographic patterns) to understand biogeographic processes can involve circularity.

Focusing on relationships of areas relegates biology to a minor consideration in biogeography. This has resulted in the unfortunate dichotomy between 'ecological' and 'historical' biogeography. A biogeography of areas also limits the information potentially available from biogeographic distributions. Choice of areas for biogeographic analysis can be problematical and analysis is sensitive to this choice.

Problems in identifying and analysing biogeographic areas are illustrated with trans-oceanic and local examples of austral biogeography.

Introduction

The concept of 'area' is central to most of the developments in biogeographic analysis of the last two decades. Data for biogeographic analysis are generally presented as taxa occupying areas (e.g. Fig. 1). Recent developments of the panbiogeographic method are apparently reliant on taxa inhabiting disjunct areas (Craw 1985; Page 1987), but it is within the vicariance cladistic school that the analysis of areas is explicitly stated as the prime object: 'The most elementary questions of historical biogeography concern areas of endemism and their interrelationships.' (Nelson and Platnick 1981: 56). However, for such an important concept, 'area of endemism' has received little critical analysis. In the major texts of vicariance cladistics operational definitions of areas of endemism are not provided. Instead, we are offered: 'fairly small areas that have a significant number of species that occur nowhere else'. (Nelson and Platnick 1981: 390); '. . . an area of endemism is delineated by more or less coincident distributions of taxa that occur nowhere else.' (Nelson and Platnick 1981: 468); '. . . areas of endemism rarely correspond exactly owing to chance and differing

1030–1887/91/010059$05.00

ecological requirements . . . if they are not exactly congruent but are largely congruent, then they may be considered equivalent for purposes of biogeographic analysis'. (Wiley 1981: 282); 'Area: A biogeographic region occupied by a monophyletic group of organisms or a species'. (Humphries and Parenti 1986: 93). Unfortunately, the terms 'fairly small', 'significant', 'more or less coincident', 'largely congruent', and what it means to occupy a region, are not further defined.

To those who work with hypothetical data in developing methods of analysis, this poses no problem. Biogeographic problems can be constructed which consist of endemic taxa distributed among disjunct areas of endemism. However, when dealing with real biogeographic data, problems do arise in the delineation of areas of endemism. Some of these problems can be illustrated even with a hypothetical example (Fig. 1).

This figure claims to show the distributions and phylogenetic relationships of a number of taxa endemic to three areas, New Zealand, Australia and Chile, but poses a number of questions about areas.

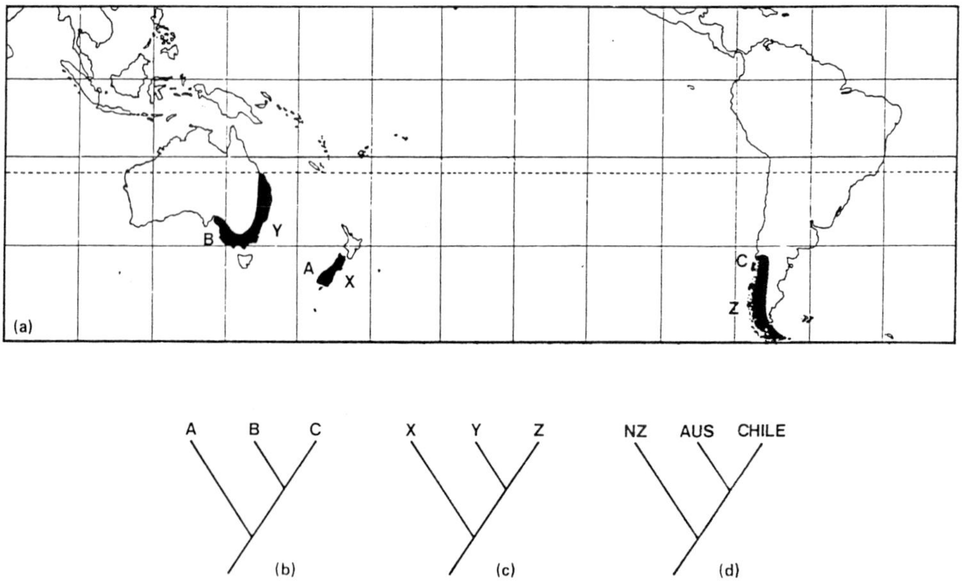

Fig. 1. (*a*) Hypothetical distributions of two groups of taxa; (*b*, *c*) cladograms for each group; (*d*) area cladogram common to both groups. From Humphries and Parenti (1986, fig. 2.8).

The areas are labelled with geo-political names. Is it appropriate to use this type of area? Are these areas 'natural' and if so, how is this established?

The map clearly shows that these taxa are *not* endemic to these countries, but to smaller parts of them. For example, the New Zealand taxa are present in South Island only. Does this matter?

Our knowledge of biogeographic distributions comes from records of taxa at locations. How do we convert these point records to a statement of area occupancy, and why should we do this if it loses information (as in this example)?

The answers to these questions depend on what we require from biogeographic analysis. If our aim is to determine the relationships of these three, predefined areas, then perhaps it is irrelevant that the taxa only occupy parts of the areas. Perhaps 'ecology' prevents them from occupying the full area. It may also be valid to ask questions about the relationships of geo-political areas, although the results might not be very informative if the areas are not natural biogeographic areas (Craw 1982).

On the other hand, our aim could be to understand the biogeography of the taxa (not the areas), and here it becomes important to know why taxa A and X are not present in North Island, or what replace taxa A and X in North Island.

These two types of questions in biogeography have been noted before: '. . . given a list of certain areas of endemism, are the interrelationships of the endemic taxa geographically nonrandom, and if so, what does the pattern (or patterns) tell us about the history of the areas and the taxa which inhabit them'? (Nelson and Platnick 1981).

While cladistic biogeographers have made some progress in answering the first part (the history of areas), the history of the taxa has received less attention, mostly amounting to the suggestion that the history of the taxa is the same as the history of the areas. This, no doubt, frustrates many biologists who look to biogeography for insights into the *biological* history of taxa, not simply the geographic constraints. If biogeographers are to maintain that the spatial aspect is not only an integral component of evolution but that spatial analysis is the key to understanding evolution (Darwin 1859: 1; and see Croizat 1964: 12; Craw and Page 1988), then we must provide something more than congruent cladograms of areas.

Historical biogeography today seems to have the emphasis on *graph* theory (cladograms and spanning trees), quite a lot on *geo*logy, but very little *bio*logy. Can we increase the biological content by considering questions about the biogeography of plants and animals, rather than areas? How might we proceed from a general area cladogram to an understanding of the history of taxa (e.g. speciation, extinction, mobilism)?

'A cladistic approach to biogeography presupposes that there are areas of endemism that are interrelated among themselves' (Nelson and Platnick 1981), and it is biological taxa (their phylogenetic relationships and occupancy of these areas) that provide the evidence for area relationships. Can this inference work both ways? After establishing a general area cladogram, can we use this to infer the history of these same taxa? This introduces circularity into the argument. If the history of areas is recorded only by vicariance of taxa, predictions of vicariance histories of these taxa are hardly novel. It would be even more problematical if the method for estimating a general area cladogram has been optimised for recovering patterns from a particular type of process, say vicariance (cf. assumptions 1 and 2 of Nelson and Platnick, 1981 and Endler 1982). One escape from this circularity is to obtain independent confirmation of area relationships from geology, but this places biogeography back in the very vulnerable position of relying on the revealed truths of geologists to 'explain' biology (see Craw and Weston 1984).

Identifying and Analysing Areas of Endemism

Problems in the 'biogeography of areas' are also evident at earlier stages in analysis. There are practical problems in actually producing area cladograms. This is evident from the repeated claims that 'the application of the cladistic biogeographic method is still in its infancy because cladistic analyses of plants and animals have only been carried out in a few groups' (Humphries and Parenti 1986: 33). This excuse is wearing a little thin after some 20 years of cladistics and 12 years of vicariance biogeography. The real problem is that to proceed with cladistic biogeography, 'at least two and preferably more phylogenetic analyses on unrelated taxa must be available for the same set of areas (Humphries and Parenti 1986: 33). Further, these phylogenetic analyses must be 'informative' with respect to these areas. This is the stumbling block: most taxa are not 'endemic' to 'areas of endemism' or are not informative (i.e. consist of two area statements, or incongruent patterns). The following examples from the biogeography of caddisflies (Trichoptera) in New Zealand will illustrate some of these problems (details of the data base in Henderson, unpublished data, 1985).

Uninformative Area Clades

After an extensive analysis of *Nothofagus* phylogeny and distribution, Humphries (1981) concurred with Patterson (1981) that '*Nothofagus* is uninformative on the interrelationships of southern hemisphere areas'. Humphries (1981) compared his partially resolved and 'uninformative' *Nothofagus* area cladogram to area cladograms based on a number of other southern hemisphere groups including the 'Hydrobiosine caddisflies (after Ross, 1956 and redrawn from Patterson 1981)'. Unfortunately, since Patterson was concerned with marsupial relationships, he 'reduced' his cladograms to ignore occurrences of taxa in areas where endemic marsupials do not occur (e.g. New Zealand). However, this reduction had deleted terms and components that might have been informative when compared to *Nothofagus* biogeography. Because of the particular target of this analysis (i.e the choice of areas), the biogeography of the Hydrobiosidae has been misrepresented.

In a recent revision of the Hydrobiosidae, Schmid (1989) offered a tentative and only partially resolved phylogeny of this family, but also provided the full character data set. I have recoded structurally related characters as unordered multistate and used Schmid's determination of character polarity to carry out a cladistic analysis using PAUP 3.0 (Swofford 1990). Extensive use of the random addition and global branch swapping options yielded two equally parsimonious trees (length = 642, Rescaled Consistency Index = 0·091). Fig. 2 shows the consensus of these two trees and the occupancy of austral land areas by the 48 genera.

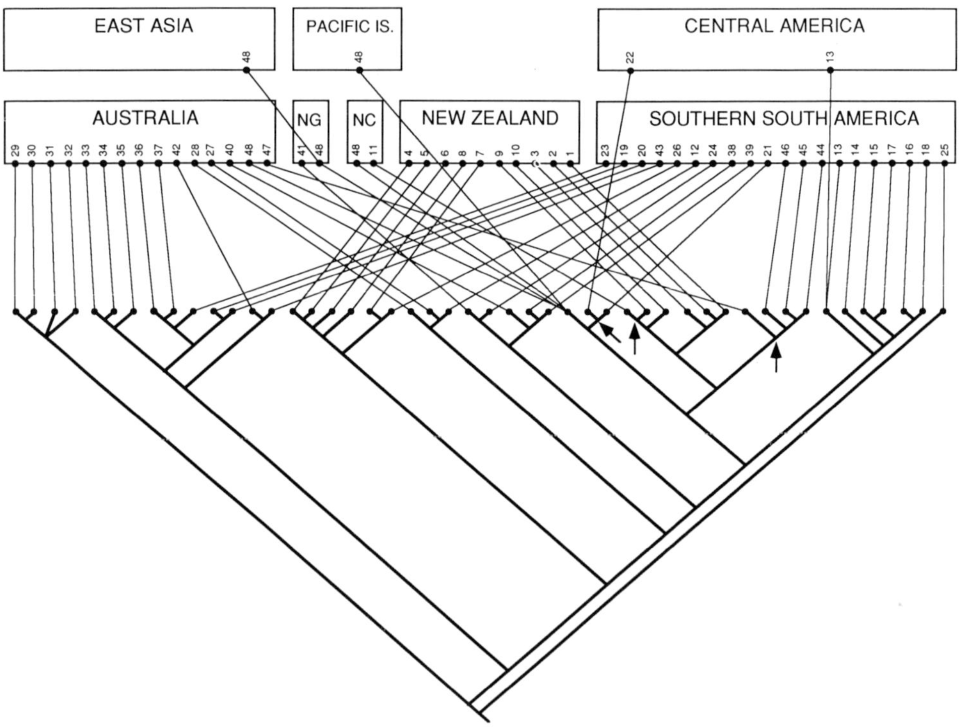

Fig. 2. Cladogram of the 48 genera of Hydrobiosidae obtained from parsimony analysis of the data of Schmid (1989). Taxon numbers refer to order in Schmid's plate 1. Informative nodes under assumption 1 (Page 1988) marked with arrows.

The Hydrobiosidae follow a familiar pattern of multiple transantartic (or transpacific) relationships (e.g. Brundin 1966; Edmunds 1975). What can this tell us about the history of the areas and taxa? Fig. 3 shows the analysis of this area cladogram under assumptions 0, 1 and 2 (Page 1988) expressed as consensus trees. Since there appears to be no consensus on which consensus method is appropriate for vicariance cladistic analysis, I have shown three commonly employed methods: strict, majority rule (Swofford 1990) and Nelson consensus (Page 1989). The results of this analysis could be described as uninformative or contradictory. If there is information on the relationships of austral land areas in the Hydrobiosidae, it is certainly a very weak signal. Note, in particular, that under assumption 2 (that preferred by Nelson and Platnick 1981) no component is consistently present in the 49 trees generated. The Nelson consensus under assumption 2 is more interesting, although the position of Australia as outgroup to New Zealand and South America finds little support in geological area cladograms (see for example, Humphries and Parenti 1986: 85).

If we now consider the biogeographic history of the Hydrobiosidae, a further problem is evident. Very few of the genera have actually contributed to the meagre results in Fig. 3. Page's (1988) algorithm for assumption 1 is useful here since the (potentially) informative nodes can easily be recognised by inspection of the matrix input to the maximum parsimony step. These are marked with arrows in Fig. 2. Only eight of the 48 genera of Hydrobiosidae have contributed to this analysis. Another way of appreciating this is to consider each of the clades encountered moving up the tree from the root. Most are one or two area clades. What have we learned of the history of these 40 taxa? Only that their history is compatible with the patterns in the left column of Fig. 3!

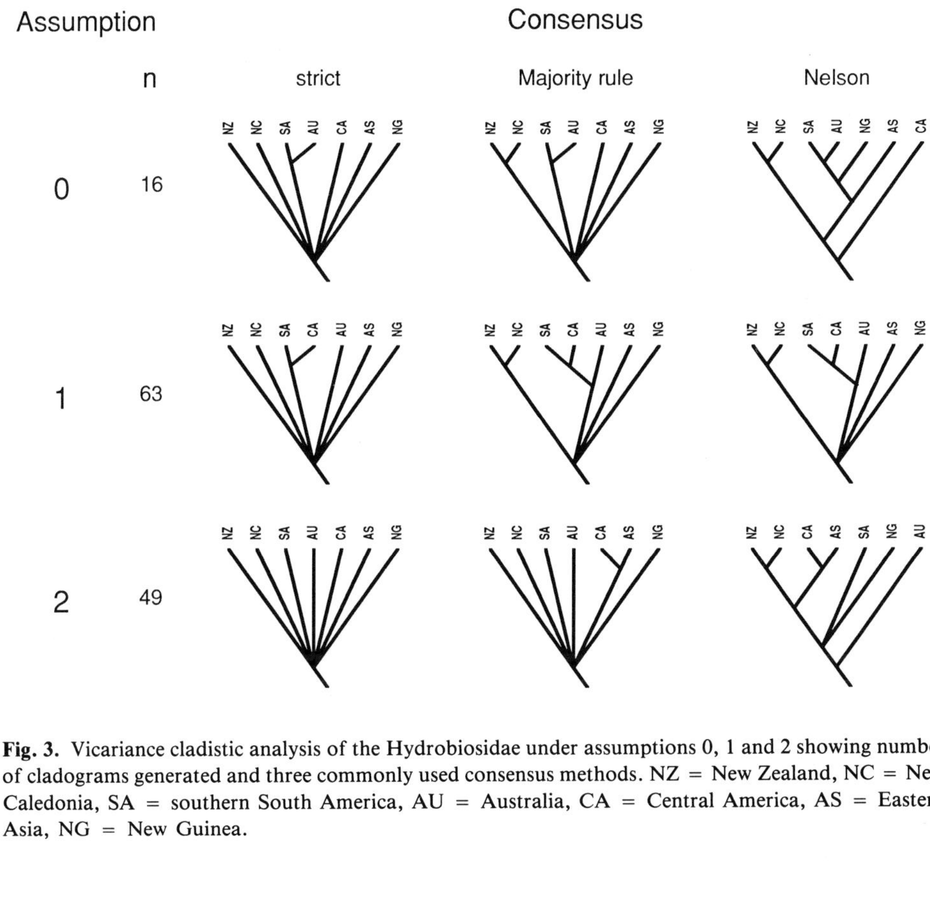

Fig. 3. Vicariance cladistic analysis of the Hydrobiosidae under assumptions 0, 1 and 2 showing number of cladograms generated and three commonly used consensus methods. NZ = New Zealand, NC = New Caledonia, SA = southern South America, AU = Australia, CA = Central America, AS = Eastern Asia, NG = New Guinea.

Non-endemic Taxa and Areas of Non-endemism

Every species is endemic to some area, but in searching for congruent patterns of area relationships, comparable (or homologous) areas for different taxa are required. There are claims that 'unproblematical areas of endemism' do exist, such as New Zealand, Hawaii, the Sahara desert (Nelson and Platnick 1981: 56). Determining the relationships of New Zealand, Hawaii and the Sahara may be an interesting problem in biogeography, but my present interest is in the biogeography of aquatic insects within New Zealand. There are patterns of distribution *within* some of these 'unproblematical' areas that require explanation. It has been suggested that there may be many local areas of endemism within New Zealand that might be analysed (Platnick and Nelson 1984).

The genus *Hydrobiosella* (Trichoptera: Philopotamidae) contains four species in New Zealand with distribution and phylogenetic relationships shown in Fig. 4 (Henderson 1983). These four species are not confined to disjunct areas of endemism, but neither are they widespread throughout New Zealand. Further, elements of this pattern will be familiar to anyone who has studied the distributions of plants or animals within New Zealand: maximum diversity in the north-west of the South Island; limits to distributions at the Manawatu gorge (Rogers 1989) and central South Island [the northern end of the 'beech gap' (Craw 1989*c*)]; and an apparent disregard for Cook Strait.

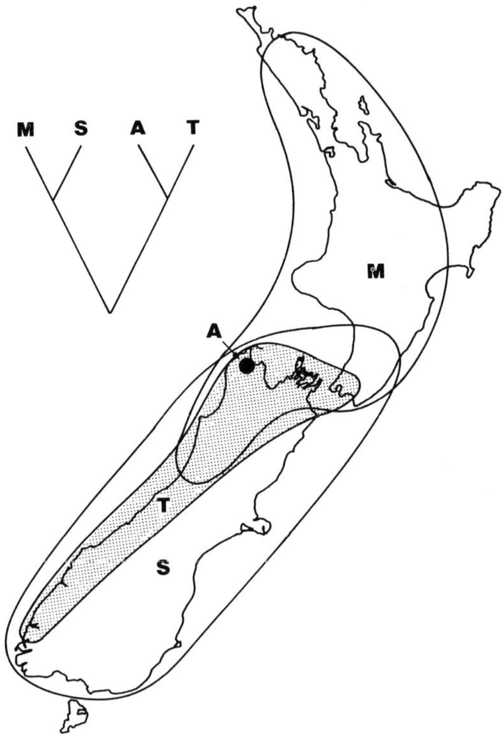

Fig. 4. Distribution and cladogram of *Hydrobiosella*. M = *H.* mixta, S = H. *stenocerca*, A = *H. aorere*, T = *H. tonela*. Northern limit of S is at Manawatu Gorge.

What areas of endemism might be appropriate here? One possible set consists of a northern, central and a southern area (Fig. 5*a*). Although no species of *Hydrobiosella* defines these areas, many other taxa do [examples from the Trichoptera (Henderson 1985) include: *Orthopsyche thomasi* (North), *Philorheithrus agilis* (Central), and *Psilochorema mataura* (South)]. However, when subjected to vicariance cladistic analysis, this pattern is uninformative because three of the four taxa are widespread. Under assumption 1 of Page (1988), no informative nodes are found, and under assumption 2, two solutions are found with no components in common (Fig. 5*a*).

Perhaps these are the wrong areas of endemism? The central area could be broken into two areas (Fig. 5*b*) and once again there are several species of Trichoptera to define these areas of endemism. Under assumption 1, *Hydrobiosella* are uninformative with respect to this scheme too, but under assumption 2, a single fully resolved solution is found (Fig. 5*b*). It may be that some distribution patterns are uninformative under either assumption, with *any* set of areas (artificial examples can be constructed). More importantly, data may only be informative when the right areas are chosen for analysis.

Unresolved Phylogenies

Another group of caddisflies, the genus *Philorheithrus*, consists of four species with distributions outlined in Fig. 6*b*. Unfortunately, phylogenetic analysis of morphological characters (Henderson 1985) yielded a completely unresolved cladogram (Fig. 6*a*). Rather than a sequential series of dichotomies defined by synapomorphies, I found combinations of characters such that no two species were more closely related than any others. Clearly, we cannot proceed with a cladistic biogeographic analysis with these data. How can we analyse the biogeographic history of *Philorheithrus* which has produced such an intriguing distribution pattern?

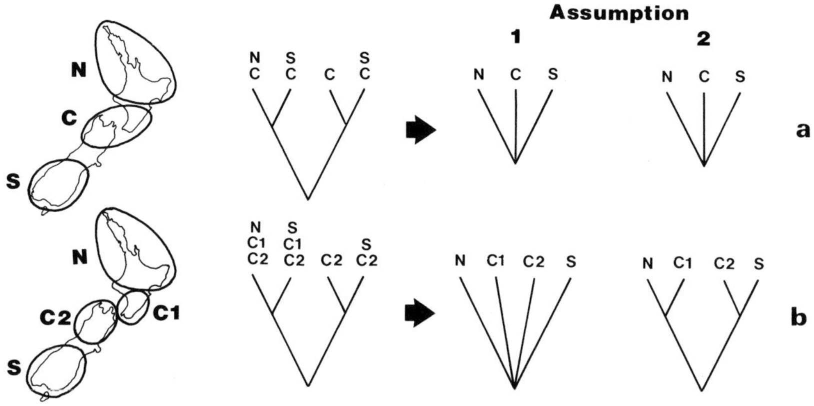

Fig. 5. (*a*) Northern (N), central (C) and southern (S) areas of endemism and area cladograms for *Hydrobiosella* under assumptions 1 and 2. (*b*) Same as (*a*) but central area treated as two areas of endemism.

Perhaps future analysis of this genus will provide a resolved phylogeny. Fig. 6*c* shows one possible outcome. It was obtained by deleting some characters with 'questionable' homologies. As with *Hydrobiosella*, it is not obvious what areas of endemism are appropriate, and similar results are achieved using the two schemes in Fig. 5. Under assumption 1, *Philorheithrus* is uninformative on both sets of areas, and under assumption 2, only the four area scheme is resolved. However, since all the evidence available now suggests that several other taxon cladograms are just as likely as Fig. 6*c*, the ambiguities multiply.

These examples have shown that some biogeographic data defy analysis by the methods of vicariance cladistics, and that some may be ambiguous. When taxa have overlapping distributions, as in these examples, the choice of areas is not obvious, although with a little thought an 'appropriate' set might be devised, i.e. a set of areas that would yield a resolved

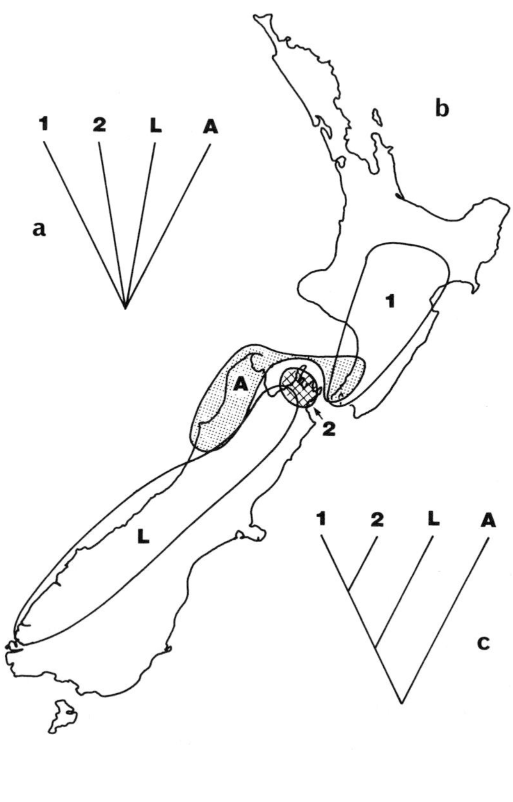

Fig. 6. (*a*) Cladogram of *Philorheithrus*; (*b*) distributions of species of *Philorheithrus*; (*c*) one possible resolved cladogram for *Philorheithrus*. L = *P. lacustris*, A = *P. agilis*, 1 and 2 = unpublished species (Henderson 1985).

area cladogram. However, vicariance cladistics takes a different approach. Once a set of areas is defined, appropriate taxa are found to test hypotheses of area relationships (e.g. Humphries 1981). Taxa such as *Hydrobiosella* and *Philorheithrus* would be dismissed as uninformative. If taxa are uninformative with respect to areas of endemism, what can we learn of the history of these taxa?

History v. Ecology

A practical problem in identifying areas of endemism is how to convert a series of point records to areas. This can be illustrated with a simple example in Fig. 7. This is the distribution of records for a species of insect in North Island, New Zealand. It could be argued that this species is found in three disjunct areas. What we know of the ecology of this species suggests otherwise. It is a caddisfly (*Hydrobiosis charadraea*) which inhabits swift mountain streams. These three areas constitute the three regions of suitable habitat so we could consider this distribution as occupying one area of endemism, southern North Island. Just as the terrestrial habitats between rivers are ignored, so we can ignore other regions with habitat unsuitable for this caddisfly. But why not draw the line further north; perhaps this species would exist there if the habitat were available? This might be important to a cladistic analysis if related species were found only to the north (in fact there is one, *Hydrobiosis budgei*). This biogeographic boundary through the centre of North Island is a standard feature of New Zealand biogeography (e.g. Gibbs 1989), but is it due to ecology, or to history? Is *H. charadraea* endemic to North Island or to southern North Island?

Fig. 7. Distribution of *Hydrobiosis charadraea* in North Island, New Zealand.

Gray (1989) has argued that the distinction between ecological and historical biogeography is a false opposition because ecology and history are reciprocally contingent. We need to know something of the ecology of a species in order to determine its area of endemism (history can be dependent on ecology), but ecology is also a result of history. Mountain dwelling caddisflies will only evolve where mountains are present. History, ecology and distribution are inextricably linked, but in order to fit distributions into a system of areas (such as New Zealand, Australia and Chile) we are forced to ignore ecology, or worse, assume that ecology will explain all the inconsistencies in distribution.

Areas in Panbiogeography

The use of minimal distance spanning trees for biogeographic analysis described by Page (1987) also requires distribution data in the form of occupancy of disjunct areas. Although this method may overcome the problems of unresolved (or unknown) phylogenetic relationships, overlapping distributions present problems. The mayfly genus *Zephlebia* (Towns 1983) exhibits an extreme form of nested, rather than vicariant distributions (Fig. 8). Applying the minimal spanning tree algorithm to these data would result in a single point somewhere within the distribution of species Zi, and no resolution of biogeographic pattern. A vicariance cladistic analysis of this example is similarly uninformative. Yet this genus has a very striking distribution pattern, elements of which are shared by many other taxa (see Climo 1989).

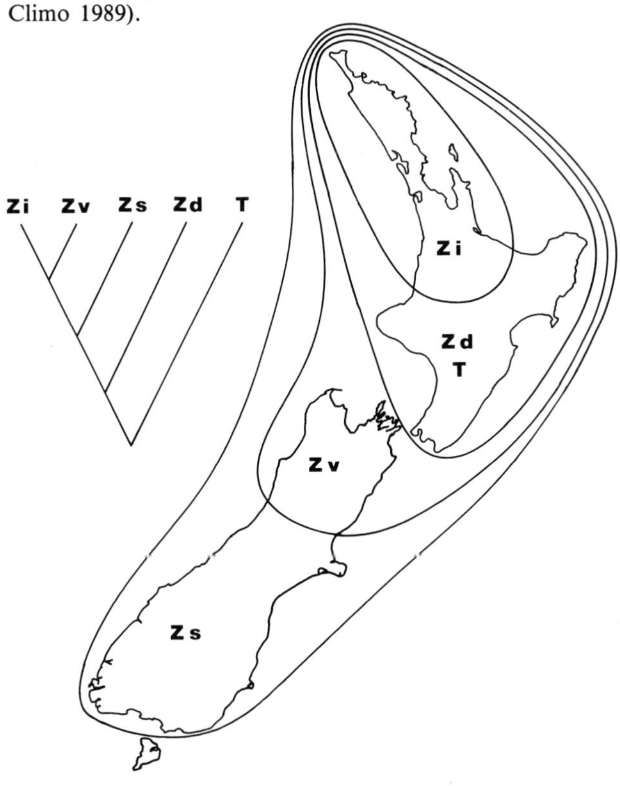

Fig. 8. Cladogram and distributions of *Zephlebia* (from Towns 1983). Zi = *Z. inconspicua*, Zv = *Z. versicolor*, Zs = *Z. spectabilis*, Zd = *Z. dentata*, T = *Zephlebia (Terama) borealis*.

In Croizat's own works there are relatively few maps showing areas; rather, he chose to encode distributions as lines (tracks). 'Geographic distribution . . . the records of occurrence at different *points* of the map' (Croizat 1964: 13, my emphasis). An example is reproduced in Fig. 9. In this case distributional ranges *are* indicated for some species (broken lines), but note that it is the tracks of dispersal (the process that gives rise to distributions) that indicate points of biogeographic significance (nodes). How did Croizat draw these tracks?

Many people have recognised the importance of Croizat's insight and have attempted to capture his method. One attempt is vicariance cladistics (Nelson and Platnick 1981) in which the track is associated with the area cladogram. Another is Page's (1987) spanning tree algorithm. Craw (1989*a*, 1989*b*) offers another approach with his clique compatibility method. Incompatible, but overlapping, tracks identify centres of incongruence, or nodes (Fig. 10). Note that while the tracks may link disjunct areas (in this case subantarctic islands), potentially any track, including those based on taxa with overlapping distributions or distributions *within an 'area'* (hatched line), can be compared.

The minimal spanning tree algorithm need not be restricted to analysing disjunct areas. Spanning trees can be used to identify species tracks and repetitive biogeographic features. The 160 species of New Zealand Trichoptera display a bewildering array of distribution patterns, the previous examples illustrate some of this variety. How can we simplify these patterns to a smaller set of key biogeographic characters? One approach is to search for distributions that are more or less the 'same' to define areas of endemism. However, the task of objectively determining the optimal position of the boundaries of these areas appears daunting due to the degree of overlap in species ranges. Further, any set of areas defined would be of limited use because most species distributions would be widespread and uninformative. The spanning tree method offers another approach.

A key concept in Croizat's panbiogeography (Croizat 1958, 1964) is the node as a centre of form-making. I have argued elsewhere (Henderson 1989) that nodes could be objectively identified by a high density of terminal vertices in minimal spanning trees of species tracks. This corresponds to Heads' (1989) first and fourth characteristics of biogeographic nodes. Because of the shape of New Zealand and the variable coverage of collecting locations, some points with high numbers of terminal vertices can be expected from random data (e.g. around the coastline generally). The analysis of Henderson (1989) can be extended to consider the distribution of nodes under a null hypothesis of random distributions. Using the same location points, numbers of species and records as in the New Zealand Trichoptera

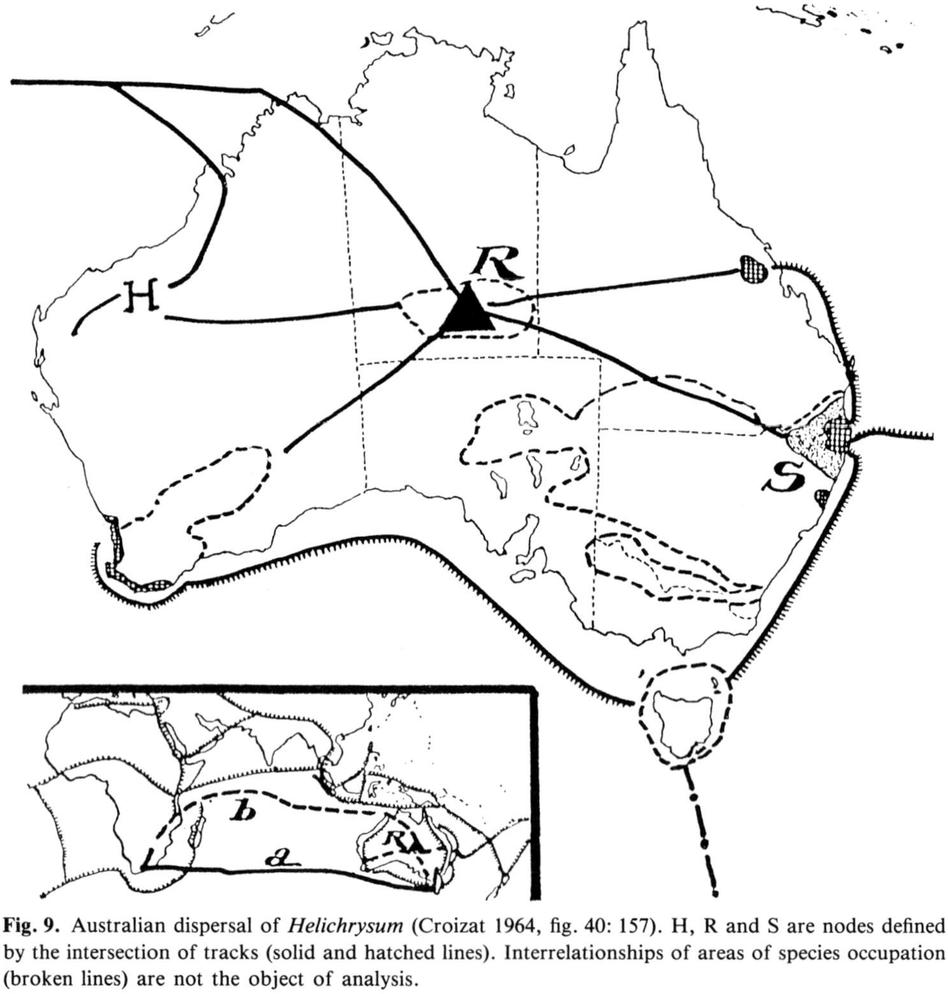

Fig. 9. Australian dispersal of *Helichrysum* (Croizat 1964, fig. 40: 157). H, R and S are nodes defined by the intersection of tracks (solid and hatched lines). Interrelationships of areas of species occupation (broken lines) are not the object of analysis.

data, a Monte Carlo approach was used to determine the statistical distribution of nodal intensity. The distribution was found to approximate the Poisson, although each point has a characteristic expected nodal intensity dependent on its geographic location. Analysis by minimal spanning trees of the real distributions identified many nodes more intense than the 5% tail of the random distribution. These are plotted in Fig. 11 as circles, with diameter proportional to the excess of terminal vertices over the mean from the random model. Points where *less* than the expected number of terminal vertices are found can also be identified. Heads (1989) argues that tracks of absence (see Craw 1989*c*) are also informative in panbiogeographic analysis and states that nodes may be marked by absences. I prefer to use the term antinode for such points (plotted as squares in Fig. 11).

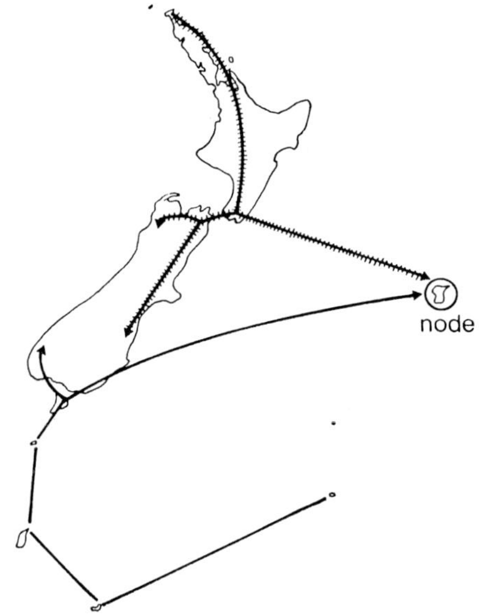

Fig. 10. Clique analysis of tracks. Overlap of incompatible tracks identifies a node at Chatham Islands. Tracks may link disjunct areas of endemism (solid lines), or summarise distribution patterns within areas (hatched lines). From Craw (1989*c*, fig. 13: 540).

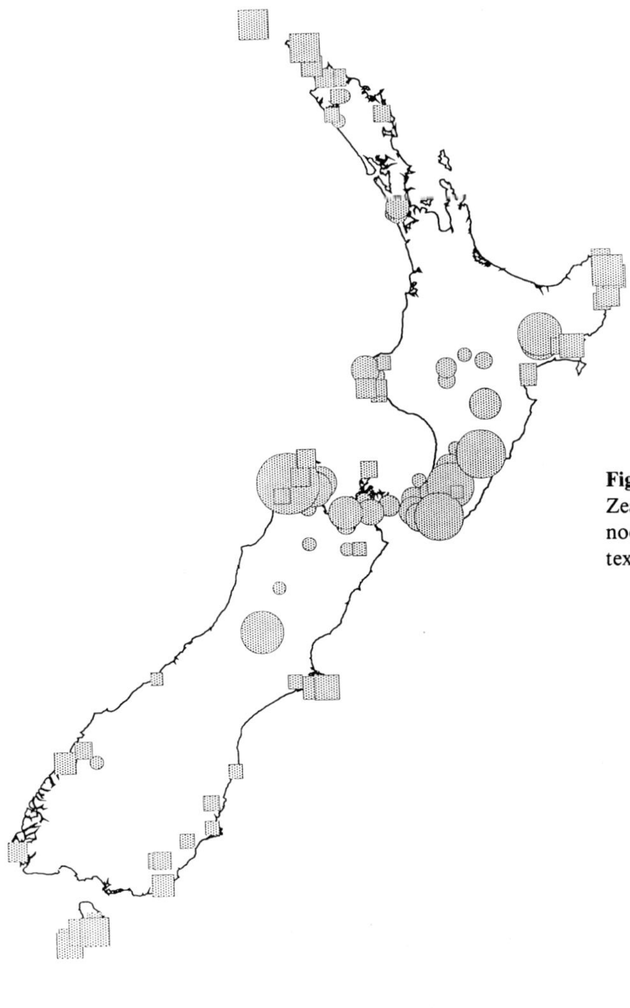

Fig. 11. Nodal analysis of New Zealand Trichoptera. Circles = nodes, squares = antinodes. See text for details.

Part of this pattern may be due to variations in sampling intensity but since much of the distribution data were obtained from workers based in Auckland (northern North Island) and Christchurch (central South Island), as well as myself based in Wellington (southern North Island), the main features of this pattern are unlikely to change with further sampling (see Henderson 1985). Most of the important nodes are clustered around Cook Strait but this is *not* predominantly due to distributions limited to one of the two main islands. Many of the nodes in the southern tip of North Island mark the *northern* limits of species and southern limits can be found in the northern parts of South Island. The north-western corner of South Island is identified as the most important node, a region noted for high diversity and endemicity (Kuschell 1975; Heads 1989). An unexpected result is the lack of nodes in the south of South Island and around Foveaux Strait. This region, like the far north of North Island, has few points of interest to Trichoptera biogeography, although the clusters of antinodes do stress the absence of many species from these points.

The achievements of the above analysis have been to establish that Trichoptera distributions are not random within New Zealand and to identify key *points* in this pattern. These nodes are biogeographic homologies and allow a classification of species tracks into groups associated with particular nodes. This classification, and hypotheses concerning the biological and geological factors responsible for the patterns, will not be attempted here. The exercise was carried out in order to demonstrate a viable alternative to the 'biogeography of areas' in which biological patterns are the focus rather than interrelationships of areas.

Conclusion

The last two decades have seen much activity in the analysis of areas of endemism whether by vicariance biogeography or panbiogeography. Much of this analysis has concerned trans-oceanic patterns. Relatively few workers have been fortunate to have data like Rosen's (1978) classic case of freshwater fish in isolated catchments, in which distributions can be resolved into largely disjunct areas of endemism *at the most local level* (Platnick and Nelson 1984). One view of cladistic methodology, whether for form or area analysis, is that the fundamental aim is to find the correct level for each piece of information in a hierarchy of clades. A character that is plesiomorphic and uninformative at one level of a phylogeny will be apomorphic at another level where it will define a monophyletic group. In vicariance cladistics, a distribution pattern will only be informative at the right hierarchical level. The distribution of the genus *Nothofagus* may be informative with respect to the relationships of southern hemisphere land masses, but the distribution of *Nothofagus truncata* cannot be. It might be informative at a more local level, if only we can find the right areas of endemism. However, if related taxa have overlapping distributions, this implies that, with respect to local areas of endemism, some taxa must be widespread and thus uninformative (Platnick and Nelson 1978). Only a tiny fraction of the potential information in biogeographic distributions may be suitable for answering questions about *any* areas, except perhaps the very smallest areas, those of zero area-points.

References

Brundin, L. (1966). Transantarctic relationships and their significance as evidenced by midges. *Kungliga Svenska Vetenskapsakademiens Handlinger* **4**(11), 1–472.

Climo, F. M. (1989). The panbiogeography of New Zealand as illuminated by the genus *Fectola* Iredale, 1915 and subfamily Rotadiscinae Pilsbury, 1927 (Mollusca: Pulmonata: Punctoidea: Charopidae). *New Zealand Journal of Zoology* **16**, 587–649.

Craw, R. C. (1982). Phylogenetics, areas, geology and the biogeography of Croizat: a radical view. *Systematic Zoology* **31**, 304–26.

Craw, R. C. (1985). Classic problems of southern hemisphere biogeography re-examined. *Sonderdruck aus. Z. f. zoo. Systematik u. Evolutionsforschung* **23**, 1–10.

Craw, R. C. (1989*a*). Quantitative panbiogeography: introduction to methods. *New Zealand Journal of Zoology* **16**, 485–94.

Craw, R. C. (1989*b*). Continuing the synthesis between panbiogeography, phylogenetic systematics and geology as illustrated by empirical studies on the biogeography of New Zealand and the Chatham Islands. *Systematic Zoology* **37**, 291–310.

Craw, R. C. (1989*c*). New Zealand biogeography: a panbiogeographic approach. *New Zealand Journal of Zoology* **16**, 527–47.

Craw, R. C., and Page, R. D. M. (1988). Panbiogeography: method and metaphor in the new biogeography. In 'Evolutionary Processes and Metaphors'. (Eds M.-W. Ho and S. W. Fox.) pp. 163–89. (John Wiley and Sons: New York.)

Craw, R. C., and Weston, P. (1984). Panbiogeography: a progressive research program. *Systematic Zoology* **33**, 1–13.

Croizat, L. (1958). 'Panbiogeography.' (Published by the author: Caracas.)

Croizat, L. (1964). 'Space, Time, Form: the Biological Synthesis.' (Published by the author: Caracas.)

Darwin, C. (1859). 'The Origin of Species.' 6th Edn. (John Murray: London.)

Edmunds, G. F. (1975). Phylogenetic biogeography of mayflies. *Annals of the Missouri Botanical Garden* **62**, 251–63.

Endler, J. A. (1982). Problems in distinguishing historical from ecological factors in biogeography. *American Zoologist* **22**, 441–52.

Gibbs, G. W. (1989). Local or global? Biogeography of some primitive Lepidoptera in New Zealand. *New Zealand Journal of Zoology* **16**, 689–98.

Gray, R. (1989). Oppositions in panbiogeography: can the conflicts between selection, constraint, ecology, and history be resolved. *New Zealand Journal of Zoology* **16**, 787–806.

Heads, M. (1989). Integrating earth and life sciences in New Zealand natural history: the parallel arcs model. *New Zealand Journal of Zoology* **16**, 549–85.

Henderson, I. M. (1983). A contribution to the systematics of New Zealand Philopotamidae (Trichoptera). *New Zealand Journal of Zoology* **10**, 163–76.,

Henderson, I. M. (1985). Systematic studies of New Zealand Trichoptera and critical analysis of systematic methods. Unpublished Ph.D. Thesis, Victoria University of Wellington.

Henderson, I. M. (1989). Quantitative panbiogeography: an investigation into concepts and methods. *New Zealand Journal of Zoology* **16**, 495–510.

Humphries, C. J. (1981). Biogeographic methods and the southern beeches (Fagaceae: *Nothofagus*). In 'Advances in Cladistics: Proceedings of the First Meeting of the Willi Hennig Society'. (Eds V. A. Funk and D. R. Brooks.) pp. 177–207. (New York Botanical Garden: New York.)

Humphries, C. J., and Parenti, L. R. (1986). 'Cladistic Biogeography.' (Clarendon Press: Oxford.)

Kuschell, G. (1975). 'Biogeography and Ecology in New Zealand.' (W. Junk: The Hague.)

Nelson, G., and Platnick, N. (1981). 'Systematics and Biogeography.' (Columbia University Press: New York.)

Page, R. D. M. (1987). Graphs and generalized tracks: quantifying Croizat's panbiogeography. *Systematic Zoology* **36**, 1–17.

Page, R. D. M. (1988). Quantitative cladistic biogeography: constructing and comparing area cladograms. *Systematic Zoology* **37**, 254–70.

Page, R. D. M. (1989). Comments on component-compatibility in historical biogeography. *Cladistics* **5**, 167–82.

Patterson, C. (1981). Methods in paleobiogeography. In 'Vicariance Biogeography: a Critique'. (Eds G. Nelson and D. Rosen.) pp. 446–89. (Columbia University Press: New York.)

Platnick, N. I., and Nelson, G. (1978). The perils of plesiomorphy: widespread taxa, dispersal, and phenetic biogeography. *Systematic Zoology* **27**, 474–7.

Platnick, N. I., and Nelson, G. (1984). Composite areas in biogeography. *Systematic Zoology* **33**, 328–35.

Rogers, G. M. (1989). The nature of the North Island floristic gap. *New Zealand Journal of Botany* **27**, 221–41.

Rosen, D. E. (1978). Vicariant patterns and historical explanations in biogeography. *Systematic Zoology* **27**, 159–88.

Schmid, F. (1989). Les Hydrobiosides (Trichoptera, Annulipalpia). *Bulletin de L'Institut Royal des Sciences Naturalles de Belgique* **59** (supplement, 1–154).

Swofford, D. L. (1990). 'PAUP: Phylogenetic Analysis using Parsimony.' Version 3.0. (Illinois Natural History Survey: Champaign.)

Towns, D. R. (1983). A revision of the genus *Zephlebia* (Ephemeroptera: Leptophlebiidae). *New Zealand Journal of Zoology* **10**, 1–52.

Wiley, E. O. (1981). 'Phylogenetics: the Theory and Practice of Phylogenetic Systematics.' (John Wiley and Sons: New York.)

Manuscript received 18 October 1990, accepted 19 March 1991

3 September 1991 *Aust. Syst. Bot.*, **4**, 73–89

Cladistic Biogeography of Afromontane Spiders

Charles E. Griswold

Department of Entomology, NHB 164, National Museum of Natural History, Smithsonian Institution, Washington, D.C. 20560, U.S.A.

Abstract

The application of cladistic data is seen as crucial to answering questions regarding the definition, mode of origin and age of historical biogeographic patterns. From the cladograms and distributional data for four groups of afromontane spiders [*Microstigmata* (Microstigmatidae), the *Moggridgea quercina* group (Migidae), and the subfamilies Vidoleini and Phyxelidini (Amaurobiidae)] a set of nine disjunct areas of endemism is defined for African and Malagasy forests. Taxon/area cladograms are combined through a parsimony method to produce a general area cladogram. General conclusions are: (1) Madagascar is related to eastern Africa and/or eastern South Africa rather than being the sister area to all of Africa; (2) eastern South Africa shows affinities with tropical Africa rather than with the nearby Cape region; (3) the Cape region of South Africa is highly distinctive; and (4) the area cladogram is hard to reconcile with historical scenarios involving primarily dispersal or Pleistocene vicariance events, and a Mesozoic origin for parts of the biogeographic pattern for afromontane spiders is possible.

Introduction

The forest biota of Africa is known for its heterogeneity. The existence of several biotic patterns has been suggested, including centres of endemism and gradients in species richness within unbroken blocks of forest, and the presence of distinct, disjunct montane and alpine forest islands surrounded by a sea of lowland forest. Since patterns of species richness and endemism in African forests were first recognised (Lönnberg 1929), their origin has frequently been attributed to the general effects of climatic changes on the distribution of forests. Acceptance of the reality of such patterns has not been unanimous, and the historical biogeography of African forests continues to be the focus of a lively debate. I perceive that three foci of controversy repeatedly emerge in biogeographic analyses of the African forest biota: (1) the reality of biogeographic patterns and definition of areas of richness and endemism; (2) the mode of origin of recognised patterns; and (3) the age of recognised patterns. Virtually all previous analyses relevant to these questions have been conducted under what I term the 'phenetic paradigm' in historical biogeography. This paper briefly summarises and critiques the phenetic paradigm and discusses why it is inadequate to address historical questions. It outlines the alternative 'cladistic paradigm' that offers greater hope of recognising and defining patterns and elucidating historical processes. Finally, as an example of the cladistic approach, it focuses on framing and answering these three questions with regard to the remarkable cool-temperate or 'afromontane' forest biota using cladistic data from several groups of spiders.

Paradigms in Historical Biogeography

Phenetic Paradigm

The majority of attempts to understand the historical biogeography of the forest biota of Africa have been undertaken under what I call the phenetic paradigm utilising lengthy

1030-1887/91/010073$05.00

tabulations of numbers of widespread species in common between the area of interest and other areas (e.g. Keay 1954; Gillet 1955; Morton 1972; Cole 1974; Hall and Medler 1975; Wild 1975; Diamond and Hamilton 1980; Struhsaker 1981; Stuart 1981; Rodgers *et al.* 1982). I refer to this approach as phenetic through analogy with phenetics in taxonomy because of the primacy given to estimates of overall similarity among the objects (taxa, areas) to be classified. Under this paradigm, data to assess the relationships among areas are the presence or absence of taxa, with widespread species considered informative and absence treated in the same way as presence. Cladistic interrelationships among taxa are usually not specified because they are not recognised as relevant. Historical relationships among areas are estimated by assuming that recency of common connection is directly proportional to the number of taxa in common.

While very useful in the description and comparison of biotas, the varied ages, ecological requirements and dispersal capacities of the individual species making up these biotas make phenetic assessments of the relationships among biotas difficult to interpret historically. Estimates of overall similarity do not accurately reflect historical relationships (Farris 1981), but are highly sensitive to the effects of local extinction or differentiation which may be unrelated to the common history of areas. The widespread taxa which provide the basic comparative data in phenetic biogeography may simply represent a failure to respond to historical events, and therefore not represent the common histories of areas. In this respect they are analogous to plesiomorphic characters in taxonomy (Nelson and Platnick 1978). The very data that may be most relevant to the common history of areas, the allopatric distributions of related lineages, are usually excluded from comparison. In short, the basic datum of phenetic biogeography, the species list, seems inadequate as a tool for historically meaningful comparisons of two or more areas.

Cladistic Paradigm

In the cladistic method (Nelson and Platnick 1981; Humphries and Parenti 1986), sister-group relationships among taxa endemic to different areas form the primary source of information about the relationships among areas. It assumes that (*a*) differentiation results primarily from the fragmentation of formerly continuous biotas through the action of barriers; (*b*) the increasing age of former connections among areas may be estimated by relationships of greater inclusivity on a cladogram of taxa endemic to those areas; and (*c*) by combining cladograms of taxa endemic to a set of areas a general cladogram for those areas may be proposed. The cladistic method has a number of advantages: (1) the independent assessment of the cladistic relationships of the taxa providing the raw data insures that we are dealing with natural groups; (2) it provides an unambiguous statement about the sequential, historical relationships among areas and, because of this, the dating of any part(s) allows relative dating of other parts; and (3) the 'signal' is less subject to being obscured by 'noise' (widespread taxa, absence of data).

Cladistics is just beginning to be applied to studies of African biogeography (e.g. Griswold 1985, 1990; Smiles 1985; Anderberg 1986; Louw 1986; Willassen and Cranston 1986; Endrödy-Younga 1988). Each of these studies deals with a single taxon, and the set of area relationships specified by each taxon has not yet been tested for generality. It is only when a set of area relationships is shown to be general (i.e. common to two or more taxa) that parsimony favours the hypothesis of common cause over unique, chance explanations, and historical biogeography becomes a predictive as well as descriptive science. As a model for the use of the cladistic paradigm in historical biogeography, I would like to focus on the remarkable 'afromontane' biota, and apply this method toward answering the three crucial questions regarding the definition, mode of origin, and age of this pattern.

Cladistic Biogeography and the Afromontane Biota

The Afromontane Biota

The existence of a distinct, homogeneous, and strikingly disjunct montane forest biota has been apparent to many workers. Among the earliest to document such as pattern were Carcasson (1964) in a study of African butterflies, and Moreau (1966) in his comprehensive study of the bird faunas of Africa. It was White (1978) who, in a study of forest trees, first used the term 'afromontane'. According to White (1983*b*) the afromontane region is an 'archipelago-like centre of endemism' occurring primarily above 2000 m altitude in the tropics but occurring at lower elevations where oceanic conditions ameliorate the climate (as in the West Usambara Mountains in Tanzania) and extending nearly to sea level in the temperate latitudes of South Africa. The afromontane archipelago is scattered from the mountains of Sierra Leone in the north-west to Somalia in the north-east, and from Sudan in the north to the Cape Province of South Africa in the South (White 1983*b*: fig. 14). Typical of the afromontane region are tree ferns (*Cyathea*), the plant family Oliniaceae, and the plant genera *Podocarpus* (Podocarpaceae), *Curtisia* (Cornaceae) and *Xymalos* (Trimeniaceae).

Defining the Afromontane Biota

Although many workers have recognised a distinct upland forest biota (Carcasson 1964; Moreau 1966; White 1978, 1983*b*; Stuart 1981; Feijen 1983; Dowsett 1986), others have suggested that there is no clear altitudinal boundary between lowland and montane biotas (Bigalke 1968; Diamond and Hamilton 1980; Stuart *et al.* 1987; Scharff 1991). Scharff (1990) investigated the vertical distribution of linyphiid spider species along an altitudinal gradient from 300 to 1850 m in the Uzungwa Mountains in south-western Tanzania. He found that whereas there were no true forest species in common between the highest (1700–1800 m) and lowest (300–500 m) parts of the forest, species from lower and higher altitudes met in a broad zone of intermixture such that it was not possible to draw any well defined borderline between a highland and lowland biota. Stuart *et al.* (1987) also found a very wide altitudinal zone of overlap between lowland and montane bird species. Delimitation of distinct biotas based upon the altitudinal distribution of single species has been unsuccessful. Perhaps the definition of an afromontane biota based on the phylogenetic relationships among its component taxa will be more successful than one based on their ecology. The question is not whether a species occurs above or below a certain altitude, but whether there is a pattern of sister-group disjunctions among isolated mountains.

Mode of Origin of the Afromontane Pattern

There is much debate regarding the relative contributions of general (i.e. vicariance) and unique (i.e. dispersal) historical factors to the creation of present-day patterns in the African forest biota. A profound influence has been attributed to the general effects of climatic changes which led to the dissection of forest into restricted patches during times of aridity and their subsequent coalescence during humid periods. Such a cyclical, climate-driven 'refugium' scenario is widely believed to have led to the complex biotic patterns currently observed within otherwise homogeneous forest (Carcasson 1964; Bigalke 1968; Grubb 1978, 1982; Diamond and Hamilton 1980; Struhsaker 1981; Stuart 1981; Rodgers *et al.* 1982; Mayr and O'Hara 1986). Whereas the refugium scenario as applied to Africa deals specifically with the origin during the Pleistocene of centres of endemism and gradients in species richness within lowland forest, conceptually it is equally applicable to the cyclical establishment and severance of contact between afromontane forest blocks. Acceptance of this model is not unanimous. Livingstone (1982) judged the palaeoecological evidence insufficient to support the concept of large forest blocks under conditions of Pleistocene aridity in East Africa, and both he and Hamilton (1976) suggested that dispersal played an important role in establishing the distributions of modern forest animals and plants.

A fresh approach that focuses on the phylogeny and distribution of component groups may illuminate this problem. Any causal explanation for biotic patterns invoking the primacy of unique events (i.e. dispersal) would not be expected to produce replicated sets of sister-group relationships among the areas making up the pattern. On the other hand, a vicariance scenario which involves sequential fission and coalescence of areas implies a specific set of hierarchic relationships among those areas which should be general. This could be tested through an area cladogram inferred from taxon cladograms. Indeed, the different phylogenetic patterns (or lack thereof) implied by different historical scenarios may be an effective and telling way to test their validity.

Age of the Patterns Within the Afromontane Biota

The presence of temperate montane forest in Africa can be documented well back into the Tertiary. Typically afromontane Monimiaceae are known from the late Cretaceous of South Africa (Mädel 1960), and fossil floras containing many taxa currently typical of afromontane forest are known from the South African Eocene (Phillips 1927; Adamson 1931). The palaeobotanical record from tropical Africa is not as complete, but Ethiopian Miocene *Podocarpus* fossils suggest widespread distribution of afromontane forest type in Africa at least by the mid-tertiary (Lemoigne and Beauchamp 1972). Forest containing many taxa which are today components of afromontane formations would seem to have been established in Africa at least at the beginning of the Tertiary (Axelrod and Raven 1978).

At least two bouts of aridification separated by several million years have been suggested, and both could have caused forest dissection leading to patterns within the forest biota. Past cycles of Pleistocene humidity and aridity (Van Zinderen Bakker and Coetzee 1988) in moist forest regions are well documented for southern, central and eastern Africa (Coetzee 1964; Morrison 1968; Kendall 1969; Van Zinderen Bakker and Coetzee 1972; Livingstone 1975; Van Zinderen Bakker 1976; Hamilton 1982; Scott 1982, 1984; Butzer 1984; Deacon *et al.* 1984; Thomas and Goudie 1984). Acceptance of the pre-eminent influence of these late Pleistocene climatic changes in shaping modern patterns in forest biota has been nearly unanimous (e.g. birds: Moreau 1966; Diamond and Hamilton 1980; Stuart 1981; Jensen and Stuart 1986; mammals: Booth 1958; Bigalke 1968; Grubb 1978, 1982; Struhsaker 1981; Rodgers *et al.* 1982; butterflies: Carcasson 1964), though some workers caution that late Quaternary events may not suffice to account for the complexity of patterns (Carcasson 1964, White 1983*a*; Griswold 1990; Scharff 1991).

Pleistocene climatic changes were not the first to affect Africa, and the possible effects of earlier events upon biotic distribution must be considered. Evidence of vegetational change during the Tertiary is sketchy, but studies from South Africa (Coetzee and Rogers 1982; Coetzee and Muller 1984) and tropical East Africa (Andrews and Van Couvering 1975; Bonnefille 1985; Retallack *et al.* 1990) suggest persistence of general forest cover into the mid-Tertiary that was then fragmented due to increasing aridity during the Miocene, presumably with profound evolutionary and biogeographical consequences (Laporte and Zihlman 1983), with the possibility of similar episodes occurring at intervals since the Cretaceous (Brain 1984).

Similar sets of evolutionary pressures (e.g. dissection of forest into isolated patches during periods of aridity) acting upon biotas under different geographic configurations would be expected to lead to different sets of biogeographic patterns. The Pleistocene desiccation took place in an Africa with modern topography, whereas the Miocene–Pliocene desiccation took place prior to or coincident with significant changes in African topography [e.g. uplift along the Natal Monocline in South Africa (King 1982); marine incursion during the Pliocene in the Uitenhage basin effectively isolating the eastern and western parts of the Cape Province of South Africa (King 1972); volcanism in East Africa which produced major lakes, e.g. Lake Victoria (Kendall 1969), and mountains, e.g. Kenya, Kilimanjaro and Meru (Grove 1983)]. Consideration of the age and mode of origin of biogeographic patterns must encompass the whole history of a biota, recognising at least the possibility of a palimpsest of ancient patterns interwoven among more recent. The application of phylogenetics to African forest biogeography will help to unravel the effects of general causes of differing ages and isolate and identify those patterns which could be attributable to stochastic (e.g. dispersal) causes.

Cladistic Biogeographic Analysis of Afromontane Spiders

Taxon Cladograms

Within this theoretical framework for the use of cladistics in historical biogeographic analysis, cladistic data from several groups of spiders can be used to define and investigate the age and mode of origin of the afromontane biota. The particular spider taxa were chosen for revision because each was largely or wholly restricted to afromontane forest and each was made up of many narrowly distributed, monophyletic groups. The spider taxa revised exhibit a variety of lifestyles, and include cursorial hunters (Microstigmatidae), fossorial ambushers (Migidae), and terrestrial and aerial web-builders (Amaurobiidae).

The family Microstigmatidae includes small to minute mygalomorphs occurring in leaf litter, beneath stones and logs, and within wet, rotting logs in moist forests. The African genus *Microstigmata* is currently known only from the eastern part of South Africa, where endemic, montane forest species in the Zuurberg Mountains, Amatola Mountains, Transkei-Natal Midlands, and Natal Drakensberg suggest origin through vicariance events in the late Miocene to early Pliocene (Griswold 1985). These area relationships are untested by phylogenies for other taxa. Based upon the occurrence of the sister-genus *Ministigmata* in South America (Raven and Platnick 1981), I suggested Cretaceous age for the genus and predicted the occurrence of this family in tropical Africa (Griswold 1985).

The Migidae are often referred to as 'tree trapdoor spiders' because many species build remarkably well camouflaged trapdoor nests on the trunks and buttresses of trees. Biogeographic data come from the revision of *Moggridgea* (Griswold 1987). The 31 *Moggridgea* species are known only from Africa and some surrounding islands, where they occur in virtually all habitats. Of interest to afromontane biogeography are six endemic, forest-dwelling species of the *quercina* group comprising *Moggridgea teresae* Griswold, *M. quercina* Simon, *M. intermedia* Hewitt, *M. microps* Hewitt, *M. verruculata* Griswold and *M. anactenidia* Griswold.

The amaurobiid subfamily Phyxelidinae includes taxa from south-east Asia, the eastern Mediterranean, Africa and Madagascar. The majority of species occur in moist forest, although some species may be found in drier habitats, including deserts. The cladogram (Griswold 1990) reveals two clades relevant to afromontane biogeography: the tribe Vidoleini, which occurs in southern and eastern Africa, and the *Lamaika* plus *Phyxelida* groups of the Phyxelidini, which occur in southern and eastern Africa and in Madagascar. It is justifiable to treat these clades as independent with regard to the information provided on the interrelationships among afromontane areas of endemism because these clades are separated on the subfamilial cladogram (Griswold 1990: fig. 13) by the genus *Namaguarachne* which occurs primarily in arid regions and is thus not relevant to analysing afromontane patterns.

Areas of Endemism

In defining areas of endemism in African and Malagasy forests the principal criteria were the smallest coincident ranges of two species and the geographic extent of forest islands. Each species was apparently restricted to forest habitats and there was detailed sister-group information relevant to the species. Distributions of all species of the four taxa were mapped. In some cases these were a point (i.e. a species might be known only from the type locality), or they might include several to many points. A species' range was considered to encompass at least all space between the points. Coincidence in ranges was tabulated. If the ranges of two species (not sister taxa) overlapped they were considered coincident, even if one was only a point. The smallest coincident ranges were taken to define an area of endemism. The areas of forest considered here are island-like, surrounded by areas of grassland or bushveld. If a species' range encompassed only part of a forest island, the whole island was defined as an area of endemism. The distributional ranges of monophyletic taxa took precedence over topographic boundaries. If a species' range encompassed more than one forest island, the area of endemism was defined to include all forest islands within the species' range. A species was considered widespread if its range included two or more of the areas defined by the smallest coincident ranges of two other species. The areas of endemism, as circumscribed on the map (Fig. 1), may contain both forest and non-forest vegetation types. As forest typically occurs as small, discrete patches, depiction of only these patches on a map covering most of Africa would be impossible. The biogeographic discussion here is relevant only to the forest habitats within the circumscribed areas of endemism.

Following are the nine areas of endemism, along with taxa that define and characterise them.

(1) Table Mountain: this includes the forests on the slopes of Table Mountain near Cape Town. While small in area, it is noted for a high rate of endemism. Defining taxa are *Moggridgea teresae* Griswold and *Malaika longipes* (Purcell).

(2) Knysna Forest: includes large areas of forest stretching from the South African coast between Mossel Bay and Cape Saint Francis onto the southern slopes of the Outeniqua and Tsitsikamma Mountains. Defining taxa are *Moggridgea intermedia* Hewitt, *Lamaika distincta* Griswold and *Matundua silvatica* (Purcell).

Areas 1 and 2 are in the 'Cape Fold Mountains' of South Africa. The oldest of these mountains were in existence in the Permian (Truswell 1977; Halbich *et al.* 1983), and are perhaps the oldest mountains in Africa. To the east of the Knysna Forest (area 2) is the region of the Algoa Bay and the Uitenhage Basin on the southern coast of South Africa. This is a region of aridity that is thought to be a barrier to the spread of forest organisms. Its significance as an interval in the distributional patterns of forest arthropods has been discussed by Griswold (1985) and Stuckenberg (1962). To the east and north of the Uitenhage Basin the topography of eastern South Africa consists of a series of step-like, north-east to south-west trending escarpments (King 1982). Moist winds move onshore from the Indian Ocean and drop their moisture as rain or mist at the bases of these escarpments, encouraging the development of forests. Three separate escarpment forest formations may be recognised at different elevations, each isolated from other forests at lower or higher elevations by intervening grassland or bushveld.

(3) Natal–Zululand Coast: comprises coastal forests stretching up the coast from the vicinity of Port St Johns in Transkei to Zululand, and extending into the interior of Zululand (montane forest at Nkandla) as well. The northern boundary of this area is not well defined for spiders

due to the scarcity of collections from Mozambique. Defining taxa are *Microstigmata zuluense* (Lawrence) and *Moggridgea microps* Hewitt. *Themacrys irrorata* Simon, *Vidole lyra* Griswold and *Xevioso amica* Griswold also characterise this area.

(4) Transkei–Natal Midlands: comprises the area of mid-elevation forests that typically occur along misty escarpments at around 800 m elevation in the midlands of the Transkei and Natal from at least Umtata in the south to at least Kranskop on the Tugela River in the north. Defining taxa are *Microstigmata lawrencei* Griswold and *Themacrys cavernicola* (Lawrence). *Vidole helicigyna* Griswold also characterises this area.

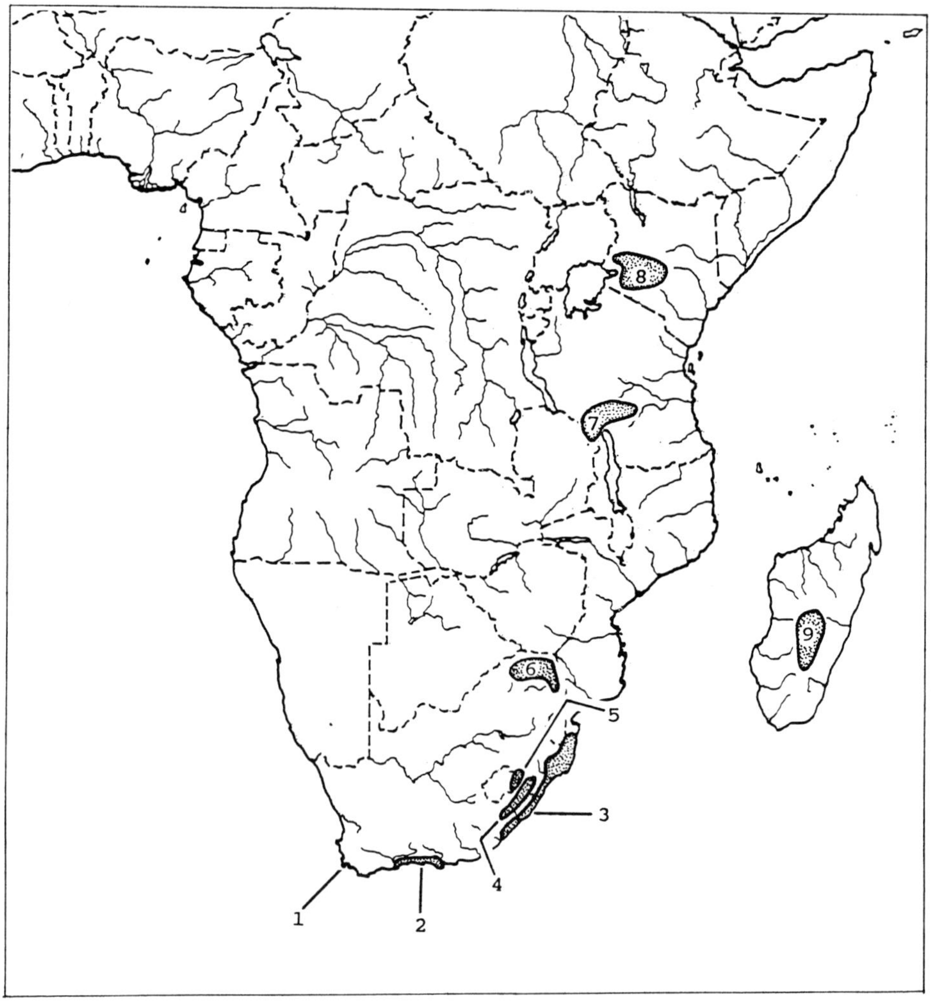

Fig. 1. Afromontane areas of endemism defined by distribution of monophyletic spider taxa: *1*, Table Mountain; *2*, Knysna Forest; *3*, Natal–Zululand Coast; *4*, Transkei–Natal Midlands; *5*, Natal Drakensberg; *6*, Transvaal Drakensberg; *7*, Eastern Arc Mountains; *8*, East African Volcanoes; *9*, Central Madagascar.

(5) Natal Drakensberg: comprises forests along the northern portion of the Drakensberg escarpment in Natal at elevations of 1000–2000 m. Defining taxa are *Microstigmata ukhahlamba* Griswold and *Themacrys ukhahlamba* Griswold.

(6) Transvaal Drakensberg: the escarpment of the Drakensberg mountains reappears in the Transvaal, where cool temperate forests occur at about 1500 m elevation along the moist eastern side. The Soutpansberg mountains in the northern Transvaal, which contain patches of montane forest, are also included in this area. The defining taxon is *Xevioso kulufa* Griswold. *Phyxelida makapanensis* Simon also characterises this area.

North of areas 1–6 lies the basin of the Limpopo River. This is another zone of aridity, comprising an ancient trough bordered on the south by Jurassic and Cretaceous faults. It has been considered a barrier to the dispersal of forest organisms that may have operated well back into the Tertiary (Stuckenberg 1962).

Two areas are defined in tropical East Africa.

(7) Eastern Arc Mountains: a series of crystalline mountains extends from south-eastern Kenya to northern Malawi. These mountains are formed along mesozoic faults (Sowerbutts 1972) and are frequently cited as being of mesozoic age. They may be the oldest mountains in East Africa. The 'eastern arc' mountains are noted for high rates of endemism and extraordinary richness (Schiotz 1981; Rodgers and Homewood 1982; Lovett 1988). Scharff (1991) found the proportion of endemism for linyphiid spiders to be much greater in the eastern arc mountains than in the volcanic mountains of Kenya and Tanzania. These mountains are thus of great biogeographic interest. For this reason I have allowed topography and geologic history to take precedence in this case over the distributions of spider taxa in defining an area of endemism. I define area 7 to include at least the uplands of the Nyika Plateau west of Lake Malawi (locality of *Phyxelida eurygyna* Griswold) and the Uzungwa Mountains north-east of Lake Malawi (locality of *Phyxelida kipia* Griswold). No taxon range defines this area. My use of the term 'eastern arc' differs from that of Lovett (1988) in that it includes Nyika in Malawi as well as areas in Tanzania.

(8) East African Volcanoes: the volcanic mountains of East Africa are far younger than the nearby eastern arc mountains, dating from the early Miocene to late Pliocene (Grove 1983). I define this area by the range of *Phyxelida pingoana* Griswold that occurs on Mount Elgon and Mount Kenya. *Phyxelida crassibursa* Griswold also characterises this area.

(9) Central Madagascar: comprises the highlands of Fianarantsoa and Tananarive Districts in the central part of Madagascar. The defining taxon is *Ambohima sublima* Griswold, and *Phyxelida fanivelona* Griswold and *P. malagasyana* Griswold also characterise this area.

Analysis

Taxon cladograms were converted to area cladograms by replacing the terminal taxon with the number for the area of endemism in which it occurred. These taxon/area cladograms were then combined using a modification of Kluge's (1988) implementation of a parsimony method for comparing area cladograms. The application of parsimony to historical biogeography treats areas as analogous to taxa and taxon cladograms as analogous to characters. The method searches for the most parsimonious hypothesis of ancestral distributions on a single cladogram for the areas of endemism assuming that the ranges of descendent sister-groups, when added together, provide the best estimate of the range of their hypothetical common ancestor. These hypothetical ancestral ranges form the characters ('syntaxa' of Kluge 1988), and are coded in an additive binary manner from the most distal to the most basal nodes on each taxon cladogram. The analysis is rooted by including a hypothetical 'outgroup' area with 0 values for all characters. This is analogous to assigning 0 values to the outgroup node in the phylogenetic analysis of taxa, and in a biogeographic analysis signifies the assumption that the set of areas are monophyletic with regard to the taxon data. The length of the resulting cladogram is measured in steps, and steps may be interpreted as vicariance events, failure to differentiate, dispersal, or extinctions.

Kluge's implementation of parsimony for historical biogeographic analysis recognises the ambiguity of area information provided by widespread taxa and deletes this potentially noisy information from consideration. Taxon cladograms were converted to reduced taxon/area cladograms by eliminating all widespread taxa and all taxa with redundant distributions if these taxa simply duplicated the hypothetical ancestral distribution specified by relatively more distal taxa. The reduced taxon/area cladograms for *Microstigmata*, the *Moggridgea quercina* group and the Phyxelidinae are depicted in Fig. 2, and syntaxa are coded in Fig. 3. Syntaxa were extracted from particular taxon cladograms: a syntaxon value of '1' was assigned for an area if that area was represented for a taxon distal to that syntaxon node, '0' was assigned if that area was represented somewhere on a given taxon cladogram but not distal to that syntaxon node, and '?' was assigned if that area was not represented on the taxon cladogram from which that syntaxon was extracted.

Results

Analysis for the nine replicated, unambiguously interrelated areas plus the hypothetical outgroup area by 17 syntaxa was done with Hennig86 (Farris 1988) using the 'ie*' routine to

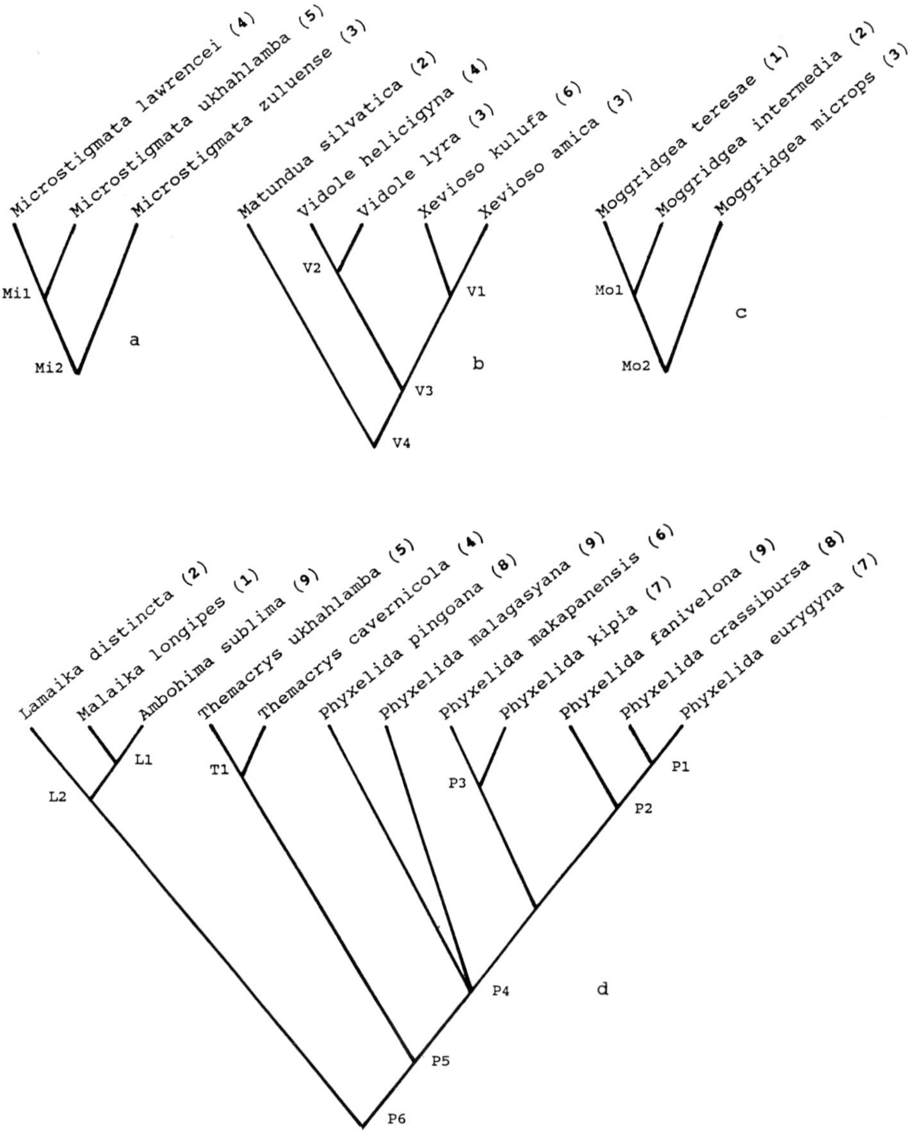

Fig. 2. Reduced taxon/area cladograms for (*a*) *Microstigmata*, (*b*), the Vidoleini, (*c*) the *Moggridgea quercina* group, and (*d*) the Phyxelidini. Numbers following terminal taxa represent areas of endemism; syntaxa are designated at nodes.

	Mo1	Mo2	Mi1	Mi2	V1	V2	V3	V4	L1	L2	T1	P1	P2	P3	P4	P5	P6
1 Table Mountain	1	1	?	?	?	?	?	?	1	1	0	0	0	0	0	0	1
2 Knysna Forest	1	1	?	?	0	0	0	1	0	1	0	0	0	0	0	0	1
3 Natal-Zululand Coast	0	1	0	1	1	1	1	1	?	?	?	?	?	?	?	?	?
4 Mid Transkei-Natal	?	?	1	1	0	1	1	1	0	0	1	0	0	0	0	1	1
5 Natal Drakensberg	?	?	1	1	?	?	?	?	0	0	1	0	0	0	0	1	1
6 Transvaal Drakensberg	?	?	?	?	1	0	1	1	0	0	0	0	0	1	1	1	1
7 Eastern Arc Mountains	?	?	?	?	?	?	?	?	0	0	0	1	1	1	1	1	1
8 East African Volcanoes	?	?	?	?	?	?	?	?	0	0	0	1	1	0	1	1	1
9 Central Madagascar	?	?	?	?	?	?	?	?	1	1	0	0	1	0	1	1	1

Fig. 3. Area by Syntaxon matrix for afromontane spiders.

give eight equally parsimonious trees of 21 steps with a consistency index of 0.80 and a retention index of 0.77. These eight trees are identical except for the placement of area 3 (Natal–Zululand Coast), which could be placed with equal parsimony at any of eight positions (Fig. 4). For graphical simplicity I present a consensus of these trees with area 3 (Natal–Zululand Coast) placed at its lowest unambiguous node (Fig. 5). Syntaxa were optimised on this tree, requiring 22 steps with a consistency index of 0.77 and a retention index of 0.72. Ambiguous optimisations were resolved to favour delayed transformations. Although this consensus tree is inferior to each of the eight trees upon which it is based, I do not believe that the precision gained by presenting all eight 21-step trees with syntaxon optimisations would be worth the space and effort. All syntaxon optimisations shown on the consensus tree (Fig. 5) were found on at least one of the 21-step trees, and the optimisations shown for 11 syntaxa (Mo1, Mo2, Mi1, V3, V4, L1, L2, T1, P3, P5, P6) were found on all eight trees.

This area cladogram (Fig. 5) has several features of interest. The basal node of the area cladogram is between the Cape region of South Africa [areas 1 (Table Mountain) and 2 (Knysna Forest)] and the remainder of Africa and Madagascar (areas 3–9). The distinctness of the Cape biota is strongly emphasised. A region south of the arid Limpopo River Basin in northern South Africa [area 6 (Transvaal Drakensberg)] shows affinities with tropical Africa [area 7 (Eastern Arc Mountains) plus area 8 (East African Volcanoes)] and Madagascar (area 9). Eastern South Africa [areas 3 (Natal–Zululand Coast), 4 (Transkei–Natal Midlands) and

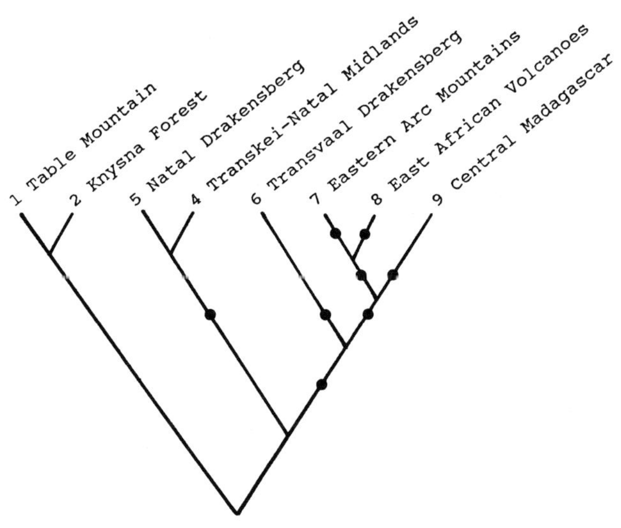

Fig. 4. Area cladogram calculated by Hennig86 showing possible positions of area 3 (Natal–Zululand Coast).

5 (Natal Drakensberg)] shows affinities with tropical Africa and Madagascar rather than with the nearby Cape region. Two upland escarpment areas 4 (Transkei–Natal Midlands) and 5 (Natal Drakensberg) in eastern South Africa are sister-areas. Perhaps the most striking aspect of this area cladogram is that Central Madagascar (area 9) is related to eastern Africa [area 7 (Eastern Arc Mountains) plus area 8 (East African Volcanoes)] rather than being the sister-area to all of Africa.

The afromontane regions of Africa do not represent a distinct biogeographic region, but rather some parts of the afromontane regions of Africa are more closely related to areas in another continent (Madagascar) than to other afromontane regions of Africa. A close relationship between East Africa and Madagascar is not confined to spiders, as a similar pattern has been suggested for several groups of plants on the basis of taxa in common (Pocs 1975; Wild 1975; Leroy 1978; Lovett 1988), and a few studies (Smiles 1985; Puff 1988) suggest explicit sister-group relationships between East African and Madagascan taxa.

The fit of the hypothetical ancestral spider distributions to this area cladogram is not perfect, and homoplasy is required for 5 of the 17 syntaxa. Kluge (1988) suggests that homoplasious syntaxa may reflect several possibilities: the terminal taxon is not a historical entity, or its sister-taxon relationships may be erroneously construed; a defined area of endemism may be a composite of areas with two or more independent histories; the geographic range of the terminal taxon may be erroneously defined; there may be unsampled taxa (uncollected taxa, or taxa that have become extinct) that are/were distributed in those areas between the apparently homoplasious occurrences of the syntaxon on the area cladogram; or dispersal may have occurred. Two instances of parallelism in syntaxa involve area 3 (Natal–Zululand Coast), suggesting that area 3 may be misdefined and have a composite history. The homoplasy for P3 between areas 6 and 7 may indicate that it was incorrect to lump Uzungwa and Nyika as area 7; or, if the area cladogram is true, dispersals may be postulated.

Homoplasies in some syntaxa (Fig. 5) may represent dispersals between those areas for which homoplasies are required (areas 3 and 6 for V1, 3 and 4 for V2, and 6 and 7 for P3). Homoplasy in syntaxon L1 (hypothetical ancestor of *Malaika longipes* and *Ambohima sublima*) might also be interpreted as dispersal of the ancestor of either one of the terminal taxa between area 1 (Table Mountain) and area 9 (Central Madagascar), but this homoplasy may be

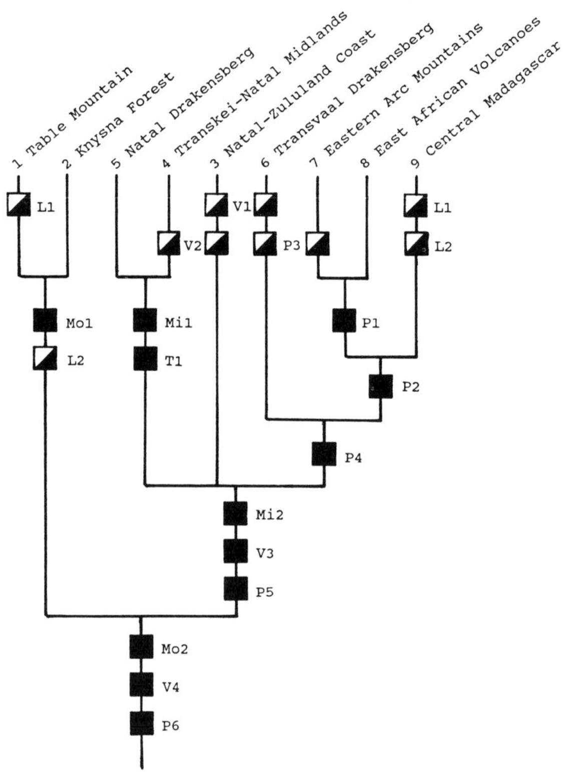

Fig. 5. Consensus area cladogram for afromontane spiders with syntaxa (hypothetical ancestral distributions) optimised. Black squares: apomorphies; half-shaded squares: homoplasies.

misleading or artefactual due to the present dearth of knowledge regarding the distribution and phylogeny of the afromontane spiders of tropical East Africa. In the case of *Ambohima*, its sister-taxon (*Kulalania antigua* Griswold; see Griswold 1990: fig. 13) also occurs in East Africa (in the volcanic Kulal Mountains), but as these mountains fall outside the area of endemism [area 8 (East African Volcanoes)], this sister-group relationship is not included in the biogeographic data set. In fact, for all Madagascan members of the spider taxa used in this analysis, the sister-group occurs in the mountains of East Africa. The homoplasy in syntaxon L2 (hypothetical ancester of *Lamaika distincta* and *Malaika longipes* + *Ambohima sublima*) is more difficult to interpret. Either the ancestor of either one of the terminal taxa dispersed between the Cape and Madagascar, or there is a set of extinct or extant but as yet undiscovered relatives of *Lamaika distincta* in areas 1 and 3–8.

Discussion

History of Afromontane Spiders

The general area cladogram calculated from the cladistic and distributional data from the Phyxelidini, Vidoleini, *Moggridgea quercina* group and *Microstigmata*, provides answers to each of the three questions posed previously concerning the afromontane biota: its reality, mode of origin and age.

If we define an afromontane biota as comprising those taxa disjunct among forest patches in temperate South Africa and on mountains in tropical Africa, the four groups of spiders treated here clearly describe such a pattern. The nine areas of endemism (Fig. 1) defined by the distributions of spider taxa coincide with several of the afromontane areas depicted by White (1983*b*: fig. 14).

The set of area relationships specified by the spider distributions may be considered to be general, suggesting origin through events affecting many components of the biota simultaneously. Little homoplasy in hypothetical ancestral distributions is necessary for these data, and it is these exceptions that may be explained as dispersals.

Finally, some nodes on the area cladogram correspond with possible barriers of known age and allow suggestions of the ages of parts of the pattern. The area cladogram is hard to reconcile with a historical scenario involving primarily Pleistocene vicariance events. A Pleistocene vicariance scenario in which forest expanded and contracted in an essentially modern Africa suggests that geographic proximity of disjuncts should be a good predictor of sister-area relationship. Such is not the case as areas 1+2 (the Cape Region of South Africa), are not the sister-areas of 3, 4 and 5 (eastern South Africa), and all of Africa (areas 1–8) is not an area distinct from Madagascar. I have previously suggested (Griswold 1985) that a species characteristic of area 4 (*Microstigmata lawrencei*) may have originated at the Miocene–Pliocene boundary and, if this is true, the age of the disjunction between areas 4 and 5 should be as old or older than this. The disjunction between area 1 (Table Mountain) and area 2 (Knysna Forest) may date from forest dissection in response to Pleistocene climatic changes documented in the southern Cape (Schalke 1973). A vicariance interpretation of the sister-area relationship between Madagascar and East Africa implies great age for the taxa involved. Rabinowitz *et al.* (1983) suggested that the separation of Africa and Madagascar began during the Jurassic (approx. 165 my) and that Madagascar had reached its present position by the mid-Cretaceous (approx. 121 my). Whereas no evidence exists to challenge this estimate for the age of the Indian Ocean between Madagascar and East Africa it is not clear that this was the only route of biotic communication between these areas.

Briggs (1989) has recently suggested possible biotic connections between Africa and Madagascar via India and an emergent Seychelles Bank that may have continued through the Late Cretaceous. Great age does not necessarily rule out the possibility of vicariance for spiders, as spiders may be extremely ancient, with fossils of modern superfamilies and families known from the Lower Cretaceous (Selden 1989, 1990). Nor is this the only case in which spider distributions may be explained by mesozoic events. Platnick (1976) presented an example in which an apparent late Cretaceous vicariant event separated sister-groups within a genus of gnaphosid spiders.

Testing Hypothesis of Biogeographic History

An area cladogram conveys an implicit prediction regarding the place of occurrence and sister-group relationships of as yet undiscovered taxa. The known distributions of *Microstigmata* and the *Moggridgea quercina* group are mapped on the area cladogram (Fig. 6). The area cladogram predicts that species of *Microstigmata* should be discovered in East Africa and Madagascar. It also predicts that members of the *Moggridgea quercina* group should be discovered in East Africa and Madagascar and that, if the area cladogram is true, these should be more closely related to one another and to species from eastern South Africa than to species from the Cape.

This method also permits us to compare alternative hypotheses. We may ask how well the spider data fit a scenario for Pleistocene vicariance of afromontane forest. Although detailed interpretations may differ regarding the timing and geographic scope of events, the prevailing opinion among palaeoecologists and biogeographers is that climatic changes during the Pleistocene, particularly changes in temperature and rainfall related to glacial advance and retreat, affected the distribution of the afromontane forest biota. Altitudinal zonation of vegetation belts on mountains moved up and down, and forest patches expanded and contracted.

Furthermore, by some means, communication among what are today disjunct patches of forest was at times much easier, and large geographic areas may have supported common afromontane biotas. At the end of the Pleistocene, or shortly after, biotic exchange among patches became unlikely, leading to the present pattern of disjunct distributions. In this scenario geographic proximity of disjuncts should be a good predictor of sister-area relationship.

There has been no explicit statement of the set of area relationships predicted by this Pleistocene scenario, but I will venture to offer one. Based upon geographic proximity of areas of endemism and the distributions of zones of aridity that may have acted as barriers, I predict the following area cladogram (Fig. 7) containing five components. *A*: southern Cape Fold

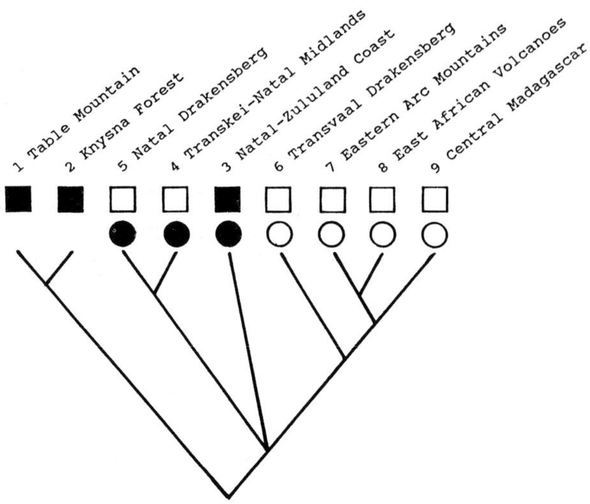

Fig. 6. Area cladogram for afromontane spiders with distributions of *Microstigmata* (circles) and the *Moggridgea quercina* group (squares) mapped on cladogram. Black symbols represent known distribution of taxon, open symbols represent predicted occurrence of taxon.

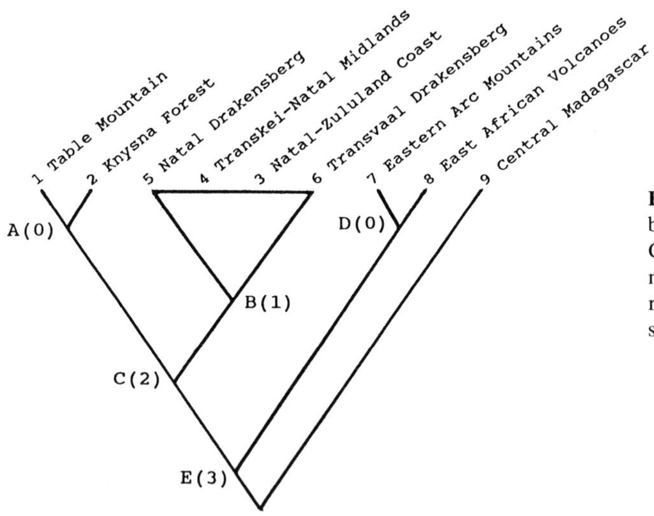

Fig. 7. Area cladogram predicted by Pleistocene vicariance scenario. Components noted by letters at nodes, along with additional steps required for spider data set to support monophyly of that node.

Mountains west of Algoa Bay and Uitenhage Basin, areas 1 and 2; *B*: eastern southern Africa east of Algoa Bay and Uitenhage Basin and south of the Limpopo, areas 3, 4, 5 and 6; *C*: Africa south of the Limpopo, areas 1–6; *D*: tropical East Africa, areas 7 and 8; and *E*: Africa, separated from Madagascar by the Indian Ocean, areas 1–8.

The parsimony of these two area cladograms can be compared quantitatively in explaining the spider distributions. I have no reason to hypothesise resolution among the areas within component B for the Pleistocene scenario, and, rather than perform tests on all 15 possible resolutions of these four areas, I have left them unresolved. This renders comparison of the two cladograms by simple calculation of their relative lengths unfair to the Pleistocene scenario, because, all other things being equal, an unresolved tree always requires more steps than a fully resolved tree. To avoid this bias, I asked what it would cost in parsimony to force a given component to come out as a monophyletic group on the area cladogram. To test this I added five dummy characters at the end of the data set, each supporting the monophyly of component A, B, C, D or E. I then tested each component individually by incrementally weighting its character through a series of analyses until monophyly of that component was achieved. Characters supporting the other components were deactivated during each analysis. I took the length of the tree for which the monophyly of a given component was achieved, and subtracted the weight of its character that was necessary to achieve that monophyly. The result was considered to be the tree length forced by the monophyly of that component. The minimum tree length for the unconstrained parsimony analysis (21 steps) was then subtracted from the minimum tree lengths required by the forced monophyly of each component. This was repeated for each hypothetical Pleistocene component and the additional steps necessary to support the monophyly of each were tabulated (Fig. 7). There is no cost in the monophyly of components A (Cape Region, areas 1 and 2) and D (East African highlands, areas 7 and 8). These components also appear on the most parsimonious area cladogram calculated from the spider data. Significant costs in parsimony are required to support B (eastern South Africa, areas 3, 4, 5 and 6) and C (South Africa, areas 1–6), and the greatest cost, 3 additional steps, is required for E (Africa, areas 1–8), i.e. to consider afromontane Africa a biogeographic region separate from Madagascar. Clearly, the Pleistocene scenario for the origin of the disjunctions in the afromontane biota is not the most parsimonious for the spider data, and the alternatives are to seek a more ancient explanation or allow the possibility of wholesale dispersal.

Conclusions

The pertinence of the cladistic paradigm in historical biogeography to defining the afromontane biota and investigating the age and mode of origin of its distributional pattern is clear. Among the points that emerge from this study is that areas that are clearly defined geographically (e.g. South Africa, Africa) were not found to be natural biogeographic regions, at least with respect to the afromontane biota. This underscores the importance of defining areas of endemism based on the distributions of monophyletic taxa rather than *a priori* based on geographic or political boundaries. The fit of the spider data to a single area cladogram is good, and it appears that a basic assumption of the parsimony approach to historical biogeography, that there is a single cladogram for the areas of interest, is not violated in this case. This may not necessarily hold for all cases (Cracraft 1988), and the danger exists that the data will be forced on to a model that is not appropriate. However, the application of cladistics to historical biogeography is still in its preliminary stages, and the notion of a single cladogram for areas of endemism if these areas are carefully defined on the basis of distributions of monophyletic groups may still serve as a reasonable null hypothesis. Dating a biogeographic pattern would seem filled with pitfalls, only one of which is the correctness of the area cladogram. Establishing the age of a taxon through fossils and of a barrier through geologic evidence are both open to wide margins of error, while establishing the concordance of a barrier with a node on an area cladogram is subjective. Geological hypotheses should not be considered more reliable than biological hypotheses. One need only consider the generations of biogeographers who felt compelled to interpret intercontinental disjunctions in the southern hemisphere in light of the hypothesis of continental stability. Still, I can think of no better way to suggest an age for a biogeographic pattern. The great ages suggested here for the spider distributions may not necessarily apply to similar disjunctions in other groups of plants and animals. Some groups that exhibit afromontane distributions may be too young to have been affected by mesozoic events. But until the phylogenetic relationships among other groups exhibiting afromontane disjunctions have been worked out and the degree to which they coincide with or contradict the area relationships specified by these spiders is understood, I believe that the potential of great age for these other groups can not be discounted out of hand. If a chance dispersal explanation is still to be preferred, it should at least be noted that it is less parsimonious than the vicariance explanation and that it presumes that a particular geological hypothesis is true. Finally, the use of parsimony to quantitatively compare alternative biogeographic hypotheses based upon how well each explains the distributions of monophyletic groups removes debates about biogeographic history from the realm of relative authority of proponents or plausibility of scenarios to a realm where alternative explanations can be objectively compared. This may be the greatest contribution of parsimony to historical biogeography.

Acknowledgments

I thank Robert Raven for inviting me to present a version of this paper at the 9th Meeting of the Willi Hennig Society in Canberra, Australia, and Pauline Ladiges and Chris Humphries for inviting me to submit the paper for publication in the symposium volume 'Austral Biogeography'. Support to carry out the basic taxonomic research upon which these conclusions are based was provided by the Natal Museum, Pietermaritzburg, the South African Council for Scientific and Industrial Research (CSIR), Kalbfleisch Fellowships at the American Museum of Natural History, New York, and Smithsonian Postdoctoral Fellowships at the National Museum of Natural History, Washington, D.C. Drs Jonathan Coddington and Nikolaj Scharff commented on a draft of the manuscript. I particularly want to thank Dr Scharff for illuminating discussions on African spiders and biogeography.

References

Adamson, R. S. (1931). Notes on some petrified wood from Banke, Namaqualand. *Transactions of the Royal Society of South Africa* **19,** 255–8.

Anderberg, A. (1986). The genus *Pegolettia* (Compositae, Inuleae). *Cladistics* **2,** 158–86.

Andrews, P., and Van Couvering, J. A. H. (1975). Palaeoenvironments in the East African Miocene. In 'Approaches to Primate Paleobiology'. (Ed. F. S. Szalay.) *Contributions to Primatology* **5,** 62–103.

Axelrod, D. I., and Raven, P. H. (1978). Late Cretaceous and Tertiary vegetation history of Africa. In 'Biogeography and Ecology of Southern Africa'. (Ed. M. J. A. Werger.) pp. 77–130. (Junk: The Hague.)

Bigalke, R. C. (1968). Evolution of mammals on southern continents. III. The contemporary mammal fauna of Africa. *Quarterly Review of Biology* **43,** 265–300.

Bonnefille, R. (1985). Evolution of the continental vegetation: the palaeobotanical record from East Africa. *South African Journal of Science* **81,** 267–70.

Booth, A. H. (1958). The zoogeography of West African primates: a review. *Bulletin de l'Institut Français d'Afrique Noire, Serie A* **20,** 587–622.

Brain, C. K. (1984). The Terminal Miocene Event: A critical environmental and evolutionary episode? In 'Late Cainozoic Palaeoclimates of the Southern Hemisphere'. (Ed. J. C. Vogel.) pp. 491–8. (Balkema: Rotterdam.)

Briggs, J. C. (1989). The historic biogeography of India: isolation or contact? *Systematic Zoology* **38,** 322–32.

Butzer, K. W. (1984). Late Quaternary environments in South Africa. 'Late Cainozoic Palaeoclimates of the Southern Hemisphere'. (Ed. J. C. Vogel.) pp. 235–64. (Balkema: Rotterdam.)

Carcasson, R. H. (1964). A preliminary survey of the zoogeography of African butterflies. *East African Wildlife Journal* **2,** 122–57.

Coetzee, J. A. (1964). Evidence for a considerable depression of the vegetation belts during Upper Pleistocene on the East African mountains. *Nature* **204,** 564–6.

Coetzee, J. A., and Muller, J. (1984). The phytogeographic significance of some extinct Gondwana pollen types from the Tertiary of the southwestern Cape. *Annals of the Missouri Botanical Garden* **71,** 1088–99.

Coetzee, J. A., and Rogers, J. (1982). Palynological and lithological evidence for the Miocene palaeoenvironment in the Saldahna region (South Africa). *Palaeogeography, Palaeoclimatology, Palaeoecology* **39,** 71–85.

Cole, N. H. A. (1974). Climate, life forms, and species distribution on the Loma Montane grassland, Sierra Leone. *Botanical Journal of the Linnean Society* **69,** 197–210.

Cracraft, J. (1988). Deep-history biogeography: retrieving the historical pattern of evolving continental biotas. *Systematic Zoology* **37,** 219–36.

Deacon, J., Lancaster, N., and Scott, L. (1984). Evidence for Late Quaternary climatic change in southern Africa: Summary of the proceedings of the SASQUA Workshop held in Johannesburg, September 1983. In 'Late Cainozoic Palaeoclimates of the Southern Hemisphere'. (Ed. J. C. Vogel.) pp. 391–404. (Balkema: Rotterdam.)

Diamond, A. W., and Hamilton, A. C. (1980). The distribution of forest passerine birds and Quaternary climatic change in tropical Africa. *Journal of Zoology, London* **191,** 379–402.

Dowsett, R. J. (1986). Origins of the high-altitude avifaunas of tropical Africa. In 'High Altitude Tropical Biogeography'. (Eds. F. Vuilleumier and M. Monastero.) pp. 557–85. (Oxford University Press: Oxford.)

Endrödy-Younga, S. (1988). Evidence for the low-altitude origin of the Cape Mountain Biome derived from the systematic revision of the genus *Colophon* Gray (Coleoptera, Lucanidea). *Annals of the South African Museum* **96,** 359–424.

Farris, J. S. (1981). Distance data in phylogenetic analysis. In 'Advances in Cladistics: Proceedings of the First Meeting of the Willi Hennig Society, New York Botanical Garden, Bronx'. (Eds. V. A. Funk and D. R. Brooks.) pp. 3–23. (New York Botanical Garden: New York.)

Farris, J. S. (1988). 'HENNIG86, Version 1.5.' Available from author at 41 Admiral St, Port Jefferson Station, New York, 11776, U.S.A.

Feijen, H. R. (1983). Systematics and phylogeny of Centrioncidae, a new afromontane family of Diptera (Schizophora). *Zoologische Verhandelingen* **202,** 1–137.

Gillet, J. B. (1955). Relation between highland floras of Ethiopia and British East Africa. *Webbia* **11,** 459–66.

Griswold, C. E. (1985). A revision of the African spiders of the family Microstigmatidae (Araneae: Mygalomorphae). *Annals of the Natal Museum* **27,** 1–37.

Griswold, C. E. (1987). The African spiders of the trap-door spider family Migidae (Araneae: Mygalomorphae) 1: the genus *Moggridgea* O. P. Cambridge, 1875. *Annals of the Natal Museum* **28,** 1–118.

Griswold, C. E. (1990). A revision and phylogenetic analysis of the spider subfamily Phyxelidinae (Araneae, Amaurobiidae). *Bulletin of the American Museum of Natural History* **196,** 1–206.

Grove, A. T. (1983). Evolution of the physical geography of the east African Rift Valley region. In 'Evolution, Time and Space: The Emergence of the Biosphere'. (Eds. R. W. Sims, J. H. Price and P. E. S. Whalley.) Systematic Association Special Vol. No. 23, pp. 115–55. (Academic Press: New York.)

Grubb, P. (1978). Patterns of speciation in African mammals. *Bulletin of the Carnegie Museum of Natural History* **6,** 152–67.

Grubb, P. (1982). Refuges and dispersal in the speciation of African forest mammals. In 'Biological Diversification in the Tropics'. (Ed. G. Prance.) pp. 537–53. (Columbia University Press: New York.)

Halbich, I. W., Fitch, F. J., and Miller, J. A. (1983). Dating the Cape orogeny. *Geological Society of South Africa Special Publication* **12,** 149–64.

Hall, J. B., and Medler, J. A. (1975). Highland vegetation in South-Eastern Nigeria and its affinities. *Vegetatio* **29,** 191–8.

Hamilton, A. C. (1976). The significance of patterns of distribution shown by forest plants and animals in tropical Africa for the reconstruction of Upper Pleistocene palaeoenvironments: a review. *Palaeoecology of Africa* **9,** 63–97.

Hamilton, A. C. (1982). 'Environmental History of East Africa.' (Academic Press: London.)

Humphries, C. J., and Parenti, L. R. (1986). 'Cladistic Biogeography.' (Clarendon Press: Oxford.)

Jensen, F. P., and Stuart, S. N. (1986). The origin and evolution of the Cameroon montane forest avifauna. In 'Conservation of Cameroon Montane Forests. Report of the ICBP Cameroon Montane Forest Survey, November 1983–April 1984'. (Ed. S. N. Stuart.) pp. 28–37. (International Council for Bird Preservation: Cambridge.)

Keay, R. W. J. (1954/1955). Montane vegetation and flora in British Cameroons. *Proceedings of the Linnean Society of London* **165,** 140–3.

Kendall, R. L. (1969). An ecological history of the Lake Victoria basin. *Ecological Monographs* **39,** 121–76.

King, L. (1972). The coastal plain of Southeast Africa: its form, deposits, and development. *Zeitschrift für Geomorphologie N. F.* **16,** 239–51.

King, L. 1982. 'The Natal Monocline Explaining the Origin and Scenery of Natal, South Africa.' 2nd ed. (University of Natal Press: Pietermaritzburg.)

Kluge, A. G. (1988). Parsimony in vicariance biogeography: A quantitative method and a greater Antillean example. *Systematic Zoology* **37,** 315–28.

Laporte, L. F., and Zihlman, A. L. (1983). Plates, climate and hominid evolution. *South African Journal of Science* **79,** 96–110.

Lemoigne, Y., and Beauchamp, J. (1972). Palaéoflores tertiares de la région de Welkite (Ethiopie, province d'Shoa). *Compte Rendu Sommaire des Séances de la Société Géologique de France* **1,** 35.

Leroy, J. F. (1978). Composition, origin, and affinities of the Madagascan vascular flora. *Annals of the Missouri Botanical Garden* **65,** 535–89.

Livingstone, D. A. (1975). Late Quaternary climatic change in Africa. *Annual Review of Ecology and Systematics* **6,** 249–80.

Livingstone, D. A. (1982). Quaternary geography of Africa and the Refuge Theory. In 'Biological Diversification in the Tropics'. (Ed. G. Prance.) pp. 523–36. (Columbia University Press: New York.)

Lönnberg, E. (1929). The development and distribution of the African fauna in connection with and depending upon climatic changes. *Archiv för Zoologi* **21A,** 1–33.

Louw, S. (1986). Revision of the Microcerinae (Coleoptera: Curculionidae) with an analysis of their phylogeny and zoogeography. *Memoirs of the National Museum, Bloemfontein* **21,** 1–331.

Lovett, J. C. (1988). Endemism and affinities of the Tanzanian montane forest flora. *Monographs in Systematic Botany from the Missouri Botanical Garden* **25,** 591–8.

Mädel, E. (1960). Monimiaceen-Hölzer aus den oberkretizischen Uzamba-Schichten von Ost Pondoland (S. Africa). *Senckenbergiana Lethaea* **41,** 331–91.

Mayr, E., and O'Hara R. J. (1986). The biogeographic evidence supporting the Pleistocene Forest refuge hypothesis. *Evolution* **40,** 55–67.

Moreau, R. E. (1966). 'The Bird Faunas of Africa and its Islands.' (Academic Press: New York.)

Morrison, M. E. S. (1968). Vegetation and climate in uplands of SW Uganda during later Pleistocene period. I. Muchoya Swamp, Kigezi District. *Journal of Ecology* **56,** 363–84.

Morton, J. K. (1972). Phytogeography of the West African mountains. In 'Taxonomy, Phytogeography, and Evolution'. (Ed. D. H. Valentine.) pp. 221–36. (Academic Press: London.)

Nelson, G., and Platnick, N. I. (1978). The perils of plesiomorphy: widespread taxa, dispersal, and phenetic biogeography. *Systematic Zoology* **27,** 474–7.

Nelson, G., and Platnick, N. I. (1981). 'Systematics and Biogeography: Cladistics and Vicariance.' (Columbia University Press: New York.)

Phillips, J. F. V. (1927). Fossil *Widdringtonia* in lignite of the Knysna Series, with a note on fossil leaves of several other species. *South African Journal of Science* **24,** 188–97.

Platnick, N. I. (1976). Drifting spiders or continents?: vicariance biogeography of the spider subfamily Laroniinae (Araneae: Gnaphosidae). *Systematic Zoology* **25,** 101–9.

Pócs, T. (1975). Affinities between the bryoflora of East Africa and Madagascar. *Boissiera* **24,** 125–8.

Puff, C. (1988). The relationships and affinities of the Madagascan *Anthospermum* species (Rubiaceae). *Monographs in Systematic Botany from the Missouri Botanical Garden* **25,** 205–13.

Rabinowitz, P. D., Coffin, M. F., and Falvey, D. (1983). The separation of Madagascar and Africa. *Science* **220,** 67–9.

Raven, R. J., and Platnick, N. I. (1981). A revision of the American spiders of the family Microstigmatidae (Araneae, Mygalomorphae). *American Museum Novitates* **2707,** 1–20.

Retallack, D., Dugas, P., and Bestland, E. A. (1990). Fossil soils and grasses of a Middle Miocene East African grassland. *Science* **247,** 1325–8.

Rodgers, W. A., and Homewood, K. M. (1982). Species richness and endemism in the Usambara mountain forests, Tanzania. *Biological Journal of the Linnean Society* **18,** 197–242.

Rodgers, W. A., Owen, C. F., and Homewood, K. M. (1982). Biogeography of East African forest mammals. *Journal of Biogeography* **9,** 41–54.

Schalke, H. J. W. G. (1973). The upper Quaternary of the Cape Flats area (Cape Province, South Africa). *Scripta Geologica* **15,** 1–57.

Scharff, N. (1990). Spiders of the family Linyphiidae from the Uzungwa mountains, Tanzania (Araneae, Linyphiidae). *Entomologica Scandinavica Supplement* **36,** 1–95.

Scharff, N. (1991). The Linyphiid Fauna of Eastern Africa (Araneae, Linyphiidae) — distributional patterns, diversity, and endemism. *Biological Journal of the Linnean Society* (in press).

Schiøtz, A. (1981). The Amphibia in the forested basement hills of Tanzania: a biogeographical indicator group. *African Journal of Ecology* **19,** 205–7.

Scott, L. (1982). A late Quaternary pollen record from the Transvaal Bushveld, South Africa. *Quaternary Research (New York)* **17,** 339–70.

Scott, L. (1984). Late Quaternary palaeoenvironments in the Transvaal on the basis of palynological evidence. In 'Late Cainozoic Palaeoclimates of the Southern Hemisphere'. (Ed. J. C. Vogel.) pp. 317–27. (Balkema: Rotterdam.)

Selden, P. (1989). Orb-web weaving spiders in the early Cretaceous. *Nature* **340,** 711–13.

Selden, P. (1990). Lower Cretaceous spiders from the Sierra De Montsech, North-East Spain. *Palaeontology* **33,** 257–95.

Smiles, R. L. (1985). Cladistics and distribution of *Euxanthe* butterflies (Nymphalidae: Charaxinae). *Journal of Natural History* **19,** 1165–89.

Sowerbutts, W. T. C. (1972). Rifting in Eastern Africa and the fragmentation of Gondwanaland. *Nature* **235,** 435–7.

Struhsaker, T. T. (1981). Forest and primate conservation in East Africa. *African Journal of Ecology* **19,** 99–114.

Stuart, S. N. (1981). A comparison of the avifaunas of seven east African forest islands. *African Journal of Ecology* **19,** 133–51.

Stuart, S. N., Jensen, F. P., and Brogger-Jensen, S. (1987). Altitudinal zonation of the avifauna in Mwanihana and Magombera forests, eastern Tanzania. *Gerfaut Revue Scientifique Belge d'Ornithology* **77,** 165–86.

Stuckenberg, B. R. (1962). The distribution of the montane palaeogenic element in the South African invertebrate fauna. *Annals of the Cape Provincial Museums* **2,** 190–205.

Thomas, D. S. G., and Goudie, A. S. (1984). Ancient ergs of the Southern Hemisphere. In 'Late Cainozoic Palaeoclimates of the Southern Hemisphere'. (Ed. J. C. Vogel.) pp. 407–18. (Balkema: Rotterdam.)

Truswell, J. F. (1977). 'The Geological Evolution of South Africa.' (Purnell and Sons: Cape Town.)

Van Zinderen Bakker, E. M. (1976). The evolution of late Quaternary palaeoclimates of Southern Africa. *Palaeoecology of Africa* **9,** 160–202.

Van Zinderen Bakker, E. M., and Coetzee, J. A. (1972). Re-appraisal of late-Quaternary climatic evidence from Tropical Africa. *Palaeoecology of Africa* **7,** 151–81.

Van Zinderen Bakker, E. M., and Coetzee, J. A. (1988). A review of Late Quaternary pollen studies in East, Central, and Southern Africa. *Review of Palaeobotany and Palynology* **55,** 155–74.

White, F. (1978). The afromontane region. In 'Biogeography and Ecology of Southern Africa'. (Ed. M. J. A. Werger.) pp. 132–43. (Junk: The Hague.)

White, F. (1983*a*). Long-distance dispersal and the origins of the afromontane flora. *Sonderband Naturwissenschaftliche Vereins in Hamburg* **7,** 87–116.

White, F. (1983*b*). 'The Vegetation of Africa: A Descriptive Memoir to Accompany the Unesco/AETFAT/UNSO Vegetation Map of Africa.' (United Nations Educational, Scientific and Cultural Organization: Paris.)

Wild. H. (1975). Phytogeography and the Gondwana position of Madagascar. *Boissiera* **24,** 107–17.

Willassen, E., and Cranston, P. S. (1986). Afrotropical montane midges (Diptera, Chironomidae, *Diamesa*). *Zoological Journal of the Linnean Society* **87,** 91–123.

Manuscript received 11 October 1990, accepted 15 January 1991

3 September 1991 *Aust. Syst. Bot.*, **4**, 91–110

A Panbiogeographic Perspective for Pre-Cretaceous Angiosperm–Lepidoptera Coevolution

J. R. Grehan

Entomology Research Laboratory, University of Vermont,
655b Spear Street, South Burlington, Vermont 05403, U.S.A.

Abstract

The coevolutionary history of Lepidoptera and angiosperms is evaluated in relation to Croizat's panbiogeographic synthesis of angiosperm evolution. The panbiogeographic method of ocean basin classification is used to identify major patterns of trans-oceanic distribution for lepidopteran families and genera (principally non-ditrysian). The Pacific basin is identified as a major evolutionary centre for several 'primitive non-ditrysian Lepidoptera, including Zeugloptera, Aglossata, Heterobathmiina, Neopsuestina, Palaephatidae, Prodoxidae, and possibly the Dacnonypha. The ditrysian Ithomiidae are similarly classified with the Pacific while the related Daniidae are identified as Indian Ocean. An Indian Ocean baseline is proposed for the Callidulidae, Tinissimae and Perissomasticini (Tineidae). A 'coevolutionary' history is supported in terms of Lepidoptera and angiosperms sharing common biogeographic (spatiotemporal) characters associated with the pre-Cretaceous tectonic history of major ocean and sea basins. The lack of congruent higher level Lepidoptera-angiosperm phylogenies emerging from systematic studies may be due to a lack of cospeciation events, but this does not exclude a close ecological and evolutionary relationship through the history of both groups.

Introduction

The larval stages of most Lepidoptera live and feed on angiosperms and the majority of adults feed on angiosperm nectar (Common 1970; Powell 1980; Shields 1988). The evolution of this ecological relationship is often attributed to a process of contemporaneous coevolutionary 'radiation' beginning in the Cretaceous and continuing through the Tertiary (Hughes 1973; Common 1975; Shields 1976; Powell 1980; Tindale 1981: 968; Mitter and Brooks 1983; Crepet 1984; Kristensen 1984). The post-Jurassic timetable rests primarily on the absence of undisputed angiosperm fossils before the Barremian Cretaceous and the premise that the earliest origin of angiospermy dates from about this time (Hughes 1973, 1977; Raven and Axelrod 1974; Wolfe *et al.* 1975; Hickey and Doyle 1977; Krassilov 1977; Melville 1983; Muller 1984; Friis *et al.* 1987; cf. Shields 1988).

The spatial (biogeographic) aspect of the post-Jurassic scenario is widely portrayed in terms of Darwinian biogeography (following Darwin 1859; Wallace 1876) where an organism or group is believed to have a localised birthplace or 'centre of origin' from which species migrate to different parts of the globe (e.g. Takhtajan 1969: 175; Axelrod 1970; Raven and Axelrod 1974; Schuster 1976). Efforts to identify the elusive angiosperm centre of origin have produced mixed results. Some investigators identified boreal zones, others the tropics (Hughes 1976) and popular locations ranged from South East Asia where 'primitive' taxa abound (Takhtajan 1969) to West Gondwana with its abundance of early angiosperm fossils (Raven and Axelrod 1974; Retallack and Dilcher 1981: 63; Muller 1984). Schuster (1976) reviewed all the main geographic candidates and decided that the original centre could not be identified with any certainty because subsequent migration had obscured its location.

1030–1887/91/010091$05.00

Botanists generally remained much more confident, however, about the migrations themselves. The question 'once evolved — in which directions did the angiosperms spread?' (Brenner 1976: 23) was reflected in a voluminous literature on migration routes all over the globe, with or without continental drift playing an active role. A picture of prolific angiosperm migration during the Cretaceous and Tertiary was built up by both phytogeographers (e.g. Raven and Axelrod 1974; Schuster 1976) and palaeobotanists (e.g. Brenner 1976; Wolfe 1981; Serlin 1982).

The paleontological argument for a post-Jurassic origin of angiospermy is tenuous because it relies on the oldest fossil record representing the actual time of phylogenetic origin rather than the minimal age of fossilisation for a recognised taxonomic entity (Croizat 1964*a*; Grande 1985). Intuitive awareness of this constraint is perhaps revealed by justifications such as the 'orderly' pattern of diversification (Doyle 1978: 373) and the 'concomitant occurrence of many angiosperm features at this time' (Muller 1984: 461). Krassilov (1977: 148) contended, however, that the concept of a 'deadline' for angiosperm origins creates an 'abominable mystery', and Hill and Crane (1982: 323) have also cautioned that an uncritical acceptance of a Barremian–Aptian boundary may lead to problematic fossils being assigned to angiosperm or gymnosperm taxa according to the time of their occurrence. Paleobotanists are continuing to entertain the possibility of a pre-Cretaceous angiosperm origin (Crane 1987) and recent phylogenetic treatments (Crane 1985; Doyle and Donoghue 1986, 1987) suggest that the Bennettitales and Gnetales, which have fossil records extending back into the Triassic, are closely related to angiosperms. One cladogram identified these as the sister group of angiosperms and therefore of the same age. However, the reconstructions were not used to suggest that extant angiosperm taxa evolved before the Cretaceous.

Recent applications of Croizat's (1952, 1958) panbiogeography have stimulated a renewed interest in the relationship between systematics and biogeography and the implications for evolutionary biology (e.g. Nelson and Platnick 1981; Nelson and Rosen 1981; Craw 1982, 1983, 1985, 1988, 1989; Platnick and Nelson 1984, 1989; Heads 1985*a*, 1985*b*; Humphries 1985; Humphries and Parenti 1986; Seberg 1986; Page 1987, 1989; Craw and Heads 1988; Craw and Page 1988; Gray 1988, 1990; Hull 1988; Humphries *et al.* 1988; Grehan and Henderson, in press see also papers in Matthews 1990). Croizat established panbiogeography as an independent and progressive research program (Craw and Weston 1984) that developed a synthesis of geological and biological relationships previously absent from evolutionary theory (Croizat 1988; Grehan 1988). The synthesis has potentially significant implications for understanding the tempo and mode of animal-plant interactions and the events responsible for establishment of specific host-plant relationships. In particular, biogeographic analysis could provide new insights into the relative age of animal-plant associations and the historical-ecological context in which they first evolved (cf. Gray 1990). This paper re-examines the evolution of angiosperm feeding Lepidoptera with emphasis on some of the spatial characteristics shared between them. I will use the term 'coevolution' to include the general sense of common historical association (biogeographic and ecological) as well as the stricter sense of cospeciation (cf. Brooks 1990).

Panbiogeographic Classificiation

From the time of Darwin (1859) and Wallace (1876) many biogeographers emphasised 'centres of origin' and the importance of the different 'means of dispersal' (walking, running, flying etc.) in relation to geological theories as the key to the origins of geographic distribution. Croizat (1952, 1958) believed this reliance on theories about migration routes from centres of origin lacked empirical biogeographic foundation. He proceeded to develop a methodological framework to test Darwinian biogeography by representing the spatiotemporal evolution of distribution in the form of line graphs or 'tracks'.

The track method provides for comparison of different distributions on a repetitive or 'statistical' basis (Croizat 1961: 1519; 1964*a*). Croizat's procedure was qualitative but a statistical framework is being developed in relation to graph theory and compatibility analyses (Page 1987; Craw 1988, 1989; Craw and Page 1988; Henderson 1990). Tracks in their simplest form are minimal spanning trees of graph theory (Nelson 1973; Page 1987) which connect the nearest disjunct localities in the order that results in a minimal overall line length. Croizat's utilisation of the track method had its predecessors but he was the first to use it in such a comprehensive and concerted manner (Craw 1988).

Croizat (1952, 1958) found that the tracks of terrestrial organisms crossed oceans and continents alike as if the sea were land. No causal relationship was apparent between the individual ecological attributes of organisms (such as 'means of dispersal') and the orientation or extent of their tracks. To provide a spatial (strictly biogeographic) criterion for comparing individual tracks, Croizat (1952) introduced the novel concept of 'baseline'. The baseline is a hypothesis suggesting which sector of the earth is most involved with the evolution of a track (Craw and Page 1988). On a global scale Croizat referred to ocean basins as principal biogeographic sectors since they represented the break or division dividing the biotas of major landmasses (Nelson 1978). Croizat suggested that migratory abilities ('means of dispersal') needed to be understood in their broad biogeographic context and in relation to the evolution of ocean basins. The relationship between baselines and tectonics has been investigated by designating the baseline as a spatial referent orienting one or more tracks to a tectonic feature such as spreading ridges, suture zones and transform faults (Croizat 1964*a*; Craw 1988; Craw and Page 1988).

Geological links for trans-oceanic disjunctions have been proposed by many biogeographers and geologists before and after Croizat but he developed the concept of ocean basin classification to suggest that major present-day land areas traditionally regarded as biogeographic units are actually biogeographic (biological and geological) composites (Craw 1982, 1983, 1984, 1985). For example, North and South America share components of the biota with their closest relatives in Africa and/or Europe while other members are most closely related to forms present in the western Pacific (Croizat 1952, 1958, 1961). Geologists recognise the presence of 'exotic' terrains in the Americas and other landmasses but there is substantial disagreement about their origin. Some geologists have suggested a Pacific basin origin (e.g. Nur and Ben-Avraham 1977, 1981; Coney *et al.* 1980; Jones *et al.* 1982; Chamberlain and Lambert 1985; Howell 1985; Coward and Ries 1986). These studies suggest a disruption of Pacific landmasses during Jurassic or earlier times. Panbiogeographic analysis represents a potentially significant resource for understanding Pacific history and has already resulted in a new global classification of the world's flora and fauna (Fig. 1) that subsumes the traditional Wallacean system (Craw 1988; Craw and Page 1988).

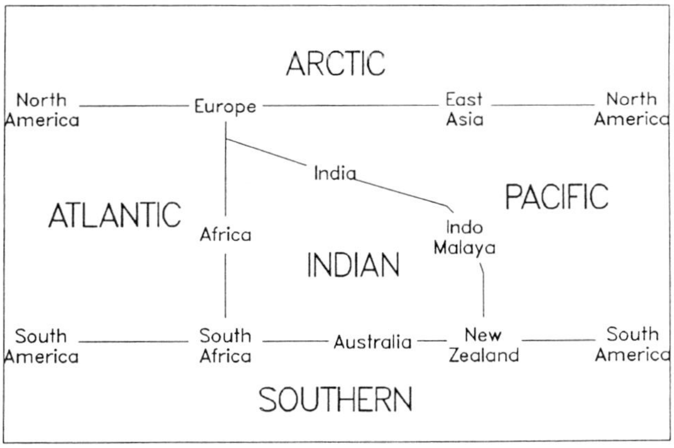

Fig. 1. Major panbiogeographic regions of the world with principal landmasses positioned at the biogeographic boundaries (Craw 1988).

Panbiogeography of Angiosperms

Croizat (1952, 1958) argued that modern trans-oceanic plant and animal distributions were at least as old as the tectonic history of the ocean and sea basins (e.g. Pacific, Atlantic, Indian, Mediterranean, Tasman, Caribbean). It is not surprising to find, therefore, that distributions of Triassic organisms such as the marine mollusc *Monotis* (Fig. 2) (Silberling 1985) have a biogeographic pattern homologous to that of extant groups (whether 'terrestrial' or 'marine' does not matter here). Furthermore, within particular landmasses the distributions of different animals and plants could be correlated with standard biogeographic boundaries or 'nodes' which have been shown to have tectonic significance (Craw and Weston 1984; Craw 1989). Croizat used numerous angiosperm examples in his panbiogeographic analyses and wrote extensively on angiosperm evolution, phylogeny and systematics. Although not acknowledged in general works on the Angiospermae, Croizat's insights have been adopted or supported by botanical specialists in a wide range of disciplines (Heads 1984). Croizat's biogeographic analyses led him to support not only a pre-Cretaceous origin for angiospermy, but also the pre-Cretaceous existence of a diverse 'angiosperm' flora ancestral to many present-day families and genera: '1) Plant forms quite proximally ancestral to the major genera of the Angiospermae now surviving were shared in common by the earth at large already in Mid-Jurassic times or about 165 million years ago; 2) It is accordingly by then that were laid down the massings of flora largely endemic today in regard of their species, and aggregates of forms of lesser order. The Cape, the highlands of Madagascar, St. Helena, Cuba and Hispaniola, the Duida-Roraima System, South China, Tenasserim, etc, etc., still contain the record clearcut of the evolution that locally followed the early settlement of angiosperms, and other, prototypes (Croizat 1964*a*: 239–40).'

Croizat's view differs from most paleontological accounts, which entertain a pre-Cretaceous origin for angiosperms, but acknowledge little (if any) differentiation of modern families and genera (e.g. Crane 1987; Doyle and Donoghue 1987: 37; Cleal 1989; Crane *et al.* 1989). However, there is an interesting agreement between panbiogeographic interpretations and recent molecular studies supporting pre-Cretaceous differentiation of angiosperms (Li *et al.* 1989; Martin *et al.* 1989). Croizat (1961, 1964*a*, 1964*b*) was inclined to regard 'pre-angiospermy' as the appropriate term to describe the evolution of many modern angiosperm features (e.g. leaves, flowers, nodes etc.) and he regarded the notion that all angiosperm

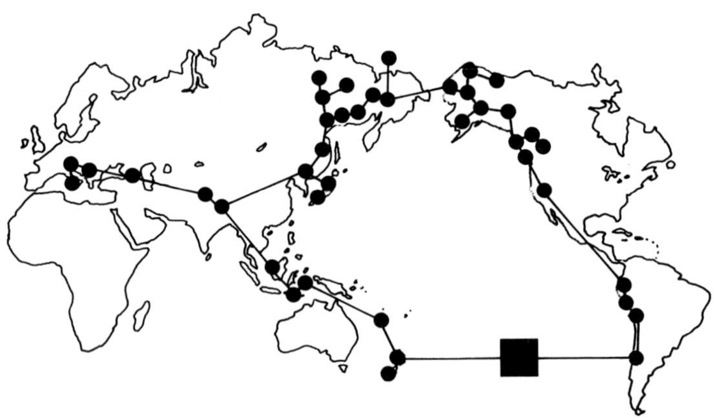

Fig. 2. Pacific baseline for the dispersal of the Triassic fossil mollusc *Monotis* (Silberling 1985). (Baseline suggested by closed square.)

families must be derivatives of a single original primitive angiosperm group, such as the 'Ranales', to be biogeographically and morphogenetically untenable: 'Had a botanist visited our world at any time between the Mid- and Late Cretaceous, he would recognize without much difficulty — in the geography then ruling — not only the Onagraceae as a family, but their major genera, whether in the New or the Old World. There is no reason why these plants are any younger than, e.g. Euphorbeaceae or Uriticaceae. . . . The notion that the modern botanical families are historically 'derived' from one another is obviously false, quite as false as is the notion that dispersal forms ranks by wild 'migrations'. Continuity is on the contrary unbroken over space, through time, by form (Croizat 1964*a*: 527).'

Croizat (1964*b*) introduced a modified 'tree' of angiosperm evolution (Fig. 3*a*) that removed what he saw as the phylogenetic and morphogenetic bottleneck present in traditional hierarchical models. The different approaches could be connected conceptually by laying the conventional tree on its side (Fig. 3*b*). This had the effect of stressing the broad basis of phylogeny, morphology and dispersal that Croizat believed to underlie the evolutionary transition from pre-angiospermy to angiospermy with respect to major angiosperm alliances (see also Croizat 1961). This proposal for angiosperm evolution in the form of a 'rhizome' opens the possibility of a much broader level of insect–plant coevolution at the boundary of angiospermy than may otherwise be allowed for in strictly hierarchical treatments of coevolution (cf. Brooks 1990).

The panbiogeographic classification identifies the Pacific as a major region for the evolution of angiosperms (e.g. *Euphrasia*, Fagaceae, Coriareaceae, *Oreobolis*, *Nertera*, *Coprosma*, Magnoliaceae, Gaultherioideae, some Monomiaceae, *Fuchsia*, *Griselinia*, *Hebe*, *Metrosiderous*, *Phrygilanthus*) (Croizat 1952, 1958, 1968; Craw 1988). Pacific distributions are sometimes characterised by a lack or absence of records from Africa, India and the eastern Americas (particularly central-eastern South America) (Fig. 4). In contrast the biogeography of other groups present in Africa, for example, is centred on the Atlantic (e.g. Turneraceae Fig. 5*a*), Indian Ocean (e.g. *Pelargonium* Fig. 5*b*) or both (e.g. Crysobalanaceae

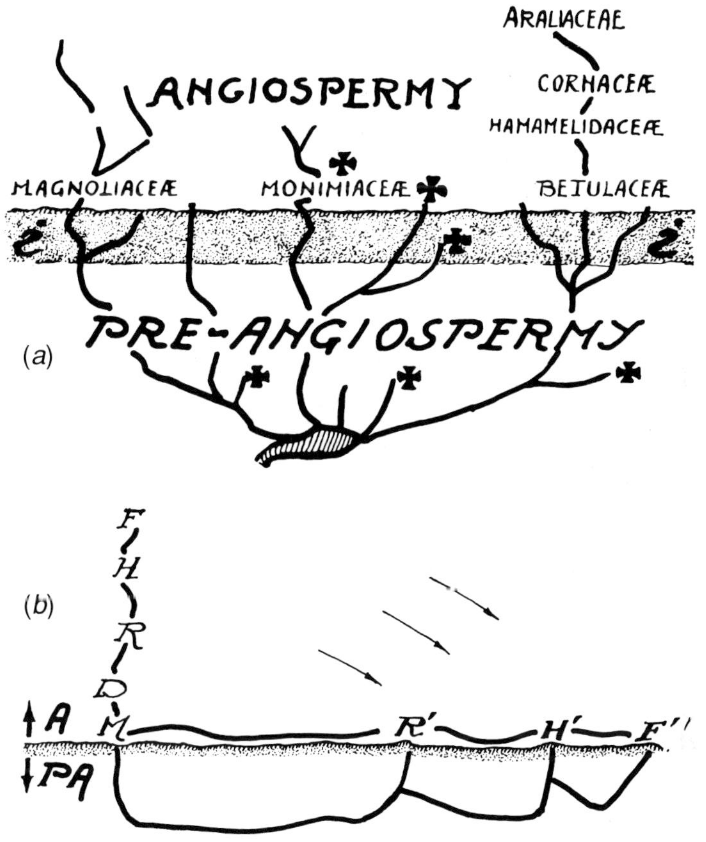

Fig. 3. Evolutionary relationship between angiospermy and pre-angiospermy proposed by Croizat (1964*b*). (*a*) Modified 'tree' of angiosperm phylogeny indicating the origin of interrelationships between extinct (crosses) and living groups and families within pre-angiospermy. The stippled region (*i*) refers to an intermediate horizon from which a number of key families emerge. (*b*) The conventional phylogenetic 'tree' turned at the Magnoliales node and reconstituted in a horizontal position with the important phylogenetic connections of various families embedded in pre-angiospermy across a broad phylogenetic and morphogenetic front.

Fig. 5*c*, Tristichaceae Fig. 5*d*). The distribution of Chrysobalanaceae borders all three major oceans, but the primary phylogenetic affinities do not appear to cross the Pacific (Prance and White 1988). A similar biogeographic (phylogenetic and distributional) orientation is found in the Tristichaceae (Croizat 1961; Cusset and Cusset 1988). These different biogeographic homologies suggest that the angiosperm families and genera involved arose not from any one restricted 'centre of origin' but over different geographic sectors of the earth now topographically represented as ocean and sea basins (cf. Craw 1983, 1988, 1989; Craw and Page 1988; Grehan 1988).

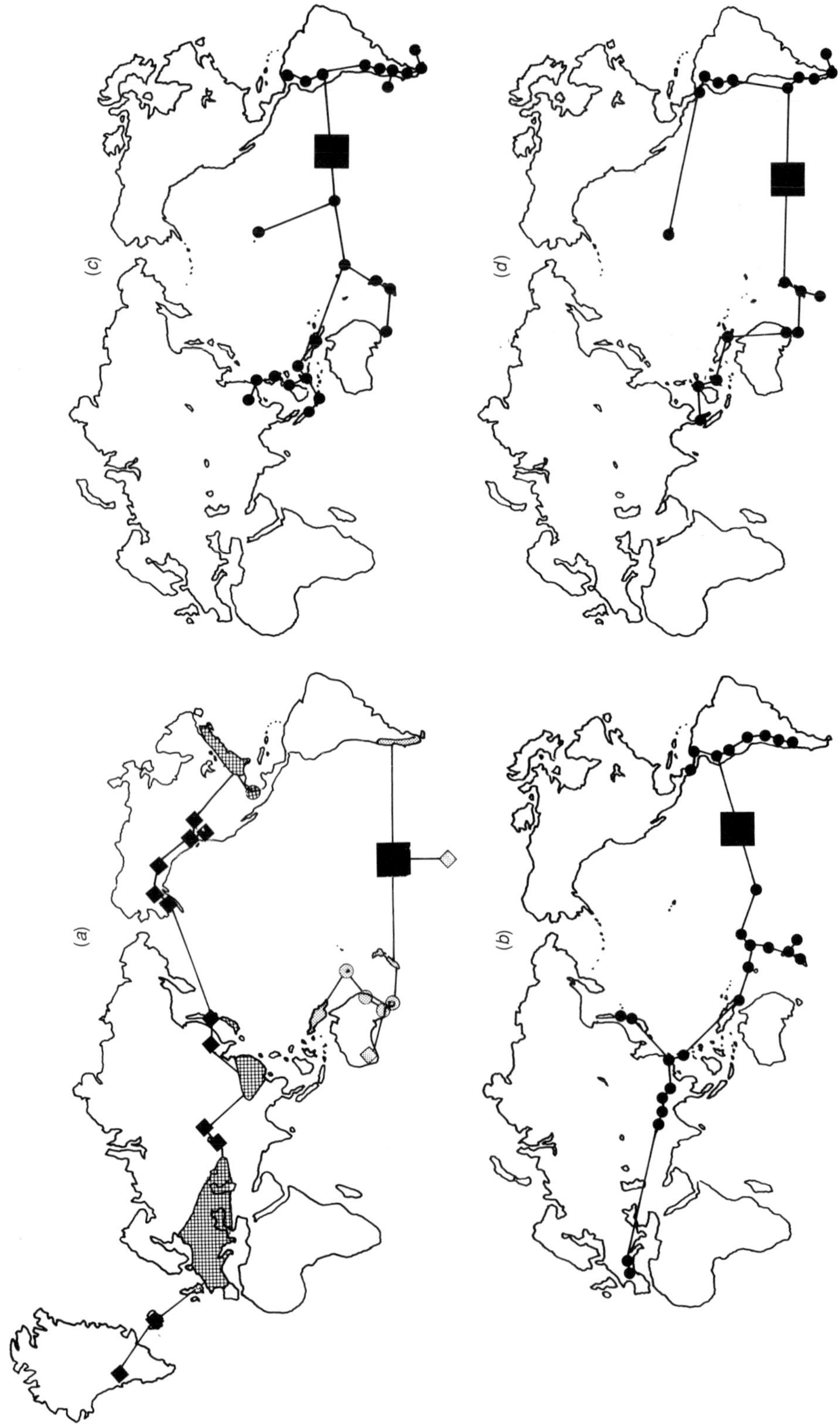

Fig. 4. Examples of Pacific baselines for angiosperm dispersal. (*a*) *Nothofagus*/*Fagus* (*Nothofagus* living, stippled, fossils beyond current range, stippled diamonds; *Fagus* living, cross-hatched, fossils beyond current range, solid diamonds); (*b*) *Coriaria*; (*c*) *Nertera*; and (*d*); *Oreobolus*. (Croizat 1968; Craw 1988.)

Lepidopteran Panbiogeography

Phylogenetic reconstructions of the major lepidopteran groups have developed considerably over the last 20 years with the discovery of new taxa, additional material of rare species and detailed character analyses focusing on the 'primitive' taxa (Kristensen 1984; Shields 1988). The higher classification of Lepidoptera (Table 1) remains in a state of flux (cf. Nielsen 1982*a*, 1982*b*; 1985*a*, 1985*b*; Davis 1986; Nielsen and Scoble 1986) and groups such as the Heterobathmiia are regarded as particularly problematic (cf. Kristensen and Nielsen 1979, 1983). A pre-Cretaceous phylogenetic origin has been traditionally accepted for the Lepidoptera, although Jurassic representatives have only recently been described (Rasnitsyn 1983; Whalley 1985). Whalley (1986: 269) predicts future confirmation of a Triassic differentiation for the Lepidoptera and Trichoptera and suggests that the present lack of higher lepidopteran Mesozoic fossils is due more to gaps in the fossil record than to a genuine absence of the groups in pre-Tertiary times. The following interpretations provide a biogeographic estimate for the age of some lepidopteran taxa by identifying the 'most probable' ocean baselines in the light of currently available geographic and taxonomic characters.

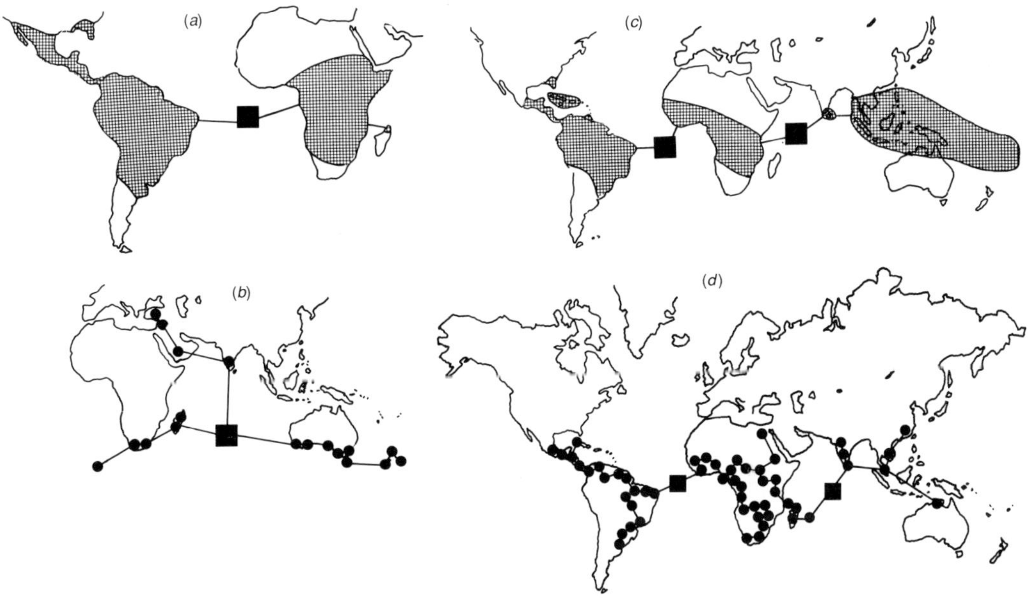

Fig. 5. Angiosperm dispersal with Atlantic–Indian Ocean baselines; (*a*) Turneraceae, Atlantic; (*b*) *Pelargonium*, Indian, (*c*) Chrysobalanaceae and (*d*) Tristichaceae with combined Atlantic–Indian Ocean baselines. (Hutchinson 1969; Cusset and Cusset 1988; Prance and White 1988).

The Micropterigidae (suborder Zeugloptera) is divided into two groups (Gibbs 1982): (*a*) 'Sabatinca' in or around the Pacific basin (Fig. 6) with outliers in South Africa and eastern North America. I interpret the localities adjacent to the Pacific as the 'main massing' to which the outliers are linked by the shortest line length because there are no Sabatinca in Europe, central Africa or the major part of eastern South America. The South African species are linked to the western Pacific to northern Australia to distinguish the pattern from the strictly 'Southern Ocean' tracks of other taxa. (*b*) 'Micropterix' ranging over parts of Eurasia from Great Britain to Japan. The Pacific orientation and main massing of the Sabatinca group supports the interpretation of the Micropterix range as an extension of the western Pacific (rather than connecting over the North Atlantic.) The overall Pacific baseline for the Micropterigidae is representative of a pattern where the Atlantic basin is not involved and the Indian Ocean is 'crossed' only as an extension of a Pacific history. Two further suborders, the Aglossata and Heterobathmiina, are distributed adjacent to or within the Pacific basin (Fig. 6).

The infra-order Dacnonypha comprises three families of uncertain affinity and perhaps doubtful monophyly (Nielsen 1982*a*; E. S. Nielsen, personal communication). The Eriocranidae occur through North America and Eurasia while the Acanthopteroctetidae are more localised in western North America and the Lophocoronidae are known only from southern Australia (Davis 1978) (Fig. 7). *If* the Dacnonypha were monophyletic a Pacific

Table 1. Systematic and biogeographic classification of 'primitive' (non-Ditrysian) Lepidoptera (systematics after Nielsen 1985)
(*Tischeriidae also included in this group by Davis 1986). Panbiogeographic classification (where sufficiently resolved) designated in parentheses

Suborder ZEUGLOPTERA		
	Micropterigidae (Pacific)	
Suborder AGLOSSATA		
	Agathiphagidae (Pacific)	
Suborder HETEROBATHMIINA		
	Heterobathmiidae (Pacific)	
Suborder GLOSSATA		
	DACNONYPHA s.lat. (Pacific?)	
		Eriocraniidae
		Lophocoronidae
		Acanthopteroctetidae
	NEOPSEUSTINA	
		Neopseustidae (Pacific)
	EXOPORIA	
	Mnesarchaeoidea	
		Mnesarchaeidae
	Hepialoidea	
		Neotheoridae
		Anomosetidae
		Prototheoridae
		Palaeosetidae
		Hepialidae (Indian / Atlantic?)
	HETERONEURA	
	NEPTICULINA*	
	Nepticuloidea	
		Nepticulidae
		Opostegidae
	Palaephatoidea	
		Palaephatidae (Pacific)
	INCURVARIINA	
		Heliozelidae
		Adelidae
		Crinopetrygidae
		Incurvariidae
		Cecidosidae (South Atlantic)
		Prodoxidae (Pacific)
	TISCHERIINA	
		Tischeriidae

baseline for the families might be inferred. The infra-order Neopsuestina (Fig. 8) is restricted to south-eastern Asia and southern South America (Davis 1975) and a minimal distance link across the Pacific shows a close geographic parallel to the angiosperm family Lardizabalaceae (Fig. 9).

Six families are recognised in the infra-order Exoporia. The largest and most widespread is the Hepialidae which ranges from boreal to sub-Antarctic latitudes and sea level to alpine habitats (Grehan 1989). Detailed intergeneric phylogenetic and distributional data for the

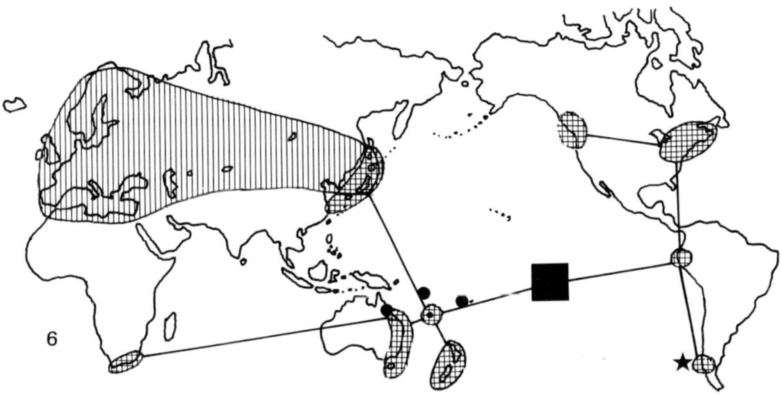

Fig. 6. Pacific dispersal of the Zeugloptera (Sabatinca group cross-hatched; Micropterix group as vertical lines), Aglossata (solid circles) and Heterobathmiina (star). (Kristensen and Nielsen 1979; Gibbs 1982.)

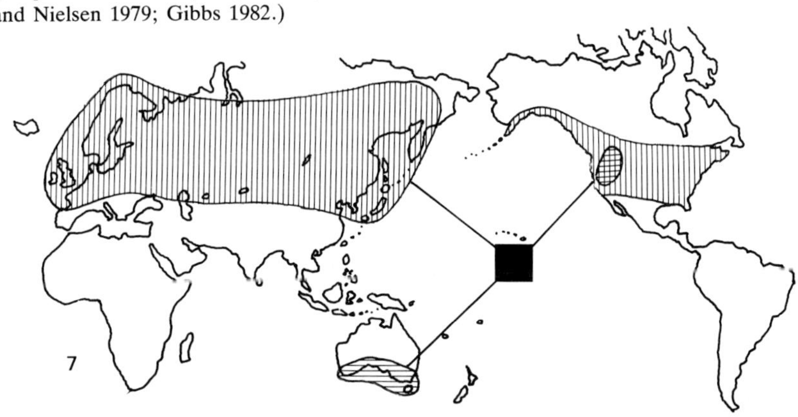

Fig. 7. Dispersal of the Dacnonypha. Eriocranidae as vertical lines, Acanthopteroctetidae cross-hatched, Lophocoronidae horizontal lines (Davis 1978).

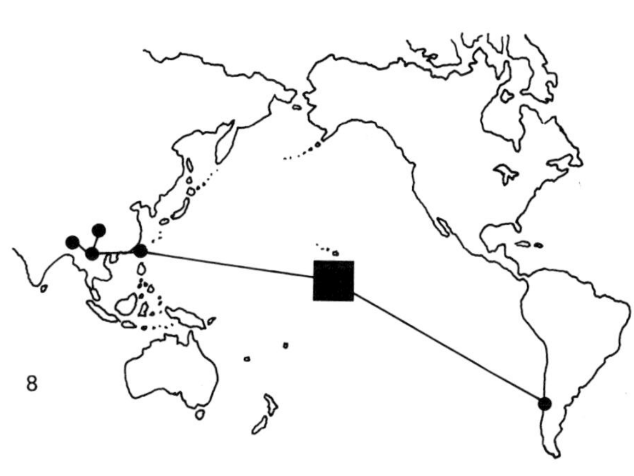

Fig. 8. Dispersal of the Neopsuestina (Davis 1975).

exoporian families are virtually non-existent (Nielsen 1988) and the major baselines remain to be determined. Trans-Pacific relationships may be applicable to some South-East Asian and South American genera (cf. Nielsen and Scoble 1983; Nielsen 1988), but the African representation of Hepialidae may also suggest that the Indian Ocean is an important biogeographic centre for the family (cf. Grehan 1987). Perhaps noteworthy in this respect is the apparent absence of Hepialidae from Madagascar which together with south-eastern Africa forms one of the world's principal biogeographic centres, the 'African Gate' (Croizat 1952). The remaining exoporian families have comparatively restricted distributions (Fig. 10). The Mnesarchaeidae are endemic to New Zealand, Prototheoridae to southern Africa, Anomosetidae to Queensland and Neotheoridae to southern Brazil (Kristensen 1978*a*, 1978*b*) while the Paleosetidae are present in South America and the western Pacific. The lack of phylogenetic resolution and the strong possibility of very incomplete distribution records precludes further determination of their baselines.

9

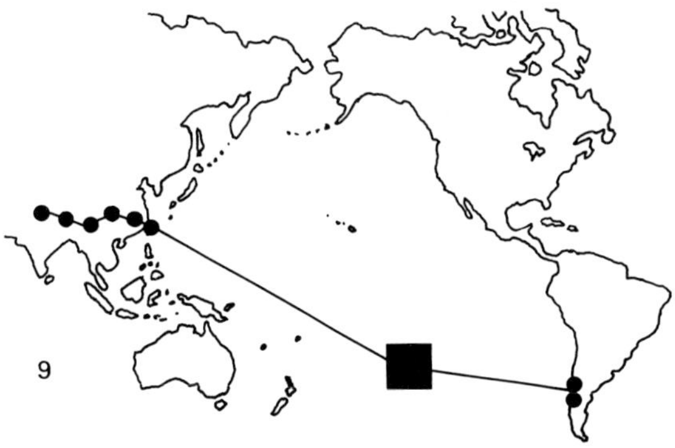

Fig. 9. Dispersal of the Lardizabalaceae (Angiospermae) (Croizat 1968).

Fig. 10. Distribution of the Exoporia (excluding Hepialidae). Mnesarchaeidae as large solid triangles (New Zealand), Prototheoridae stippled (South Africa), Anomosetidae as small solid triangle (offset Queensland), Neotheoridae as solid diamond, Paleosetidae as solid circles (South America, Australia, South-East Asia) (Kristensen 1978*a*, 1978*b*).

The non-Ditrysian Heteroneura comprise three main groups: Nepticulina, Incurvariina and Tischeriina (Table 1). Two families of the Nepticulina, the Nepticulidae and Opostegidae, are widespread (Scoble 1983; Nielsen 1985*b*). Distribution records of the Nepticulidae (Van Nieukerken 1986) may suggest major biogeographic connections between southern Africa and Eurasia/North America (Fig. 11). This connection is one of the oldest standard tracks for global distribution (cf. Croizat 1958, 1961) but with the current taxonomic and distributional information I am not confident of the ocean basin orientation (although the Mediterranean with its Rift Valley and Tethyan extensions could be important). The Nepticulid family Palaephatidae is represented in Australia and southern South America (Fig. 12) (Davis 1986; Nielsen 1987). A Pacific baseline is suggested by minimal distance and this southern Pacific range may be geographically comparable to angiosperm tracks such as *Aristotelia* and *Nothofagus*.

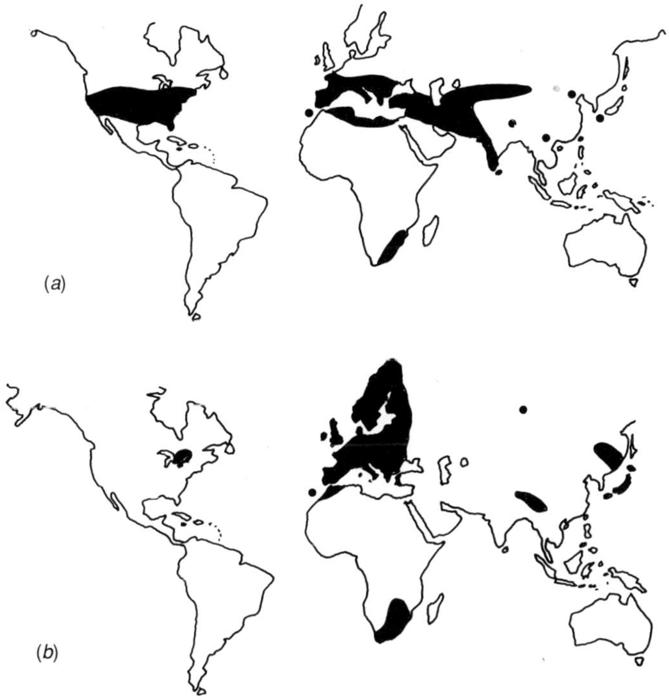

Fig. 11. Distribution of (*a*) *Acalyptris* and (*b*) *Ectodemia* (Nepticulidae Van Nieukerken 1986).

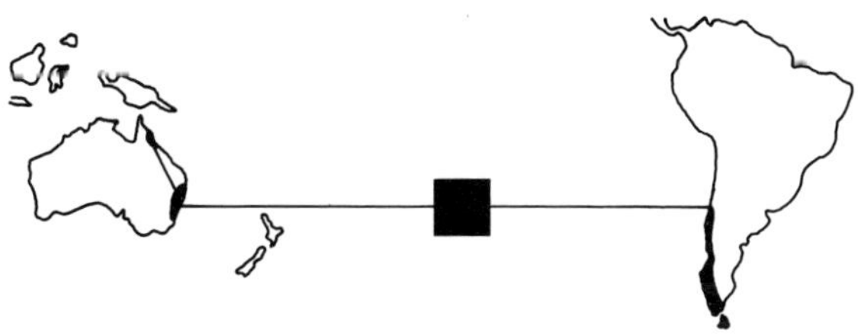

Fig. 12. Dispersal of the Palaephatidae (Davis 1986; Nielsen 1987).

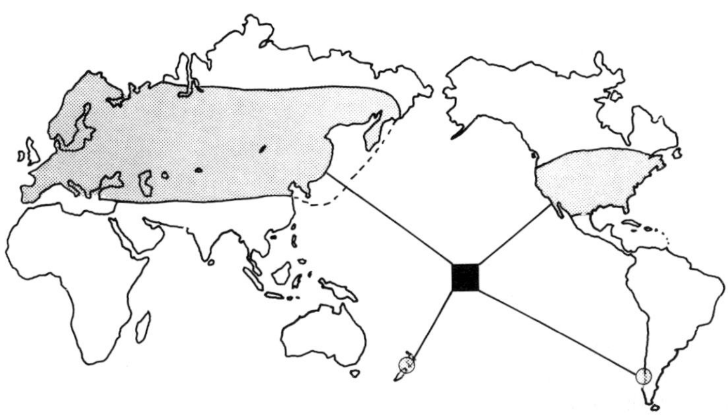

Fig. 13. Dispersal of the Prodoxidae (Davis 1967; Nielsen 1982*b*; Nielsen and Davis 1985; Dugdale 1989).

I am unable to present a detailed summary at this time of the Tischeriina or most of the Incurvariina. The incurvariid family Cecidosidae in South America and South Africa can be linked by minimal distance to a south Atlantic baseline. In contrast the incurvariid family Prodoxidae may be oriented to a Pacifc baseline (Fig. 13) connecting localities in New Zealand, North America and the southern Andes of South America (Davis 1967; Nielsen and Davis 1985; Dugdale 1989). The Pacific baseline of the Prodoxidae may be extended to include the related genus *Lampronia* distributed through North America and Eurasia (Nielsen 1982*b*).

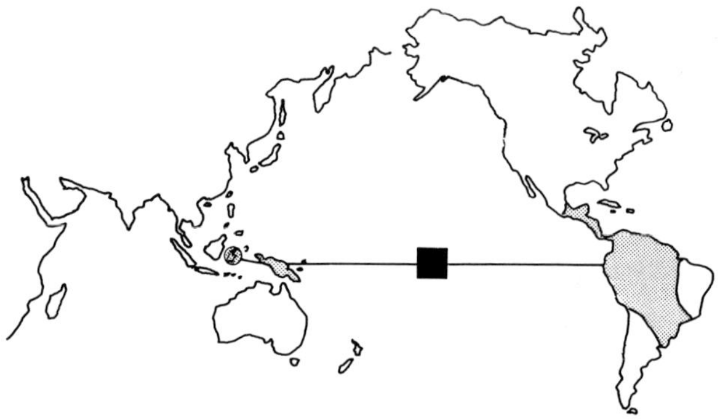

Fig. 14. Dispersal of Ithomiinae (Lepidoptera).

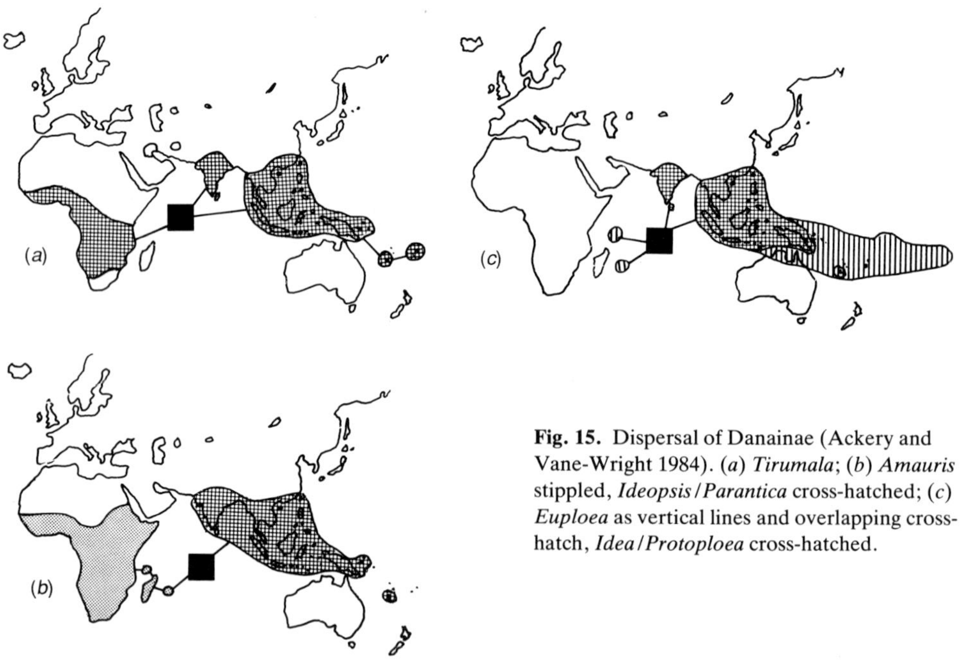

Fig. 15. Dispersal of Danainae (Ackery and Vane-Wright 1984). (*a*) *Tirumala*; (*b*) *Amauris* stippled, *Ideopsis*/*Parantica* cross-hatched; (*c*) *Euploea* as vertical lines and overlapping cross-hatch, *Idea*/*Protoploea* cross-hatched.

The suborder Ditrysia contains over 90% of all Lepidoptera species and it would therefore be desirable to have a detailed biogeographic analysis comparable to the non-ditrysian groups presented here. Observations by Shields (1976, 1979, 1983) suggest that even the 'higher' Lepidoptera have different ocean basin relationships. Shields (1976) adopted Croizat's base-line method (without formal application, although Shields was aware of panbiogeography and corresponded with Croizat on Lepidoptera–angiosperm origins; Shields 1988: 256) and supported a Jurassic–early Cretaceous butterfly radiation involving a Pacific history for groups present in the Neotropics and Indo-Australia but absent in Africa, such as the Ithomiinae (Fig. 14) and Morphinae-Amathusiinae (Shields 1976: 134).

A systematic study of the Danainae by Ackery and Vane-Wright (1984) offers distributional and phylogenetic information suitable for the *beginning* of a biogeographic analysis. Unfortunately, the records are only in the form of generalised tables so the maps presented here are not definitive. An Indian Ocean baseline may be assigned by minimal distance to three clades: (*a*) *Tirumala* (Fig. 15*a*), (*b*) *Amauris, Ideopsis, Parantica* (Fig. 15*b*) and (*c*) *Euploea, Idea, Protoploea* (Fig. 15*c*). Group (*c*) may be related to the American genera *Anetia* and *Lycorea* but the biogeographic connection (not to mention the phylogenetic link, J. S. Dugdale, personal communication) needs further analysis.

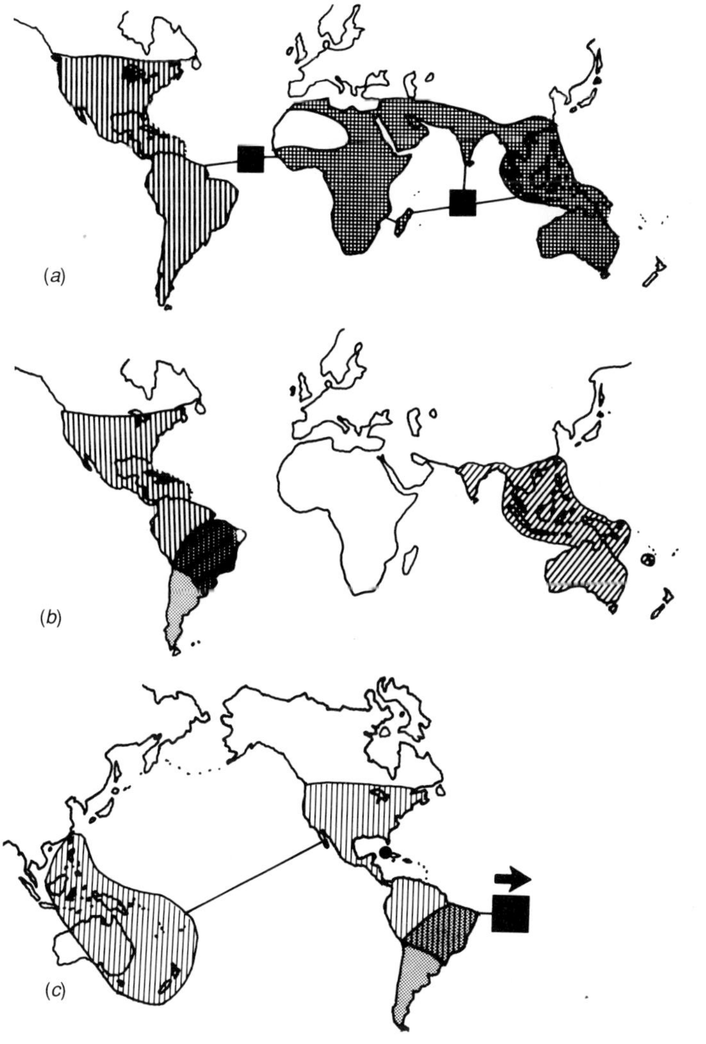

Fig. 16. Dispersal of Danaeinae. (*a*) *D.* (*Anosa*) *chrysippus* cross-hatched, *D. gilippus* vertical lines; (*b*) *D.* (*Anosa*) *eresimus* vertical lines, *D. plexaure* stippled, *Danaus* (*Salatra*) diagonal lines; (*c*) *D.* (*Danaus*) *cleophile* as solid circle, *D. plexippus* vertical lines, *D. erippus* stippled. Primary Atlantic orientation suggested by arrowed baseline.

The genus *Danaus* is divided into three subgenera. An Atlantic/Indian Ocean baseline is suggested for the *D.* (*Anosa*) group (Fig. 16*a*, *b*) with three species in the Americas (*D. ceresimus*, *D. plexaure*, *D. gallipus*) and *D. chrysippus* in and around the Indian Ocean (Fig. 16*a*). The *D.* (*Salatra*) group lies between the Indian and Pacific Oceans from India/southern China to Australia/New Caledonia (Fig. 16*b*) while *D.* (*Danaus*) is represented by three species in the Americas with one of these also present in the western Pacific (*D. chrysippus*) (Fig. 16*c*). I suggest that the western Pacific range is treated as a Pacific extension of the American distribution, i.e. the primary connection for this species is not across the Indian/Atlantic basins. In summary, the biogeography of the Danainae is principally Indian Ocean with a major extension across the Atlantic.

An Indian Ocean baseline is also proposed for the Tinissinae (Fig. 17) (Robinson 1976) and Perissomaticini (Fig. 18) (Peterson 1988) of the Tineidae, and the Callidulidae (Fig. 19) (Minet 1989). These distributions differ considerably in geographic outline, but their main massings share close geographic proximity to the Indian Ocean. The distributions of the Tinissinae and Callidulidae, in particular, are both absent from the Flores/Soemba node, a major biogeographic boundary in the East Indies (Croizat 1958). The dispersal of Callidulidae parallels angiosperms such as *Sonneratia* (Croizat 1968) and *Nepenthes* (Croizat 1958) that also bypass Flores/Soemba and reach Madagascar to the west. The Indian Ocean baseline of the Callidulidae agrees with Minet's (1989) linking the history of the Indian Ocean to the evolution of this family.

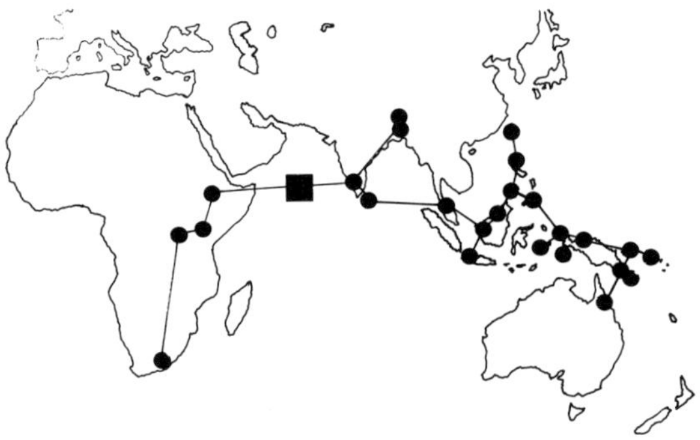

Fig. 17. Dispersal of the Tinissinae, Tineidae (Robinson 1976).

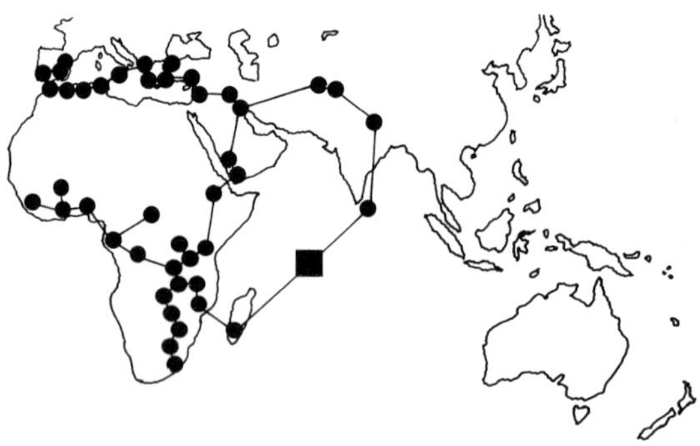

Fig. 18. Dispersal of the Perissomasticini, Tineidae (Peterson 1988).

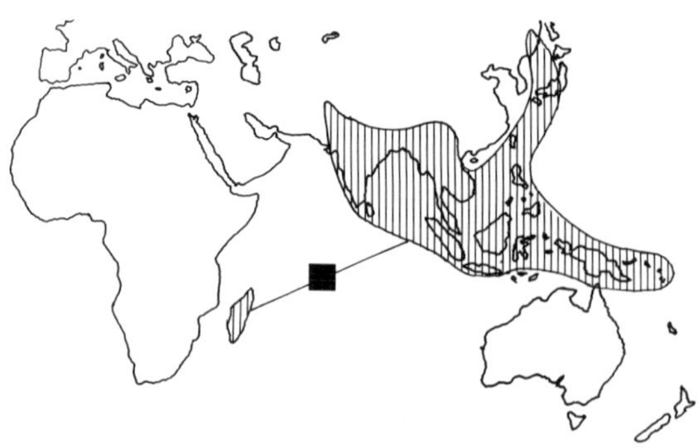

Fig. 19. Dispersal of the Callidulidae (Minet 1989).

Discussion

Non-biogeographic attempts to synthesise the evolutionary associations between angiosperms and Lepidoptera by Powell (1980) and Mitter and Brooks (1983) have met with limited success. An evaluation of feeding in the 'microlepidoptera' led Powell (1980) to offer only a 'vague picture' in which external feeding Ditrysia or their immediate ancestors had evolved by the mid-Cretaceous, possibly as fungus-feeding, detritus-feeding or early angiosperm-feeding forms. The lack of correlation between lepidopteran and angiosperm phylogenies led Powell (1980) to suggest that the principal evolutionary relationship occurred through development of specialised larval feeding niches or horizons within communities rather than along botanical 'evolutionary lines'. The ancestral branches of Lepidoptera were considered to be extinct leaving over an 'anomalous guild of relics' associated with certain lower plants. Mitter and Brooks (1983) also examined the phylogenetic patterns in Lepidoptera and their associated host plants as an estimator of mutual evolutionary history. The paucity of cladistic congruence between Lepidoptera and host-plant groups was attributed to host transfers having been so pervasive that the present-day associations have no significant joint history, or because extinction has obliterated the original pattern (essentially the same view as that of Powell 1980).

In this paper I have presented biogeographic evidence (i.e. spatial homologies) to suggest (*a*) The Pacific basin represents a major centre of evolution for both angiospermy and some of the Lepidoptera (principally non-ditrysian) and (*b*) both groups had differentiated into modern families or genera before Cretaceous. This proposed biogeographic history antedates the earliest fossil records and the difference may be attributed to either (*a*) a fallacious biogeography or (*b*) an extensive separation between phylogenetic origins and the earliest formal appearance of taxa in the fossil record. Croizat (1961, 1964*a*, 1964*b*) argued that different angiosperm groups had each evolved from 'pre-angiosperm' ancestors over a broad front of differentiation (this may be comparable to a 'parallel' development of characters in conventional systematics). Croizat's model is not necessarily incompatible with 'monophyly' as a clade where angiosperms remain each other's closest relatives (see also Heads 1984, 1990*b*). Hill and Crane (1982: 323) alluded to this possibility by acknowledging that it is unlikely that the synapomorphies defining the Angiospermae were all acquired at the same time. Doyle and Donoghue (1986: 384) similarly suggest the 'possibility' that angiosperm autapomorphies did not evolve *until* the Cretaceous. This continuing uncertainty over angiosperm evolution points to the importance of integrating current systematic studies of angiosperms with the biogeographic and morphogenetic analyses of angiosperm structure by Croizat (1961).

My biogeographic interpretations support some of the conclusions of Powell (1980) and Mitter and Brooks (1983) by suggesting an extensive history of close ecological association between Lepidoptera and angiosperms. There is, however, a major contrast concerning (*a*) the extended time-scale, (*b*) the proposal that the pre-Cretaceous diversity of host-plant associations is not extinct and (*c*) that the modern complexity of lepidopteran host-plant relationships reflects the original situation rather than later pervasive host transfers. I suggest that it is probable for modern angiosperm host-plant relationships of Lepidoptera (and many other insects) to have been established at a 'pre-angiosperm' level of morphological evolution. In this context the apparent lack of congruence between the higher Lepidoptera and their host-plant phylogenies (Powell 1980; Mitter and Brooks 1983) could be treated as a 'plesiomorphic' character rather than the result of subsequent extinction, migration or absence of a common history, i.e. the very early stages of lepidopteran host-plant relationships were characterised by broad host-plant associations with angiosperms or 'pre-angiosperms'. I do not imply that lepidopteran host-plant relationships had remained unchanged since the Jurassic, by major evolutionary events leading to the establishment of angiosperm associations by many Lepidoptera and angiosperm lineages (family and genus) had already occurred by the close of this period.

Studies by Craw (1982, 1983) have shown that systematic congruence of form characters alone is not necessarily sufficient to denote a common biogeographic history. I suggest that this also applies for coevolutionary studies of host-plant relationships since all aspects of the evolutionary process occur through space/time (Craw 1988). Identification of the ocean baseline orientation of animals and their host plants represents, therefore, an essential step in coevolutionary studies of the present biogeographic and systematic structure of ecological communities and regional biotas. It is no longer appropriate to place lepidopteran or angiosperm evolution in the traditional context of Wallacean regions that are only artefacts of current geopolitical boundaries (cf. Craw and Page 1988; Grehan 1988).

The ocean basin classification method provides a biogeographic estimate of the times of divergence between host-plant associations and also the incorporation of plants into the host range of an insect. For example, *Aenetus virescens* (Hepialidae) may have an Indian Ocean baseline, but the host range includes both Pacific and Indian/Atlantic angiosperms (Grehan

1984) brought together during the Rangatata fusion of Pacific and Gondwanic biotas (cf. Craw 1982, 1989, 1990; Heads 1990*a*). A panbiogeographic approach was recently adopted for coevolutionary studies by Humphries *et al.* (1986) for the evolution of *Nothofagus* and its parasites by proposing a Pacific baseline in a similar manner to Croizat's (1952) analysis of *Nothofagus* (Craw 1989). The division of angiosperms into different ocean basin affinities is well established through the extensive analyses by Croizat and in subsequent works by other authors, but the outline of Lepidoptera-angiosperm evolution suggested in this paper needs to be examined in detail for the major families and genera of the Lepidoptera and their host plants to develop a comprehensive evolutionary and ecological model of their coevolution.

Acknowledgments

Helpful discussion and comments on the contents of this paper were contributed by A. Drinnan, P. R. Crane, R. C. Craw, J. S. Dugdale, G. W. Gibbs, N. P. Kristensen, E. S. Nielsen, M. J. Scoble, D. L. Wagner, P. Weston and P. Whalley.

References

Ackery, P. R., and Vane-Wright, R. I. (1984). 'Milkweed Butterflies: their Cladistics and Biology.' (Cornell University Press: Ithaca, New York.)

Axelrod, D. I. (1970). Mesozoic paleogeography and early angiosperm history. *Botanical Review* **36**, 277-320.

Brenner, G. J. (1976). Middle Cretaceous floral provinces and early migrations of angiosperms. In 'Origin and Early Evolution of Angiosperms'. (Ed. C. B. Beck.) p. 23-47. (Columbia University Press: New York.)

Brooks, D. R. (1990). Parsimony analysis in historical biogeography and coevolution: methodological and theoretical update. *Systematic Zoology* **39**, 14-30.

Chamberlain, V. E., and Lambert, R. St J. (1985). Cordilleria: a newly defined Canadian microcontinent. *Nature* **314**, 707-13.

Cleal C. J. (1989). Angiosperm phylogeny: evolution in hidden forests. *Nature* **339**, 16.

Common, I. F. B. (1970). Lepidoptera. In 'The Insects of Australia'. (Ed. D. F. Waterhouse.) pp. 765-866. (Melbourne University Press: Melbourne.)

Common, I. F. B. (1975). Evolution and classification of the Lepidoptera. *Annual Review of Entomology* **20**, 183-201.

Coney, P. J., Jones, D. L., and Monger, J. W. H. (1980). Cordilleran suspect terranes. *Nature* **288**, 328-33.

Coward, M. P., and Ries, A. C. (1986). 'Collision Tectonics.' (Blackwell Scientific Publications: Oxford.)

Crane, P. R. (1985). Phylogenetic analysis of seed plants and the origin of angiosperms. *Annals of the Missouri Botanical Garden* **72**, 716-93.

Crane, P. R. (1987). The leaf venation and reproductive structures of a late Triassic angiosperm, *Sanmiguelia lewisii* (review). *Taxon* **36**, 778-9.

Crane, P. R., Doyle, J. A., and Friis, E. M. (1989). Angiosperm origins. *Nature* **342**, 131.

Craw, R. C. (1982). Phylogenetics, areas, geology and the biogeography of Croizat: a radical view. *Systematic Zoology* **32**, 304-16.

Craw, R. C. (1983). Panbiogeography and vicariance cladistics: are they truly different? *Systematic Zoology* **32**, 431-7.

Craw, R. C. (1984). Biogeography and biogeographical principles. *New Zealand Entomologist* **8**, 49-52.

Craw, R. C. (1985). Classic problems of southern hemisphere biogeography re-examined. *Zeitschrift fuer Zoologische Systematik und Evolutionsforschung* **23**, 1-10.

Craw, R. C. (1988). Panbiogeography: method and synthesis on biogeography. In 'Biogeographic Analysis: Methods, Patterns, Processes'. (Eds A. A. Myers and P. S. Giller.) pp. 405-35. (Chapman and Hall: London.)

Craw, R. C. (1989). Continuing the synthesis between panbiogeography, phylogenetic systematics and geology as illustrated by empirical studies on the biogeography of New Zealand and the Chatham islands. *Systematic Zoology* **37**, 291-310.

Craw, R. C. (1990). New Zealand biogeography: a panbiogeographic approach. *New Zealand Journal of Zoology* **16**, 527-47.

Craw, R. C., and Heads, M. J. (1988). Reading Croizat: on the edge of biology. *Rivista di Biologia Forum* **81**, 499-532.

Craw, R. C., and Page, R. (1988). Panbiogeography: method and metaphor in the new biogeography. In 'Process and Metaphor in the Evolutionary Paradigm'. (Eds M. W. Ho and S. Fox.) (John Wiley and Sons Ltd: Chichester, England.)

Craw, R. C., and Weston, P. (1984). Panbiogeography: a progressive research programme? *Systematic Zoology* **33**, 1-13.

Crepet, W. L. (1984). Ancient flowers for the faithful. *Natural History* **93**, 38–44.

Croizat, L. (1952). 'Manual of Phytogeography.' (Junk: The Hague.)

Croizat, L. (1958). 'Panbiogeography.' (Published by the author: Caracas.)

Croizat, L. (1961). '*Principia Botanica*.' (Published by the author: Caracas.)

Croizat, L. (1964*a*). 'Space, Time, Form: the Biological Synthesis.' (Published by the author: Caracas.)

Croizat, L. (1964*b*). Thoughts on high systematics, phylogeny and floral morphogeny, with a note on the origin of the Angiospermae. *Candollea* **19**, 17–96.

Croizat, L. (1968). The biogeography of the tropical lands and islands east of Suez-Madagascar: with particular reference to the dispersal and form-making of *Ficus* L., and different other vegetal and animal groups. *Attidel' Istituto Botanico della Univiversita e Laboratorio Crittogamico Pavia*, Ser. 6, **4**, 1–400.

Croizat, L. (1988). Out of nothing, nothing comes: Interview with J. Baskin. *Rivista di Biologia Forum* **81**, 589–611.

Cusset, G., and Cusset, C. (1988). Etudes sur les *Podostemopsida*. 11. Répartition et évolution des *Tristichaceae*. *Adansonia* **3**, 223–62.

Darwin, C. (1859). 'On the Origin of Species by Natural Selection.' (John Murray: London.)

Davis, D. R. (1967). A revision of the moths of the subfamily Prodoxinae (Lepidoptera: Incurvariidae). United States National Museum Bulletin 255.

Davis, D. R. (1975). Systematics and zoogeography of the family Neopseustidae with the proposal of a new subfamily (Lepidoptera: Neopseustoidea). *Smithsonian Contributions to Zoology* **210**, 1–27.

Davis, D. R. (1978). A revision of the North American moths of the superfamily Eriocranioidea with the proposal of a new family Acanthopteroctetidae (Lepidoptera). *Smithsonian Contributions to Zoology* **251**, 1–131.

Davis, D. R. (1986). A new family of monotrysian moths from austral South America (Lepidoptera: Palaephatidae) with a phylogenetic review of the Monotrysia. *Smithsonian Contributions to Zoology* **434**, 1-202.

Doyle, J. A. (1978). Origin of angiosperms. *Annual Review of Ecology and Systematics* **9**, 365–92.

Doyle, J. A., and Donoghue, M. J. (1986). Seed plant phylogeny and the origin of angiosperms: an experimental cladistic approach. *Botanical Review* **52**, 321-431.

Doyle, J. A., and Donoghue, M. J. (1987). The origin of angiosperms. In 'The Origins of Angiosperms and their Biological Consequences'. (Eds E. M. Friis, W. G. Chaloner and P. R. Crane.) pp. 17–49. (Cambridge University Press: Cambridge.)

Dugdale, J. S. (1989). A preliminary note on Incurvarioidea in New Zealand. *New Zealand Journal of Zoology* **16**, 277–81.

Friis, E. M., Chaloner, W. G., and Crane, P. R. (Eds) (1987). 'The Origins of Angiosperms and their Biological Consequences.' (Cambridge University Press: Cambridge.)

Gibbs, G. W. (1982). Evolution of Micropterigidae (Lepidoptera) in the SW Pacific. *Geo Journal* **7**, 505–10.

Grande, L. (1985). The use of paleontology in systematics and biogeography, and a time control refinement for historical biogeography. *Paleobiology* **11**, 234–43.

Gray, R. D. (1988). Metaphors and methods: behavioural ecology, panbiogeography and the evolving synthesis. In 'Process and Metaphor in the Evolutionary Paradigm'. (Eds M. W. Ho and S. Fox.) pp. 209–42. (John Wiley and Sons Ltd: Chichester, England.)

Gray, R. D. (1990). Oppositions in panbiogeography: can the conflicts between selection, constraint, ecology, and history be resolved. *New Zealand Journal of Zoology* **16**, 787–806.

Grehan, J. R. (1984). The host range of *Aenetus virescens* (Lepidoptera: Hepialidae) and its evolution. *New Zealand Entomology* **8**, 51–64.

Grehan, J. R. (1987). Evolution of arboreal tunnelling by the larvae of *Aenetus* (Lepidoptera: Hepialidae). *New Zealand Journal of Zoology* **14**, 441–62.

Grehan, J. R. (1988). Panbiogeography: evolution in space and time. *Rivista di Biologia Forum* **81**, 469–85.

Grehan, J. R. (1989). Larval feeding habits of the Hepialidae. *Journal of Natural History* **23**, 803–24.

Grehan, J. R., and Henderson, I. M. (in press). Panbiogeography. In 'Vertebrate Zoogeography and Evolution in Australasia'. (Eds M. Archer, J. Long and G. Clayton.) (Hesperian Press: Marrickville, N.S.W.)

Heads, M. J. (1984). *Principia Botanica*: Croizat's contribution to botany. *Tuatara* **27**, 26–48.
Heads, M. J. (1985*a*). On the nature of ancestors. *Systematic Zoology* **34**, 205–15.
Heads, M. J. (1985*b*). Biogeographic analysis of *Nothofagus* (Fagaceae). *Taxon* **34**, 474–92.
Heads, M. J. (1990*a*). Integrating earth and life sciences in New Zealand natural history: the parallel arcs model. *New Zealand Journal of Zoology* **16**, 549–85.
Heads, M. J. (1990*b*). Mesozoic tectonics and the deconstruction of biogeography: a new model of Australasian biology. *Journal of Biogeography* **17**, 223–5.
Henderson, I. M. (1990). Quantitative panbiogeography: an investigation into concepts and methods. *New Zealand Journal of Zoology* **16**, 495–510.
Hickey, L. J., and Doyle, A. J. (1977). Early Cretaceous fossil evidence for angiosperm evolution. *Botanical Review* **43**, 3–104.
Hill, C. R., and Crane, P. R. (1982). Evolutionary cladistics and the origin of angiosperms. In 'Problems of Phylogenetic Reconstruction'. (Eds K. A. Joysey and A. E. Friday.) pp. 269–361. (Academic Press: London and New York.)
Howell, D. G. (1985). 'Tectonostratigraphic Terranes of the Circum-Pacific Region.' (Circum-Pacific Council for Energy and Mineral Resources: Houston, Texas, U.S.A.)
Hughes, N. F. (1973). Cretaceous paleobotanic problems. In 'Origin and Early Evolution of Angiosperms'. (Ed. C. Beck.) pp. 11–46. (Columbia University Press: New York.)
Hughes, N. F. (1976). 'Paleobiology of Angiosperm Origins.' (Cambridge University Press: Cambridge.)
Hughes, N. F. (1977). Paleo-succession of earliest angiosperm evolution. *Botanical Review* **43**, 105–27.
Hull, D. L. (1988). 'Science as a Process: An Evolutionary Account of the Social and Conceptual Development of Science.' (University of Chicago Press: Chicago.)
Humphries, C. J. (1985). Temperate biogeography and an intemperate botanist. *Taxon* **34**, 480–92.
Humphries, C. J., Cox, J. M., and Nielsen, E. S. (1986). *Nothofagus* and its parasites: a cladistic approach to coevolution. In 'Coevolution and Systematics'. (Eds A. R. Stone and D. L. Duckworth.) pp. 55–76. (Clarendon Press: Oxford.)
Humphries, C. J., and Parenti, L. (1986). 'Cladistic Biogeography.' (Clarendon Press: Oxford.)
Humphries, C. J., Ladiges, P. Y., Roos, M., and Zandee, M. (1988). Cladistic biogeography. In 'Analytical Biogeography: an Integrated Approach to the Study of Animal and Plant Distribution'. (Eds A. A. Myers and P. S. Giller.) pp. 371–404. (Chapman and Hall Ltd: London.)
Hutchinson, J. (1969). 'Evolution and Phylogeny of Flowering Plants.' (Academic Press: London and New York.)
Jones, D. L., Cox, A., Coney, P., and Beck, M. (1982). The growth of western North America. *Scientific American* **247**, 50–64.
Krassilov, V. A. (1977). The origin of angiosperms. *Botanical Review* **43**, 143–76.
Kristensen, N. P. (1978*a*). Observations on *Anomoses hylecoetes* (Anomosetidae), with a key to the hepialoid families (Insecta: Lepidoptera). *Steenstrupia* **5**, 1–19.
Kristensen, N. P. (1978*b*). A new familia of Hepialoidea from South America, with remarks on the phylogeny of the subordo Exoporia (Lepidoptera). *Entomologica Germanica* **4**, 272–94.
Kristensen, N. P. (1984). Studies on the morphology and systematics of primitive Lepidoptera (Insects). *Steenstrupia* **10**, 141–91.
Kristensen, N. P., and Nielsen, E. S. (1979). A new subfamily of micropterigid moths from South America. A contribution to the morphology of the Micropterigidae, with a generic catalogue of the family (Lepidoptera: Zeugloptera). *Steenstrupia* **5**, 69–147.
Kristensen, N. P., and Nielsen, E. S. (1983). The *Heterobathmia* life history elucidated: Immature stages contradict assignment to suborder Zeugloptera (Insecta: Lepidoptera). *Zeitschrift fuer Zoologische Systematik und Evolutionsforschung* **21**, 101–24.
Li, Wen-Hsiung, Wolfe, K. H., and Sharp, P. M. (1989). Angiosperm origins. *Nature* **342**, 131–2.
Martin, W., Gierl, A., and Saedler, H. (1989). Molecular evidence for pre-Cretaceous angiosperm origins. *Nature* **339**, 46–8.
Matthews, C. (1990). (Ed.) Panbiogeography special issue. *New Zealand Journal of Zoology* **16**, i–iv, 471–806.
Melville, R. (1983). Glossopteridae, Angiospermidae and the evidence for angiosperm origin. *Journal of the Linnean Society of London, Botany* **86**, 279–323.
Minet, J. (1989). Nouvelles frontières, géographiques et taxonomiques, pour la famille des Callidulidae (Lepidoptera, Calliduloidea). *Nouvelle Revue Entomologie* (*N.S.*) **4**, 351–68.
Mitter, C., and Brooks, D. R. (1983). Phylogenetic aspects of coevolution. In 'Coevolution'. (Eds D. J. Futuyma and M. Slatkin.) (Sinauer Associates Inc.: Sunderland, Massachusetts.)
Muller, J. (1984). Significance of fossil pollen for angiosperm history. *Annals of the Missouri Botanical Garden* **71**, 419–43.

Nelson, G. (1973). Comments on Leon Croizat's biogeography. *Systematic Zoology* **22**, 312–20.

Nelson, G. (1978). From Candolle to Croizat: comments on the history of biogeography. *Journal of Historical Biology* **11**, 269–305.

Nelson, G., and Platnick, N. (1981). 'Systematics and Biogeography: Cladistics and Vicariance.' (Columbia University Press: New York.)

Nelson, G., and Rosen, D. (1981). (Eds) 'Vicariance Biogeography: A Critique.' (Columbia University Press: New York.)

Nielsen, E. S. (1982*a*). Review of the higher classification of the lepidoptera, with special reference to lower heteroneurans. *Tyo to Ga* **33**, 98–101. (Transactions of the Lepidopterological Society of Japan.)

Nielsen, E. S. (1982*b*). Incurvariidae and Prodoxidae from the Himalayan area (Lepidoptera: Incurvarioidea). *Insecta Matsummurana* **26**, 187–200.

Nielsen, E. S. (1985*a*). The monotrysian heteroneuran phylogeny puzzle: a possible solution (Lepidoptera). Proceedings of the 3rd Congress of European Lepidoptera, Cambridge 1982.

Nielsen, E. S. (1985*b*). Primitive (non-ditrysian) Lepidoptera of the Andes: diversity, distribution, biology and phylogenetic relationships. *Journal of Research on the Lepidoptera* **1**, 1–16.

Nielsen, E. S. (1987). The recently discovered primitive (non-Ditrysian) family Palaephatidae (Lepidoptera) in Australia. *Invertebrate Taxonomy* **1**, 201–29.

Nielsen, E. S. (1988). The peculiar Asian ghost moth genus *Bipectilus* Chu & Wang: taxonomy and systematic position (Lepidoptera: Hepialidae s.str.). *Systematic Entomology* **13**, 171–95.

Nielsen, E. S., and Davis, D. R. (1985). The first southern hemisphere prodoxid and the phylogeny of the Incurvarioidea (Lepidoptera). *Systematic Entomology* **10**, 307–22.

Nielsen, E. S., and Scoble, M. J. (1983). Ghost moths of southern South America (Lepidoptera: Hepialidae). *Entomonograph* **4**, 1–192.

Nielsen, E. S., and Scoble, M. J. (1986) *Afrotheora*, a new genus of primitive Hepialidae from Africa (Lepidoptera: Hepialoidea). *Entomologica Scandinavica* **17**, 29–54.

Nur, A., and Ben-Avraham, Z. (1977). Lost Pacifica continent. *Nature* **280**, 41–3.

Nur, A., and Ben-Avraham, Z. (1981). Lost Pacifica continent: a mobilistic interpretation. In 'Vicariance Biogeography: a Critique'. (Eds G. Nelson and D. E. Rosen.) pp. 341–58. (Columbia University Press: New York.)

Page, R. D. M. (1987). Graphs and generalised tracks: quantifying Croizat's panbiogeography. *Systematic Zoology* **36**, 1 17.

Page, R. D. M. (1989). Quantitative cladistic biogeography: constructing and comparing area cladograms. *Systematic Zoology* **37**, 254–70.

Peterson, G. (1988). Revision der Perissomasticini (Lepidopetera: Tineidae). *Beitraege zur Entomologie Berlin* **38**, 3–64.

Platnick, N., and Nelson, G. (1984). Composite areas in vicariance biogeography. *Systematic Zoology* **33**, 328–35.

Platnick, N., and Nelson, G. (1989). Spanning tree biogeography: shortcut, detour or dead-end? *Systematic Zoology* **37**, 410–19.

Powell, J. A. (1980). Evolution of larval feeding preferences in microlepidoptera. *Annual Review of Entomology* **25**, 133–59.

Prance, G. T., and White, F. (1988). The genera of Chrysobalanaceae: a study in practical and theoretical taxonomy and its relevance to evolutionary biology. *Philosophical Transactions of the Royal Society of London B* **320**, 1–184.

Rasnitsyn, A. P. (1983). Pervaya nakhodka babochki yurskogo vozrasta. *Doklady Akademii Nauk SSSR* **269**, 467–71.

Raven, P. H., and Axelrod, D. I. (1974). Angiosperm biogeography and past continental movements. *Annals of the Missouri Botanical Garden* **61**, 439–673.

Retallack, G., and Dilcher, D. L. (1981). Arguments for a glossopterid ancestry of angiosperms. *Paleobiology* **7**, 54–67.

Robinson, G. S. (1976). A taxonomic revision of the Tinissinae of the world (Lepidoptera: Tineidae). *Bulletin of the British Museum* (*Natural History*), Entomology **32**, 255–300.

Schuster, R. M. (1976). Plate tectonics and its bearing on the geographical origin and dispersal of angiosperms. In 'Origin and Early Evolution of Angiosperms'. (Ed. C. B. Beck.) pp. 48–138. (Columbia University Press.)

Scoble, M. J. (1983). A revised cladistic classification of the Nepticulidae (Lepidoptera). *Transvaal Museum Monograph* **2**, 1–105.

Seberg, O. (1986). A critique of the theory and methods of panbiogeography. *Systematic Zoology* **35**, 369–80.

Serlin, B. S. (1982). An early Cretaceous fossil flora from northwest Texas: its composition and implications. *Palaeontographia Italica Abteilung B* **182**, 52–86.

Shields, O. (1976). Fossil butterflies and the evolution of Lepidoptera. *Journal of Research on the Lepidoptera* **15**, 132–43.

Shields, O. (1979). Butterfly distribution and continental drift between the Americas, the Caribbean and Africa. *Journal of Natural History* **13**, 221–50.

Shields, O. (1983). Trans-Pacific biotic links that suggest earth expansion. In 'Expanding Earth Symposium'. (Ed. S. W. Carey.) pp. 199–205. (University of Tasmania: Hobart.)

Shields, O. (1988). Mesozoic history and neontology of Lepidoptera in relation to Trichoptera, Mecoptera, and Angiosperms. *Journal of Paleontology* **62**, 251–8.

Silberling, N. J. (1985). Biogeographic significance of the Upper Triassic bivalve *Monotis* in circum-Pacific accreted terranes. In 'Tectonostratigraphic Terranes of the Circum-Pacific Region'. (Ed. D. G. Howell.) pp. 63–70. (Circum-Pacific Council: Houston, Texas.)

Takhtajan, A. (1969). 'Flowering Plants: Origin and Dispersal.' (Smithsonian Institution Press: Washington.)

Tindale, N. B. (1981). The origin of Lepidoptera relative to Australia. In 'Ecological Biogeography of Australia'. (Ed. A. Keast.) pp. 356–77. (Junk: The Hague.)

Van Nieukerken, E. J. (1986). Systematics and phylogeny of holarctic genera of Nepticulidae (Lepidoptera, Heteroneura: Monotrysia). *Zoologische Verhandelingen Rijksmuseum van Natuurlijke Historie* (Leiden) **236**, 3–93.

Wallace, A. R. (1876). 'The Geographical Distribution of Animals.' (MacMillan: London.)

Whalley, P. (1986). A review of the current fossil evidence of Lepidoptera in the Mesozoic. *Biological Journal of the Linnean Society* **28**, 253–71.

Whalley, P. (1986). A review of the current fossil evidence of Lepidoptera in the Mesozoic. *Biological Journal of the Linnaens Society* **28**, 253–71.

Wolfe, J. A. (1981). Vicariance biogeography of angiosperms in relation to paleobotanical data. In 'Vicariance Biogeography: A Critique'. (Eds G. Nelson and D. Rosen.) (Columbia University Press: New York.)

Wolfe, J. A., Doyle, J. A., and Page, V. M. (1975). The bases of angiosperm phylogeny: paleobotany. *Annals of the Missouri Botanical Garden* **62**, 801–24.

Manuscript received 17 January 1990, accepted 17 December 1990

3 September 1991 *Aust. Syst. Bot.*, **4**, 111–16

Application of Evidence from Molecular Biology to the Biogeography of Angiosperms

P. G. Martin and J. M. Dowd

Department of Botany, University of Adelaide,
Box 498, G.P.O., Adelaide, S.A. 5001, Australia.

Abstract

Previously published sequence data from species, whose ancestors are thought to have been separated during the break-up of Gondwanaland, yield mean number of inferred nucleotide differences (i.n.d.) for two comparisons, Africa *versus* Australia and New Zealand *versus* Australia. In the geophysical literature, the dating of these separations appears to be uncontroversial, viz. 130 Ma and 80 Ma respectively. The mean rates for the two comparisons are one i.n.d. in 11·7 Ma and 11·9 Ma. The similarity of these estimates encouraged us to deduce the dates of other events for which we present new data. The deepest division detected within the angiosperms corresponds to 244 Ma (Triassic). This is compared with recent equivalent datings. The division between representatives of northern primitive families (Magnoliaceae, Calycanthaceae) and southern primitive families (Winteraceae, Idiospermataceae) corresponds to 162 Ma (Upper Jurassic). This is discussed briefly in relation to geophysical information. Although there is considerable scope for error in this approach, it is considered to be worth pursuing in the light of the imminence of more and longer sequences from nucleic acids and of the interest and difficulty of such biogeographical problems.

Introduction

Through the derivation of phylogenetic trees from macromolecular sequence data, molecular biology can elucidate the course that evolution has taken. Under certain circumstances, molecular data can also yield rates of evolution and divergence times which, in turn, can be used in biogeographical studies. For example, Parks and Wendel (1990) have deduced that *Liriodendron* species from China and North America diverged at 11–14 Ma.

This study derives from a survey of the dicotyledons (Martin and Dowd 1990, 1991) in which, for more than 300 dicots, 20 monocots and five gymnosperms, the N-terminal 40 amino acid sequences of the small subunit of rubisco were determined. After conversion to inferred nucleotide sequences using the genetic code parsimoniously, there were 71 variable sites; comparison of species yielded inferred nucleotide differences, henceforth i.n.d. A phylogenetic tree was computed and the ancestral sequence of the base of the angiosperm tree derived. The family nearest the base, i.e. closest to the gymnosperms, was Schisandraceae. This result, which our data indicated unequivocally, is at variance with that of Donoghue and Doyle (1989), but more in line with that of Sporne (1980), whose index of advancement placed Schisandraceae closer to the base of the dicotyledons than all but two of the 26 primitive dicot families studied by Donoghue and Doyle (1989). In our tree, the monocots arose from within the dicot tree near Calycanthaceae and Aristolochiaceae.

1030-1887/91/010111$05.00

Rates of Evolution

When deriving the phylogenetic tree, we grouped the 121 dicot families into 26 major groups which had been derived from a consensus of the phylogenetic systems of Cronquist (1981), Dahlgren (1983), Takhtajan (1983) and Thorne (1983). For every species, the number of differences (i.n.d.) between it and the computed base of the angiosperm tree was counted. Although the mean value of the slowest evolving group (14·1) was not dramatically different from the fastest (19·6), analysis of variance (Table 1*a*) showed that the differences between groups were significant. This test, which is a form of the relative rate test of Wu and Li (1985), indicated that rates of evolution vary and therefore cannot be applied widely. The taxa that will be considered below were 11 species of *Nothofagus*, 14 species of *Solanum*, 12 species from Proteaceae and 11 species from Magnoliiflorae. These belonged to four of our groups, all of which evolved more slowly than average (their mean was 14·8) and analysis of variance (Table 1*b*) showed no significant difference among these groups. This indicates that errors due to different rates of evolution can be ignored for this limited group.

In order to estimate rates of evolution, we have derived times from the breakup of Gondwanaland. Although the sequence of events is not completely certain, there are two separation times whose dates are uncontroversial in the geophysical literature. In a review, Scotese *et al.* (1988) gave the date of separation of Africa from East Gondwanaland as 130 Ma, quoting Rabinowitz and Labrecque (1979), and for the isolation of New Zealand as 80 Ma, quoting Weissel and Hayes (1977).

Table 1. Analyses of variance of distances (inferred nucleotide differences) from derived angiosperm ancestral sequence to extant species

	Sum of squares	Degree of freedom	Variance	Ratio	*P*
		(*a*) 26 groups of dicots			
Between groups	1534·9	25	61·40	10·39	< 0·001
Within groups	1431·2	276	5·91		
Total	2966·1	301			
		(*b*) Four groups which contributed to clock			
Between groups	38·0	3	12·67	1·13	> 0·20
Within groups	492·0	44	11·18		
Total	530·0	47			

Martin and Dowd (1988) have published sequence data that can be related to these two dates, the assumption being that when Africa separated from East Gondwanaland, Australian members of Proteaceae and *Solanum* were isolated from African relatives and when New Zealand was isolated its representatives in *Nothofagus*, Proteaceae and Winteraceae were separated from relatives in Australia. The mean of 15 comparisons between pairs from Australia and Africa was 11·10 ± 0·67 i.n.d. (mean ± s.e.), which gives a rate of 1 i.n.d. in 11·7 Ma. The mean of 10 comparisons between Australia and New Zealand had a mean of 6·70 ± 0·56, which gives a rate of 1 i.n.d. in 11·9 Ma. The very good agreement between these two estimates encourages us to use their mean (1 i.n.d. in 11·8 Ma) to deduce the dates of other putative separations for which we have corresponding molecular data.

Hooker's South Temperate Flora

The flora of southern South America, i.e. the southern parts of Chile and Argentina, includes components of the south temperate flora first described by Hooker (1851). We have insufficient appropriate comparisons between Africa and South America, or between Africa and New Zealand, but we have published (Martin and Dowd 1988) 15 comparisons between South America and Australia (Proteaceae, Winteraceae, *Nothofagus*) which had a mean of 5·93 ± 0·48 i.n.d. New data for the comparison New Zealand–South America are given in Table 2 and the mean is 5·31 ± 0·33 i.n.d. Thus, if we apply the rate of 1 i.n.d. in 11·8 Ma, times of 70 Ma and 63 Ma (both near the Cretaceous–Tertiary boundary) can be deduced for the migration barrier between South America and Australia, and South America and New Zealand respectively. We do not wish to compare these estimates with geophysical data because the intervening West Antarctic archipelago consists of at least five crustal blocks (Storey *et al.*, 1988) whose movements are uncertain.

Two Biogeographical Groups of Primitive Dicotyledons

Schuster (1972) mapped the present distributions of what he considered to be the dicot families that are 'by and large the most primitive and conservative extant Angiospermae'. In his figs 29–31, Schuster showed a bi-hemispheric distribution with very little overlap across Wallace's line. He drew the conclusion that the two groups had been in their present ranges when the Australian plate collided with Indonesia in the Miocene and had migrated little since. Martin (1977) made the obvious deduction that if the angiosperms are monophyletic the two groups must have been separated much earlier. It is the date of this putative separation that we now consider.

Table 2. Inferred nucleotide differences between New Zealand and South American species

A. Nothofagus species				
New Zealand	South America			
	betuloides	*procera*	*antarctica*	*pumilio*
fusca	7	5	5	4
menziesii	5	6	5	3
solandri	6	6	6	4
B. Winteraceae				
Drimys winteri (S. Am.) — *Pseudowintera colorata* (N.Z.)				7

Table 3. Inferred nucleotide differences between primitive angiosperms from either side of Wallace's line, as described by Schuster (1972)

Northern group	Southern group			
		Winteraceae		Idiospermataceae
	Drimys winteri	*Pseudowintera colorata*	*Tasmannia purpurascens*	*Idiospermum australiense*
Calycanthaceae				
Calycanthus floridus	13	14	14	10
Chimonanthus praecox	14	15	17	13
Magnoliaceae				
Liriodendron tulipiferum	18	13	15	14
Magnolia soulangiana	14	12	13	14
Michelia figo	14	11	13	13
Schisandraceae				
Kadsura japonica	21	21	23	22
Schisandra rubriflora	21	19	21	18

During our survey of the dicots, we studied some of the families mapped by Schuster (1972). In the northern group we sampled Calycanthaceae, Magnoliaceae and Schisandraceae and in the southern group, Winteraceae. To the latter we have added the monospecific *Idiospermum* which was not described in 1972, but since it is either placed in a family on its own or in Calycanthaceae, is clearly appropriate here. The differences are given in Table 3. It is clear that comparisons involving Schisandraceae (mean 20·7) are greater than the others (mean 13·7) and this difference is significant ($P < \cdot 001$). Because Schisandraceae is the family closest to the base of our phylogenetic tree, we interpret this as meaning that the earliest divergence we have observed corresponds to $20{\cdot}7 \times 11{\cdot}8 = 244$ Ma. The average difference between the rest of the northern group and the southern group is $13{\cdot}7 \times 11{\cdot}8 = 162$ Ma.

Discussion

This approach to the study of plant evolution and biogeography is of a pioneering nature with, we believe, the potential for expansion and greater precision when data from nucleic acid sequencing become available. While this is the spirit in which we present the above, we also want to record the sources of error that we recognise. We are completely dependent on the earth sciences for estimates of dates of continental separations and must assume that intercontinental migration and gene flow stop at those times. The choice of species for intercontinental comparisons may be faulty in that some long distance dispersal, subsequent to drift, may have occurred in some taxonomic groups. There is also an assumption of a linear relationship between time and the number of differences, when theory predicts that there may be saturation with number of differences per unit time decreasing with time (Li *et al.* 1985). Finally, it is always precarious to extrapolate beyond the range of the data as we have done in the previous paragraph. These are formidable potential sources of error, but we do not foresee a more profitable experimental and predictive approach to the problems at the interface between biogeography and macroevolution. It is therefore instructive to see how realistic our deductions appear to be.

Our deduced age of the angiosperms is 244 Ma, which is Triassic. Donoghue and Doyle (1989) have carried out careful cladistic studies that use 54 morphological characters and these indicate that the sister-group of the angiosperms includes extant Gnetales and Triassic fossil Bennettitales; from this they have deduced that the angiosperm lineage was distinct in the Triassic. The same study confirmed that the monocots arose from within the dicots and this is relevant to another recent approach, the molecular sequence studies of Wolfe *et al.* (1989). In this study, the age of the monocot–dicot divergence was estimated as 200 Ma with an uncertainty of about 40 Ma; because the monocots arose from within the dicots, this is probably an underestimate for the age of the angiosperms. Thus, despite all the possible sources of error, our own estimate agrees with these other studies.

It is difficult to find a biological comparison against which to test our deduction that the northern and southern groups of primitive angiosperms diverged at 162 Ma (upper Jurassic). However, it is possible to examine it in relation to what is known about geological movements at that time. Upper Jurassic magnetic lineations, which are oriented approximately ENE.–WSW. and with a northerly origin, are located in two oceanic areas near South-East Asia: the Argo lineations (Veevers *et al.* 1985) are between the north-western coast of Australia and Sumba; the Japanese lineations (Larson and Chase 1972; Larson and Pitman 1972) are near the Mariana trench. The latter are associated with the Japanese ridge postulated by Larson and Pitman (1972), whose fig. 7 suggests that the ridge may also be connected with the subsequently discovered Argo lineations. (In a personal communication, R. Larson has confirmed that this is a reasonable hypothesis.) Thus, at about the right time, there was a ridge system that could have effected a division between land masses carrying the northern and southern groups of primitive angiosperms in the general region where they subsequently came together again at Wallace's line.

The fact that we have dated the north–south separation following a hypothesis set in the South-East Asian region does not exclude the possibility that it occurred elsewhere. Raven and Axelrod (1974) have suggested that the angiosperms originated in West Gondwanaland (Africa plus South America), although Takhtajan (1987) has argued on botanical grounds against this and in favour of a South-East Asian origin. The papers of Zhao *et al.* (1990), Metcalfe (1990) and Audley-Charles (1991) suggest that in the early Mesozoic there was no shortage of terranes and isolated landmasses between Australia and Asia including the South China and North China blocks that are so rich in both extant and fossil angiosperms.

To establish firmly the ideas that we present will require more data and more sophisticated analyses, but the main point we make is that sequence data have the potential to elucidate biogeographical problems.

Acknowledgment

This work has been supported by the Australian Research Council.

References

Audley-Charles, M. G. (1991). Tectonics of the New Guinea Area. *Annual Review of Earth and Planetary Science* **19,** 17–41.

Cronquist, A. (1981). 'An Integrated System of Classification of Flowering Plants.' (Columbia University Press: New York.)

Dahlgren, R. (1983). General aspects of angiosperm evolution and macrosystematics. *Nordic Journal of Botany* **3,** 119–49.

Donoghue, M. J., and Doyle, J. A. (1989). Phylogenetic studies of seed plants and angiosperms based on morphological characters. In 'Nobel Symposium 70, The Hierarchy of Life'. (Eds B. Fernholm, K. Bremer and H. Jornvall.) pp. 181–193. (Elsevier: Amsterdam.)

Hooker, J. D. (1851). 'The Botany of the Antarctic Voyage of H. M. Discovery Ships Erebus and Terror in the Years 1839–1843. II. Flora Novae-Zelandiae. Part 1. Flowering Plants. Introductory Essay.' (Lovell Reeve: London.)

Larson, R. L., and Chase, C. G. (1972). Late Mesozoic evolution of the western Pacific Ocean. *Geological Society of America Bulletin* **83,** 3627–44.

Larson, R. L., and Pitman, W. C. (1972). World-wide correlation of mesozoic magnetic anomalies, and its implications. *Geological Society of America Bulletin* **83**, 3645–62.

Li, W.-H., Luo, C.-C., and Wu, C.-I. (1985). Evolution of DNA Sequences. In 'Molecular Evolutionary Genetics'. (Ed. R. J. McIntyre.) pp. 1–94. (Plenum Press: New York.)

Martin, P. G. (1977). Marsupial biogeography and plate tectonics. In 'The Biology of Marsupials'. (Eds B. Stonehouse and D. Gilmore.) pp. 97–115. (Macmillan: London.)

Martin, P. G., and Dowd, J. M. (1988). A molecular evolutionary clock for angiosperms. *Taxon* **37,** 364–77.

Martin, P. G., and Dowd, J. M. (1990). A protein sequence study of the dicotyledons and its relevance to the evolution of the legumes and nitrogen fixation. *Australian Systematic Botany* **3,** 91–100.

Martin, P. G., and Dowd, J. M. (1991). Studies of angiosperm phylogeny using protein sequences. *Annals of the Missouri Botanical Garden* **78**(2) (in press).

Metcalfe, I. (1990). Allochthonous terrane processes in South-East Asia. *Philosophical Transactions of the Royal Society of London A* **331,** 625–40.

Parks, C. R., and Wendel, J. F. (1990). Molecular divergence between Asian and North American species of *Liriodendron* (Magnoliaceae) with implications for interpretation of fossil floras. *American Journal of Botany* **77,** 1243–56.

Rabinowitz, P. D., and Labrecque, J. L. (1979). The Mesozoic South Atlantic and the evolution of its continental margins. *Journal of Geophysical Research* **84,** 5973–6002.

Raven, P. H., and Axelrod, D. I. (1974). Angiosperm biogeography and past continental movement. *Annals of the Missouri Botanical Garden* **61,** 539–673.

Schuster, R. M. (1972). Continental movements, 'Wallace's line' and Indomalayan–Australasian dispersal of land plants: some eclectic concepts. *Botanical Review* **38,** 3–86.

Scotese, C. R., Gahagan, L. M., and Larson, R. L. (1988). Plate tectonic reconstruction of the Cretaceous ocean basins. *Tectonophysics* **155,** 27–48.

Sporne, K. R. (1980). A re-investigation of character correlations among dicotyledons. *New Phytologist* **85,** 419–49.

Storey, B. C., Dalziel, I. W. D., Garrett, S. W., Grunow, A. M., Pankhurst, R. J., and Vennum, W. R. (1988). West Antarctica in Gondwanaland: crustal blocks, reconstruction and breakup processes. *Tectonophysics* **155,** 381–90.

Takhtajan, A. (1983). The systematic arrangement of dicotyledonous families. In 'Anatomy of the Dicotyledons'. Vol. II, 2nd ed. (Eds C. R. Metcalfe and L. Chalk.) pp. 180–201. (Clarendon Press: Oxford.)

Takhtajan, A. (1987). Flowering plant origin and dispersal: the cradle of the angiosperms re-visited. In 'Biogeography of the Malay Archipelago'. (Ed. T. C. Whitmore.) pp. 5–25. (Oxford University Press: Oxford.)

Thorne, R. F. (1983). Proposed new realignments in the angiosperms. *Nordic Journal of Botany* **3,** 85–117.

Veevers, J. J., Tayton, J. W., and Johnson, B. D. (1985). Prominent magnetic anomaly along the continent–ocean boundary between the northwestern margin of Australia (Exmouth and Scott Plateaus) and the Argo Abyssal Plain. *Earth and Planetary Science Letters* **72,** 415–26.

Weissel, J. K., and Hayes, D. E. (1977). Evolution of the Tasman sea re-appraised. *Earth and Planetary Science Letters* **36,** 77–84.

Wolfe, K. H., Guoy, M., Yang, Y.-W., Sharp, P. M. and Li, W.-H. (1989). Date of the monocot–dicot divergence estimated from chloroplast DNA sequence data. *Proceedings of the National Academy of Sciences U.S.A.* **86,** 6201–5.

Wu, C.-I., and Li, W.-H. (1985). Evidence for higher rates of nucleotide substitution in rodents than in man. *Proceedings of the National Academy of Science U.S.A.* **82,** 1741–5.

Zhao, X., Coe, R. S., Zhou, Y., Wu, H., and Wang, J. (1990). New palaeomagnetic results from northern China: collision and suturing with Siberia and Kazakhstan. *Tectonophysics* **181,** 43–81.

Manuscript received 3 September 1990, accepted 13 March 1991

3 September 1991 *Aust. Syst. Bot.*, **4**, 117–26

A Comparative Review of Cladistic Approaches to Historical Biogeography of Southern South America

J. V. Crisci,[AB] *M. M. Cigliano,*[AC] *J. J. Morrone*[AC] *and S. Roig-Juñent*[AC]

[A]Laboratorio de Sistemática y Biología Evolutiva (LASBE),
[B]División Plantas Vasculares,
[C]División Entomología,
Museo de La Plata, Paseo del Bosque, 1900 La Plata, Argentina.

Abstract

Cladistic historical biogeographic studies on the austral continents are briefly reviewed with special reference to southern South America. The biogeography of marsupials, southern beeches and the relationships of New Zealand and southern South America are compared. No general pattern of interrelationships is common to all the analyses. Differences in delimitation of areas, selection of taxa and techniques applied are discussed as possible causes of incongruities. The comparative review of these empirical studies reveals that, although it is valid to investigate the existence of a unique pattern of interrelationships among areas of endemism, this cannot be considered as an *a priori* assumption in the analyses.

Introduction

More than 10 years have passed since the first symposium on Vicariance Biogeography was held at the American Museum of Natural History, in New York. In the summary conclusions, Gareth Nelson (1981:525) stated:

'Vicariance biogeography is not an attempt to explain everything, or even the greater part, of the geographical distribution of plants and animals, for its focus is allopatric differentiation as manifested by the phenomenon of endemism . . . . It begins by asking one question: Is there a cladogram of areas of endemism? Or, in other words: Are areas of endemism interrelated among themselves in a way analogous to the interrelationships of the species of a certain group of organisms?'.

Under this working model several historical biogeographic analyses on the austral continents have been carried out. We briefly review the current state of knowledge on patterns of relationships among austral regions, as revealed by four different cladistic approaches. In order to achieve this objective we compare Patterson's (1981) study on historical biogeography of marsupials, Humphries' (1981) analysis of the biogeography of southern beeches, Craw's (1989) study of the biogeographic classification of New Zealand and Crisci *et al.*'s (1991) analysis of the relationships of southern South America.

Historical Biogeography of Marsupials

Patterson (1981) intended to explain the biogeographic history of marsupial mammals, which are mainly distributed in the southern hemisphere (South America, Australia and New Guinea) with one species in North America, and fossil records from Europe. Patterson (1981) applied the reduced area cladogram method (Rosen 1978).

To compare the marsupial area cladogram, Patterson (1981) chose five vertebrate taxa (Table 1) of the same age as the marsupials. Based on these data, four reduced area cladograms were obtained. These were consistent with the marsupial area cladogram and were subsequently

1030-1887/91/010117$05.00

Table 1. List of taxa used in the reviewed studies

Taxa	Patterson (1981)	Humphries (1981)	Craw (1989), analysis 1	Crisci *et al.* (1991)
Fungi				
Cyttaria				+
Plants				
Cupressaceae			+	
Crinodendron–Dobouzetia				+
Peripentadenia				
Embothriinae				+
Gevuinae	+	+		
Macadamiinae	+	+		
Nothofagus–Fagus	+	+	+	+
Negria–Drepanthus		+		+
Oreomyrrhis				+
Drapetes		+		+
Drymis	+	+		+
Aristotelia–Vallea				+
Leptocarpus–Calopsis		+		
Lindenia		+		
Phyllacne		+		
Donatiae		+		
Nicotiana		+		
Lagenifera–Solenogyne		+		
Oreobolus			+	

Taxa	Patterson (1981)	Humphries (1981)	Craw (1989), analysis 1	Crisci *et al.* (1991)
Animals				
Molluscs				
Bullimulidae			+	
Vertebrates				
Osteoglossum–Scleropages	+	+		
Hylidae	+	+		
Chelidae	+			
Ratites	+	+	+	
Galliformes	+	+		
Marsupialia	+	+		
Insects				
Hydrobiosinae	+	+		
Siphlonuridae				+
Eriococcus–Madarococcus				+
Podonominae	+			+
Diamesinae	+	+	+	+
Rhinotoridae	+			
Chalcosyrphus	+	+		
Sciadoceridae	+	+		
Pericompsus	+			
Anisotarsus	+	+		
Metallicina				+
Pseudopsinae				+
Oxelytrum–Ptomaphila				+

combined by Seberg (1986) into a general area cladogram (Fig. 1), where South America is closely related to North America + Europe and New Guinea is related to Australia.

In a second step of the analysis, Patterson selected eight insect and four plant taxa (Table 1), which provided seven different reduced area cladograms. The patterns specified by these area cladograms could not be combined into any single general area cladogram; the Sciadoceridae cladogram was unique and the other 11 reduced area cladograms supported either the vertebrate general area cladogram or a new one (Fig. 2), where South America is related to Australia and New Guinea instead of being related to North America + Europe. Patterson then added the areas that he had previously omitted from the reduced area cladograms, such as south-eastern Asia and Africa.

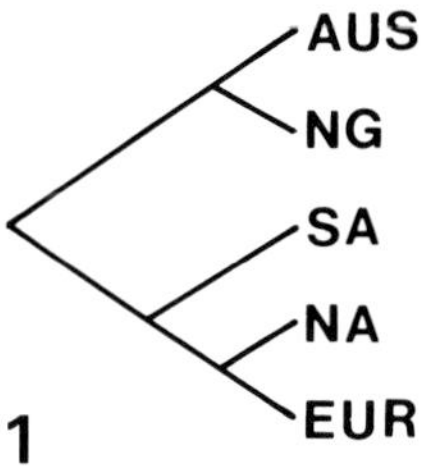

1

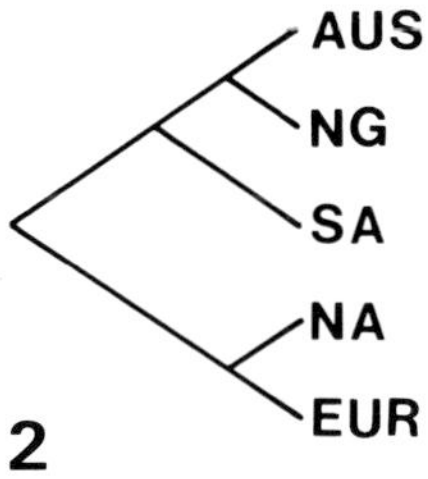

2

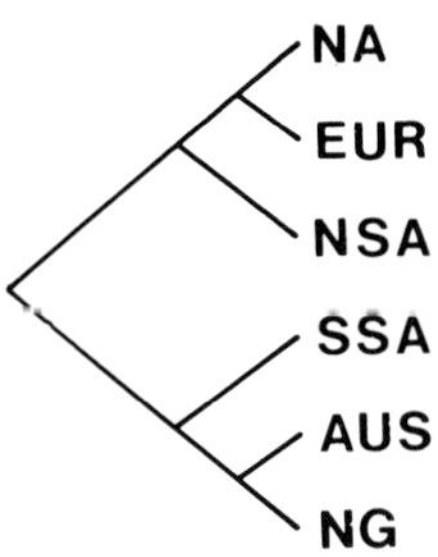

3

4

Figs 1 and 2. General area cladograms in Seberg (1986) based on Patterson's (1981) analysis.
Fig. 3. General area cladogram of Parenti (1981) based on Patterson's (1981) analysis.
Fig. 4. General area cladogram after Humphries (1981).
AUS, Australia; EUR, Europe; NA, North America; NC, New Caledonia; NG, New Guinea; NSA, Northern South America; NZ, New Zealand; SA, South America; SSA, Southern South America; TAS, Tasmania.

Patterson (1981) concluded that the taxa showing congruent Australasian/South American patterns displayed two different patterns beyond these areas, one with younger relationships extending into northern areas and the other with older relationships extending into Africa.

As a summary of Patterson's analysis, Parenti (1981) presented a general area cladogram (Fig. 3) where South America is divided into two areas, northern South America and southern South America. Patterson's data set was also used by Wiley (1988) in a biogeographic parsimony analysis (Wiley 1988; Fig. 13). Based on a data matrix of 5 areas × 122 components, PAUP analysis (Swofford 1985) produced a single tree, identical with that shown in Fig. 1.

Historical Biogeography of Southern Beeches

In a study of the historical biogeography of the southern beeches *Nothofagus*, Humphries (1981) presented a general area cladogram as a summary of the historical relationships of land masses inhabited by austral and boreal groups, following Rosen's (1978) reduced area cladogram technique. Humphries based his study on the reduced area cladograms of 22 plant and animal taxa (Table 1), with endemic representatives in at least three of the areas occupied by *Nothofagus*.

Different patterns of area relationships, particularly with regard to the position of South America, were obtained. The general area cladogram presented (Fig. 4) showed an unresolved position for South America, which could be either the sister group to North America + Europe or to the monophyletic group containing New Zealand, Australia, Tasmania, New Guinea and New Caledonia.

Humphries (1981) hypothesised that the two positions for South America supported the hypothesis that it is a composite area and should not be treated as a single area of endemism.

Biogeographic Classification of New Zealand

Craw (1989) conducted three cladistic biogeographic analyses based on Gondwanic and circum-Antarctic taxa to appraise varied views on the biogeographic classification and relationships of New Zealand. The analyses were performed using the Wagner parsimony algorithm of the computational program PAUP 2.4.1 (Swofford 1985).

Analysis 1. Craw applied a quantitative parsimony analysis (Wiley 1988; Fig. 13) to six original taxon cladograms (Table 1), considering South America, New Zealand, Australia + Tasmania, New Guinea, New Caledonia and Africa as the study areas.

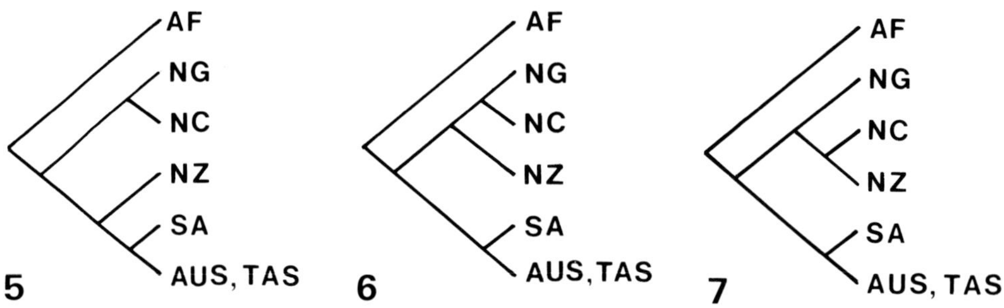

Figs 5–7. General area cladograms based on a quantitative parsimony analysis (Craw 1989, analysis 1). AF, Africa; AUS, Australia; NC, New Caledonia; NG, New Guinea; NZ, New Zealand; SA, South America; TAS, Tasmania.

The Wagner parsimony analysis of the data matrix (6 areas × 27 components) resulted in three general area cladograms (Figs 5–7), which agree in showing South America to be most closely related to Aústralia + Tasmania.

Analysis 2. Craw undertook a parsimony analysis of endemicity (Rosen 1988; Fig. 8) based on the presence/absence of 73 plant and animal families and subfamilies. He analysed the relationships among Australia + Tasmania, New Zealand, South America, Africa and the Oriental region.

Fig. 8. Flow chart of the steps in the parsimony analysis of endemicity.

The parsimony analysis of endemicity led to three equally parsimonious general area cladograms that were combined by Craw (1989) into a strict consensus tree (Fig. 9). According to the information that can be extracted from the general area cladograms, Australia + Tasmania and New Zealand are most closely related to South America, and the relationships of Africa and the Oriental region are unresolved. By deleting the plant data, one fully resolved general area cladogram was obtained (Fig. 10), with Africa and the Oriental region constituting a monophyletic group.

Analysis 3. Another parsimony analysis of endemicity was applied, to 31 genera of the circum-Antarctic biota, considering New Zealand mainland, New Zealand subantarctic islands, Indian Ocean subantarctic islands, Tierra del Fuego, Islas Malvinas (Falkland Islands) and Australia + Tasmania.

This analysis resulted in two equally parsimonious general area cladograms (Figs 11, 12), showing a close relationship between the New Zealand mainland and Tierra del Fuego and the Islas Malvinas.

Craw (1989) concluded that his three analyses supported three alternative views of the biogeographic relationship of New Zealand to Australia and South America.

Composite Origin of South American Biota

Crisci *et al.* (1991) tried to test the hypothesis of a hybrid origin for South America (Humphries 1981; Parenti 1981; Humphries and Parenti 1986). They conducted a study in which northern South America and southern South America were treated as independent units. To determine the areas which exhibit affinities with southern South America, a non-exhaustive search of distributional patterns of several plant and animal taxa was undertaken. Australia, New Guinea, New Caledonia, Tasmania, New Zealand, South Africa and North America were selected. Cladistic analyses were available for only 17 taxa of those showing austral patterns of relationships (Table 1). Three different techniques were applied: biogeographic parsimony analysis (Wiley 1988), component analysis (Nelson and Platnick 1981) and quantification of component analysis (Humphries *et al.* 1988). The main steps conducted under each technique are summarised in flow charts (Figs 13–15). Assumptions 1 and 2 (Nelson and Platnick 1981) and intersections (Nelson 1984; Page 1988) were performed using Component version 1.5 (Page 1989). Wagner parsimony analyses were performed using Hennig86 version 1.5 (Farris 1988), applying the implicit enumeration option for calculating trees (Farris 1989). When the parsimony analyses yielded more than two trees a Nelson consensus tree (Nelson 1979) of Hennig86 was calculated to be used as an illustration.

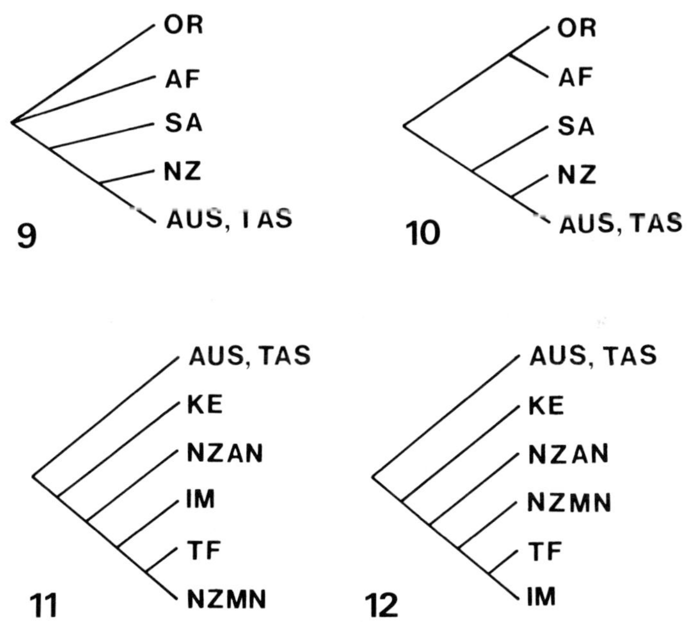

Figs 9 and 10. General area cladograms based on a parsimony analysis of endemicity for 73 plant and animal taxa (Craw 1989, analysis 2). *Fig. 9.* Consensus area cladogram based on plant and animal taxa. *Fig. 10.* General area cladogram for animal taxa only. Abbreviations as in Figs 5–7; AF, Africa; OR, Oriental Region.

Figs 11 and 12. General area cladograms based on a parsimony analysis of endemicity for circum-Antarctic land areas (Craw 1989, analysis 3). AUS, Australia; IM, Islas Malvinas; KE, Indian Ocean subantarctic islands, Kerguelen Province; NZAN, New Zealand subantarctic islands; NZMN, New Zealand mainland; TAS, Tasmania; TF, Tierra del Fuego.

Analysis 1. Biogeographic parsimony analysis (Fig. 13). The data matrix (10 areas × 93 components) yielded two general area cladograms (Figs 16, 17). These cladograms show North America and northern South America as sister areas, and southern South America included in a monophyletic group with Australia, Tasmania, New Guinea and New Caledonia, with New Zealand as the sister area. Both trees differ only in the position of the pair North America–northern South America. Either North America–northern South America is sister to South Africa or is the sister to the rest of the areas, exclusive of South Africa.

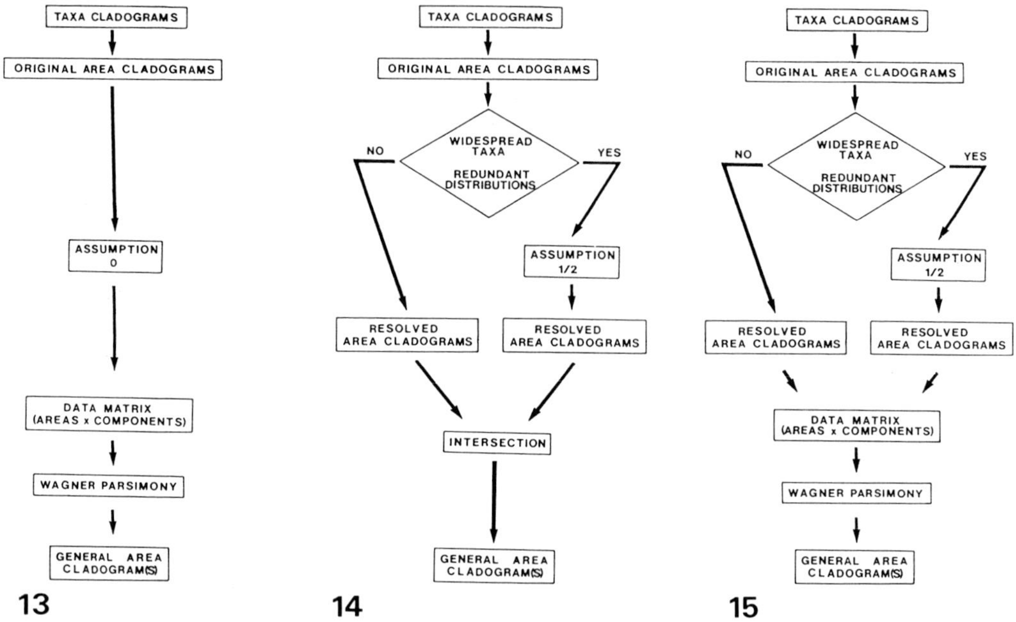

Figs 13–15. Flow charts for the steps in cladistic biogeographic techniques applied by Crisci *et al.* (1991). *Fig. 13.* Biogeographic parsimony analysis. *Fig. 14.* Component analysis. *Fig. 15.* Quantification of component analysis.

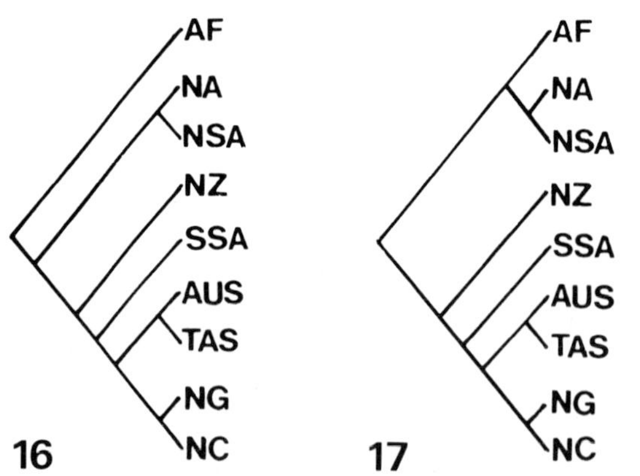

Figs 16 and 17. General area cladograms based on biogeographic parsimony analysis (after Crisci *et al.* 1991). AF, South Africa; AUS, Australia; NA, North America; NC, New Caledonia; NG, New Guinea; NSA, northern South America; NZ, New Zealand; SSA, southern South America; TAS, Tasmania.

Analysis 2. Component analysis (Fig. 14). Eleven out of the 17 original area cladograms showed widespread taxa and/or redundant distributions; assumptions 1 and 2 were applied, producing a variable number of resolved area cladograms. No general area cladogram common to the 17 sets of area cladograms derived under assumption 1 could be found. The same was true for the analysis under assumption 2. Even though not supported by the 17 sets of trees, searches for a tree shared by a smaller number of sets of area cladograms were performed. As a result of this procedure the largest number of sets able to share one or more cladograms was 10. Under assumption 1, two general area cladograms were obtained (Figs 18, 19) which differed in the basal branch, which may be either South Africa or the pair North America–northern South America. The southern temperate areas (southern South America, Australia, Tasmania, New Zealand, New Guinea and New Caledonia) constitute a monophyletic group, where southern South America is the sister area of the other five.

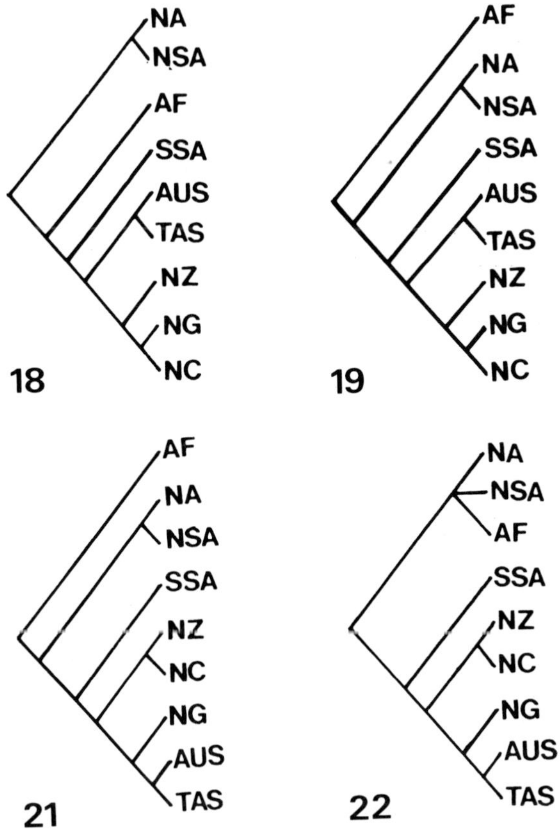

Figs 18–22. General area cladograms based on component analysis (after Crisci *et al.* 1991). Abbreviations as in Figs 16, 17.

Under assumption 2, four general area cladograms were found through the intersection of two groups of trees derived from 10 taxa each. One of these replicates the second general area cladogram obtained under assumption 1 (Fig. 19); another is consistent with both general area cladograms obtained under assumption 1 (Fig. 20); the remaining two are consistent with the others obtained under assumption 2, with the exception of the placement of New Guinea (Figs 21, 22).

Analysis 3. Quantification of component analysis (Fig. 15). Two data matrices (areas × components) were extracted from the sets of resolved area cladograms obtained under assumptions 1 or 2, and from the original area cladograms that showed neither widespread taxa nor redundant distributions. Under assumption 1, the parsimony analysis of the data matrix (10 areas × 102 components) produced nine general area cladograms. The Nelson consensus tree (Fig. 23) for these general area cladograms shows the pairs of sister areas: Australia–Tasmania, New Guinea–New Caledonia, southern South America–New Zealand, and northern South America–North America.

Under assumption 2, the analysis of the data matrix (10 areas × 172 components) led to only one general area cladogram (Fig. 24). This cladogram shows northern South America, North America and South Africa as a monophyletic group in which the first two are sister areas. The southern temperate areas constitute a monophyletic group in which southern South America is the sister area to Australia.

Crisci *et al.*'s (1991) three analyses support the hybrid origin of the South American biota, with northern South America most closely related to North America and southern South America always constituting a monophyletic group together with Australia, Tasmania, New Guinea, New Caledonia and New Zealand, reflecting the existence of an ancient austral biota. Four conflicting hypotheses of area relationships concerning southern South America and the other austral areas suggest that southern South America may be a composite area in itself.

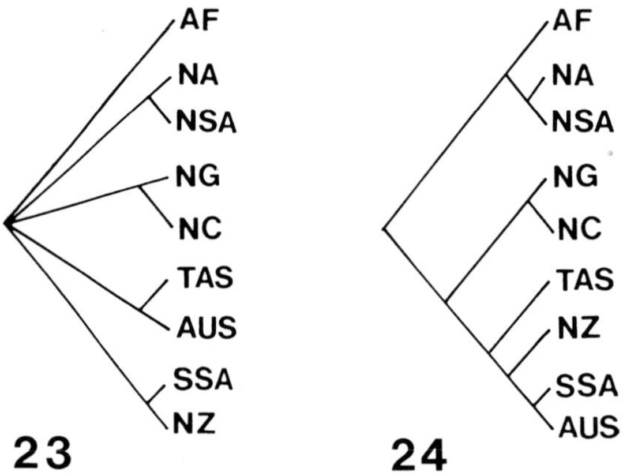

Figs 23 and 24. General area cladograms based on quantification of component analysis (after Crisci *et al.* 1991). *Fig. 23.* Nelson consensus tree of the nine general area cladograms obtained under assumption 1. *Fig. 24.* General area cladogram obtained under assumption 2. Abbreviations as in Figs 16, 17.

Discussion

Comparison of the results obtained under the different studies reviewed seems to show no single pattern of interrelationships of southern South America with the other austral areas. The incongruities among the different general area cladograms may be due to at least three methodological causes.

Areas. The lack of a universal methodology to define areas of endemism is one possible cause that leads to incongruities in empirical studies. Although we reviewed only those papers dealing with austral regions, we found that the areas of study are not equivalent. Basically two approaches for the delimitation of South America were followed in the papers reviewed. South America was considered as a single area by Patterson (1981), Humphries (1981) and Craw (1989, analyses 1 and 2) or treated as a composite (Craw 1989, analysis 3; Crisci *et al.* 1991). Craw (1989) considered the subantarctic portion of South America, including only Islas Malvinas and Tierra del Fuego in his analysis 3, whereas Crisci *et al.* (1991) divided South America into northern and southern South America.

The comparison of descriptive biogeographic schemes also presents problems. When searching for a unique descriptive biogeographic scheme for austral biota, we could not find one. Depending on the criteria for delimiting different areas, whether based on flora, fauna or ecology, the areas are defined in different ways. Recent examples of different points of view can be found in Fleming (1987) and Udvardy (1987) on the Antarctica biogeographical realm.

Taxa. Another cause of incongruity may be due to different sets of taxa used in different studies. From the 39 taxa listed in Table 1, only two (*Nothofagus* and Diamesinae) are common to all the data sets compared, thus making comparison more difficult.

The consequences of choosing different sets of taxa are clearly seen in most of the studies reviewed. Patterson (1981) found that through selection of taxa two different patterns of area relationships could be obtained. A secondary selection from the original taxa in Craw's (1989) second analysis and that of Crisci *et al.*'s (1991) application of component analysis allowed one to obtain resolved results from unresolved cladograms.

Differences in selection of taxa in the studies reviewed suggest that, if we accept the analogy between systematic and biogeographic analyses, a comparable criterion to homology is still lacking in biogeography. At present, the solely dubious criterion to choose taxa is related to the availability of cladistic analyses. Furthermore, the unreliability of taxon cladograms may bring noise to the system (Seberg 1986).

Techniques. A third cause of incongruity is due to different techniques. This is clearly shown in Crisci *et al.* (1991), where different results were obtained through the application of three techniques on the same data.

Finally, even though there were accepted universal methodologies for delimitation of natural areas, for selection of taxa and agreement on the most appropriate technique, the real world could also generate apparent incongruities. Cracraft (1988) has already stated that area-hybridisation, disappearance of barriers, origination of new barriers within areas, randomisation effects of population dispersion, and extinction, are some of the processes which threaten the integrity of areas and cause complex historical patterns.

Ten years after Nelson's (1981:525) question, the empirical studies seems to reveal that, although it is valid to investigate the existence of a unique pattern of interrelationships among areas of endemism, it cannot be accepted as an *a priori* assumption of the analysis.

Acknowledgments

We are grateful to Pauline Ladiges for inviting us to participate in the symposium. This research was supported by grant 3966-88 of the National Geographic Society. The continuous support of the 'Consejo Nacional de Investigaciones Científicas y Técnicas (CONICET)', Argentina, to which the authors belong, is gratefully acknowledged.

References

Cracraft, J. (1988). Deep-history biogeography: retrieving the historical pattern of evolving continental biotas. *Systematic Zoology* **37,** 221–36.

Craw, R. (1989). New Zealand biogeography: A panbiogeographic approach. *New Zealand Journal of Zoology* **16,** 527–47.

Crisci, J. V., Cigliano, M. M., Morrone, J. J., and Roig-Juñent, S. (1991). Historical biogeography of southern South America. *Systematic Zoology* **40** (in press).

Farris, J. S. (1988). 'Hennig86 Reference. Version 1.5.' (Published by the author.)

Farris, J. S. (1989). The retention index and the rescaled consistency index. *Cladistics* **5,** 417–19.

Fleming, C. A. (1987). Comments on Udvardy's biogeographical realm Antarctica. *Journal of the Royal Society of New Zealand* **17,** 141–68.

Humphries, C. J. (1981). Biogeographical methods and the southern beeches (Fagaceae: *Nothofagus*). In 'Advances in Cladistics, 1, Proceedings of the First Meeting of the Willi Hennig Society'. (Eds V. A. Funk and D. R. Brooks.) pp. 177–207. (New York Botanical Gardens: Bronx.)

Humphries, C. J., Ladiges, P. Y., Roos, M., and Zandee, M. (1988). Cladistic biogeography. In 'Analytical Biogeography: an Integrated Approach to the Study of Animal and Plant Distributions'. (Eds A. A. Myers and P. S. Giller.) pp. 371–404. (Chapman and Hall: London.)

Humphries, C. J., and Parenti, L. R. (1986). 'Cladistic Biogeography.' (Clarendon Press: Oxford.)

Nelson, G. (1979). Cladistic analysis and synthesis: principles and definitions, with a historical note on Adanson's Familles des Plantes (1763–64). *Systematic Zoology* **28,** 1–21.

Nelson, G. (1981). Summary. In 'Vicariance Biogeography: a Critique'. (Eds G. Nelson and D. E. Rosen.) pp. 524–37. (Columbia University Press: New York.)

Nelson, G. (1984). Cladistics and biogeography. In 'Cladistics: Perspectives on the Reconstruction of Evolutionary History'. (Eds T. Duncan and T. F. Stuessy.) pp. 273–93. (Columbia University Press: New York.)

Nelson, G., and Platnick, N. I. (1981). 'Systematics and Biogeography: Cladistics and Vicariance.' (Columbia University Press: New York.)

Page, R. D. M. (1988). Quantitative cladistic biogeography: constructing and comparing area cladograms. *Systematic Zoology* **37,** 254–70.

Page, R. D. M. (1989). 'Component User's Manual. Release 1.5.' (Published by the author.)

Parenti, L. R. (1981). Discussion. In 'Vicariance Biogeography: a Critique. (Eds G. Nelson and D. E. Rosen.) pp. 490–7. (Columbia University Press: New York.)

Patterson, C. (1981). Methods of paleobiogeography. In 'Vicariance Biogeography: a Critique'. (Eds G. Nelson and D. E. Rosen.) pp. 446–89. (Columbia University Press: New York.)

Rosen, B. R. (1988). From fossils to earth history: applied historical biogeography. In 'Analytical Biogeography: an Integrated Approach to the Study of Animal and Plant Distributions'. (Eds A. A. Myers and P. S. Giller.) pp. 437–81. (Chapman & Hall: London.)

Rosen, D. E. (1978). Vicariant patterns and historical explanation in biogeography. *Systematic Zoology* **27,** 1–16.

Seberg, O. (1986). Taxonomy, phylogeny, and biogeography of the genus *Oreobolus* R. Br. (Cyperaceae), with comments on the biogeography of the South Pacific continents. *Botanical Journal of the Linnean Society* **96,** 119–95.
Swofford, D. (1985). PAUP '(Phylogenetic Analysis Using Parsimony). Version 2.4' (Illinois Natural History Survey: Champaign, Ill.)
Udvardy, M. D. F. (1987). The biogeographical realm Antarctica: a proposal. *Journal of the Royal Society of New Zealand* **17,** 187–94.
Wiley, E. O. (1988). Parsimony analysis and vicariance biogeography. *Systematic Zoology* **37,** 271–90.

Manuscript received 3 September 1990, accepted 26 February 1991

3 September 1991 *Aust. Syst. Bot.*. **4**, 127–36

Biogeographic Congruence in the South Pacific

O. Seberg

Botanical Laboratory, Copenhagen University,
Gothersgade 140, DK-1123 Copenhagen K, Denmark.

Abstract

Ever since J. D. Hooker's famous 'Introductory Essay' to Flora Novæ-Zelandiæ, a classical problem in biogeography has been to give a casual explanation of southern hemisphere distribution patterns. An attempt is made to see whether the cladograms for the circum-Pacific areas (South America, New Zealand, Tasmania and Australia) are congruent. The area cladograms are derived from *Nothofagus* (Fagacae), Embothriinae (Protaceae), *Oreobolus* (Cyperaceae), *Cyttaria* (Helotiales) and Eriococcidae (Homoptera). The resulting general area cladogram showing southern South America as the sister-area to New Zealand, south-eastern Australia and Tasmania, and Tasmania plus south-eastern Australia as sister-areas to New Zealand are compared with different geological hypotheses for the area. The biological area cladogram is shown to be congruent with widely different geological hypotheses.

Introduction

To give a causal explanation of austral distribution patterns — patterns involving widely separated, temperate parts of the circum-Pacific continents, extending both up along the Andes and into the highlands of New Guinea and New Caledonia — is one of the classic problems in historical biogeography (see e.g. Hooker 1853; Darlington 1965; Brundin 1966).

It is very difficult to give a reliable estimate of the number of taxa sharing this 'generalized track' (Croizat *et al.* 1974). Although there is no obvious correlation between taxonomic rank and distribution pattern and many traditionally defined taxa are not monophyletic, no less than 50 genera of flowering plants include a group of species that has an austral distribution pattern (Thorne 1972). The pattern occurs also in a multitude of different groups: bryophytes, lichens, flies, dragonflies, butterflies, birds, mammals etc. Looking through the distribution maps and tables etc. in the following papers: Fulford (1951), Darlington (1957), Steenis (1963), Steenis and Balgooy (1966), Schuster (1969, 1972, 1979), Cracraft (1973, 1974, 1975), Keast (1973), Good (1974), Balgooy (1975, 1984), Baumann (1988), Engel (1990) provides an impression of the generality of this 'generalised track'.

Perhaps the most famous botanical example showing this pattern is *Nothofagus* Bl. (Fagaceae), which for more than a century has caught the attention of systematists, paleontol–ogists, and biogeographers alike (e.g. Steenis 1971; Schlinger 1974; Humphries 1981, 1983, this issue; Hill 1983; Tanai 1986; Philipson and Philipson 1988). It has even been postulated that *Nothofagus* holds the key to an understanding of how the pattern evolved (Steenis 1971).

1030-1887/91/010127$05.00

The breakthrough of cladistic biogeography (Platnick and Nelson 1978; Nelson and Platnick 1981; Humphries and Parenti 1986) has lifted historical biogeography out of the narrative and into the analytical phase (Ball 1975). Once a recurring pattern has been realised, e.g. as a 'generalised track', the basic strength of cladistic biogeography is that it does not attempt to give one explanation to each pattern (i.e. for each area cladogram), but searches for repetitions of the pattern in order to see whether it is general or not, and hence requires a general explanation.

The construction of individual area cladograms is rather time-consuming. This is probably the main reason why so little work has been done on genera other than *Nothofagus*. However, recent work by Humphries *et al.* (1986) on both *Nothofagus* and some of its parasites [*Cyttaria* Berkeley (Helotiales), *Heterobathmia* Kristensen & Nielsen (Lepidoptera) and Eriococcidae (Homoptera)], by Weston and Crisp (1987) on Embothriinae (Proteaceae), by Seberg (1988) on *Oreobolus* R. Br. (Cyperaceae) and by Crisci *et al.* (1988) on *Cyttaria* has to some extent changed this state of affairs.

Therefore, it seems pertinent to see to what extent the individual area cladograms based upon these taxa have components in common.

Methods

Whenever possible, data matrices were either rerun or checked using the computerised parsimony program Hennig86 (version 1.5), of J. S. Farris (1988). The program was run using the exhaustive search option *ie**, which ensures that all equally parsimonious cladograms are found. This procedure was occasionally rather time-consuming. The running of the entire *Oreobolus*-matrix (see below) lasted more than 2½ days on an Olivetti M24SP, using an Intel 8086-processor. When several equally parsimonious trees were found, they were treated separately in the biogeographic analyses.

Biogeographic analyses were undertaken on the program COMPONENT (release 1.5) (Page 1989). In all cases the reduced area cladograms (see below) were used as input to the subprogram BUILD using the option MAKE TREE and Assumption 2. Shared trees were found using the output from the above runs as input to the subprogram COMPARE and the option COMPARE ALL.

Phylogeny of the Involved Taxa

It is a serious impediment to most biogeographic analysis that the underlying systematic hypotheses frequently are founded on different systematic philosophies and therefore have different basic assumptions. Even within a relatively new paradigm, as phylogenetic systematics or cladistics, there is not just one method (see Meacham 1980; Felsentein 1982; Zandee 1987).

The taxon cladograms in the four previously mentioned studies are all constructed using the same basic principles (outgroup comparison etc.) in cladogram construction. Thus they are immediately comparable.

Only an out-dated (Humphries *et al.* 1986: 61–62) data matrix can be reconstructed for *Nothofagus* (Humphries 1981: 194, Table 4 and 199, fig. 15) and a data-matrix has never been published for Eriococcidae (Humphries *et al.* 1986) or Humphries *et al.*'s study of *Cyttaria* (1986). Crisci *et al.* (1988) have, however, recently published a matrix for *Cyttaria*. Data-matrices for Embothriinae (Weston and Crisp 1987), *Oreobolus* (Seberg 1988), and for a subset of *Oreobolus*, the neotropical species (Mora-Osejo 1987), have also been published.

Accepting the original character polarities, data-matrices for Embothriinae (10 OTUs, 26 characters; Weston and Crisp 1987: 27, table 2), the full *Oreobolus*-matrix (18 OTUs, 17 characters; Seberg 1988: 136, table 2) and the *Cyttaria*-matrix (11 OTUs, 18 characters; Crisci *et al.* 1988: 282, table 4) have been used as input to Hennig86 (Farris 1988). In the character analysis of Embothriinae, *Lomatia* R. Br. (Lomatiinae) functioned as outgroup (Weston and Crisp 1987: 21). *Lomatia* is not included in the published matrix, but was added to it as an ancestor. Character 26, however, was considered an apomorphy (Weston and Crisp 1987: 27–28). The decision to treat character 26 as an apomorphy for *Lomatia* (and for Embothriinae, except *Embothrium coccineum* Forster & Forster f.) was made by arguments irrelevent to the cladistic analysis. For Embothriinae 10 equally parsimonious trees were found (length 28, consistency index 0.92). Weston and Crisp (1987: 26, fig. 3), published only one tree (length 30, consistency index 0.87) as their 'preferred, most parsimonious cladogram' based on 'choosing one of the suite of equally parsimonious cladograms as a preferred solution' (Weston and Crisp 1987: 24). Thus, in the present investigation all 10 equally parsimonious trees are used for further analysis. A summary of the 10 trees is shown as a strict consensus tree in Fig. 1. The individual trees are found in Appendix 1.

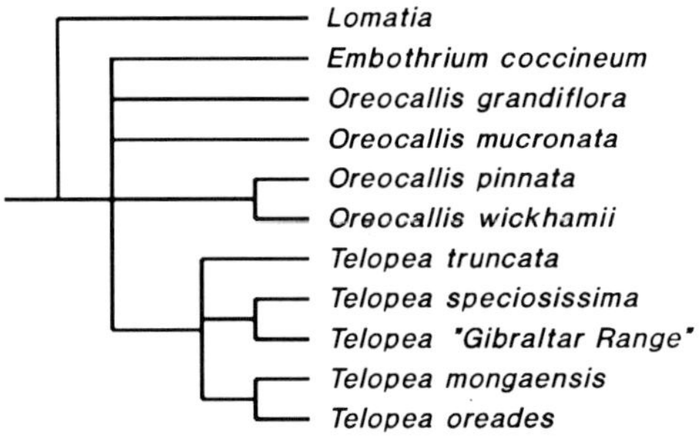

Fig. 1. Strict consensus tree for the 10 equally parsimonious cladograms for Embothriinae. For the individual cladograms see Appendix 1.

For the *Oreobolus* data the number of trees included in the group of equally parsimonious trees was raised from the original 3 (length 93, consistency index 0.46; Seberg 1988) to 10 (length 93; consistency index 0.46). A summary of the 10 trees is shown as a strict consensus tree in Fig. 2. The individual trees are found in Appendix 2. Despite the fact that seven more equally parsimonious taxon cladograms have been found for *Oreobolus*, none of the previously published taxonomic conclusions have to be changed (Seberg 1988). In all trees, *Schoenoides oligocephalus* (originally included within *Oreobolus*) appears as the sister-group of *Oreobolus*. All three pairs of subspecies either remain each other's closest relatives (e.g. *O. obtusangulus* ssp. *unispicus* and *O. obtusangulus* ssp. *obtusangulus*) or, in the case of *O. pumilio* ssp. *pumilio* and *O. pumilio* ssp. *clemensiae*, this conclusion remains among the possibilities. More importantly, at least with respect to the present investigation, *O. distichus* (from mainland Australia and Tasmania) remains the sister-species of *O. goeppingeri* (from Central and northern South America) on all cladograms and this clade is invariably embedded in a larger clade, that otherwise exclusively consists of Central and South American species.

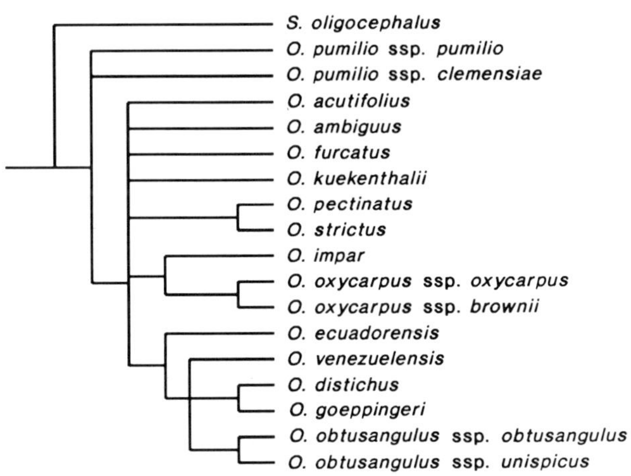

Fig. 2. Strict consensus tree for the 10 equally parsimonious cladograms for *Oreobolus* and *Schoenoides*. For the individual cladograms see Appendix 2.

Mora-Osejo (1987) has recently analysed and published a considerably smaller data-matrix (6 OTUs, 10 characters) including only the neotropical species of *Oreobolus*. With *O. pumilio* as out-group, this matrix was subjected to a cladistic parsimony algorithm (Kluge and Farris 1969). However, although most of the characters are either multistate or continuous, all are arbitrarily treated as binary (Mora-Osejo 1987: 160). Even accepting this rather dubious character treatment and despite a fortunate choice of out-group two very important facts have been neglected in the analysis: although treated as monophyletic, no attempt has been made to justify the neotropical *Oreobolus*-species as a clade within *Oreobolus*, which they apparently are not (Seberg 1988); and the Hypothetical Taxonomic Units (HTUs) on the initial tree have not been optimised (Farris 1970). It seems futile to analyse, and even less sensible to make biogeographic hypotheses, about non-monophyletic groups (Croizat *et al.* 1974; Nelson and Platnick 1981; Humphries and Parenti 1986). With Mora-Osjo's matrix as input to Hennig86, not one but three equally parsimonious trees were found for the data-set (length 13, consistency index 0.76). A strict consensus tree of the three trees has only one informative component, which erroneously (Seberg 1988) indicates that *O. ecuadorensis* is the sister-species of *O. goeppingeri*. Neither successive weighting nor the addition of an ancestor to the matrix could improve this state of affairs. Accordingly, Mora-Osejo's treatment is abandoned for the present purpose.

For the *Cyttaria* data the conclusions of Crisci *et al.* (1988: 283–84, figs 1–2) have been verified, as only two trees (length 40, consistency index 0.70) were found.

As no OTU-character matrices have been published for either *Nothofagus* or Ericoccidae, the taxon cladograms published by Humphries *et al.* (1986: 61, fig. 4.4 and 64, fig. 4.7) must be taken at their face value. The *Cyttaria* data from Humphries *et al.* (1986) are ignored in favour of the more recent data of Crisci *et al.* (1988).

As has been stressed by both Farris (1969) and more recently Carpenter (1988), a consensus tree is a conservative estimate of relationship. Successive weighting (Farris 1969; Carpenter 1988) is one way of reducing the number of equally parsimonious trees. This appears to be a reasonable approach, because fewer and bolder hypotheses about relationship are presented, which in turn require fewer tests. Cladograms obtained by successive weighting can also be used as input for the construction of area cladograms. However, if the number of equally parsimonious taxon cladograms, and hence area cladograms, is reasonably small, one can test each individual area cladogram for congruence with the area cladograms obtained from every other group of taxon cladograms. The latter approach has been adopted here.

Area Cladograms

Three of the five taxa in question have four (or more) areas in common, corresponding to four areas of endemism (Table 1): (1) southern South America (sSAm), roughly defined as the southern part of Chile and Argentina (Patagonia) plus Tierra del Fuego, (2) New Zealand (NZ), (3) south-eastern Australia (AUS), and (4) Tasmania (TAS). A more detailed picture of the distribution of the various species of *Nothofagus* may be found in Humphries (1981: 200, fig. 15), of the species of Embothriinae in Weston and Crisp (1987: 21–22 and 30, fig. 5), of the species of *Oreobolus* in Seberg (1988: 142, fig. 8), and of the species of *Cyttaria in Crisci et al.* (1988: 280, table 1). The distribution of the species of Eriococcidae may be reconstructed from the distribution of their hosts in Humphries *et al.* (1986: 64, fig. 4.7).

Table 1. Distribution of the five taxa studied

NC, New Caledonia; NG, New Guinea; AUS, south-eastern Australia; TAS, Tasmania, NZ, New Zealand; HAW, Hawaii + Tahiti; sSAm, southern South America; and nSAm, northern South America. *Note that Emborthriinae also occurs in northern Queensland

Taxon	Areas							
Cyttaria			AUS	TAS	NZ		sSAm	
Eriococcidae			AUS		NZ		sSAm	
Oreobolus		NG	AUS	TAS	NZ	HAW	sSAm	nSAm
Nothofagus	NC	NG	AUS	TAS	NZ		sSAm	
Embothriinae		NG	AUS*	TAS			sSAm	nSAm

Individual taxon cladograms were transformed into area cladograms by substituting the terminal taxa with their distribution areas. Individual area cladograms may be reconstructed from Appendices 1 and 2 and Figs 1 and 2, and from Humphries *et al.* (1986) and Crisci *et al.* (1988).

Whenever widespread and redundant distributions are involved, it may be difficult to judge whether or not the area cladograms derived from one of the taxa share components with the area cladograms derived from one or more of the other taxa.

Even a superficial examination of the area cladograms reveals that, if there is only one area relationship for the areas involved, some information is either missing, redundant or ambiguous (e.g. the same areas occur several times in each area cladogram or individual components may be missing from an area cladogram).

Part of the information on the individual area cladograms, e.g. the distribution data for the species of *Oreobolus* in northern South America, is at current level of analysis too detailed to be analysed, and hence this information must be further reduced. Fortunately, these distribution areas are nearly congruent (Seberg 1988), hence they are treated as a single area (nSAm). Additionally, some of the area cladograms include components involving only areas unique to them (e.g. Hawaii + Tahiti on the area cladograms derived from *Oreobolus*). These components are also excluded from further consideration.

Given that there is one area relationship for the areas: sSAm, NZ, AUS and TAS (= defining component = sSAm + NZ + AUS + TAS), the question I would now ask is what are the relationships between the four areas?

In order to find any shared components between the area cladograms based upon Embothriinae, *Nothofagus, Oreobolus, Cyttaria* and Eriococcidae, all were reduced to include only the areas defined by the defining component.

The reduced area cladograms for *Nothofagus,* Embothriinae, *Oreobolus*, Eriococcidae and *Cyttaria* were used as input to Page's program COMPONENT. In the area cladograms derived from *Oreobolus,* for example, the problem of widespread taxa remains. Widespread taxa, e.g. *Oreobolus distichus* occurring in both mainland Australia and Tasmania, may be analysed under one of three assumptions (Nelson and Platnick 1981; Humphries and Parenti 1986; Zandee and Ross 1987; Wiley 1988*a*, 1988*b*):

Assumption 0: The best estimate of ancestral distributions resides in the original phylogeny of the group (*viz.* the components defined by the widespread taxon are real).

Assumption 1: Whatever is true of a widespread taxon in one part of its range is also true of the taxon in other parts of its range (*viz.* all identifiable components are meaningful).

Assumption 2: Whatever is true of a widespread taxon in one parts of its range might not be true of the taxon in other parts of its range (*viz.* some of the components might be false).

An excellent critique of premises of Assumption 0 may be found in Page (1990). Assumption 2 is considered the most realistic (Nelson and Platnick 1981; Humphries and Parenti 1986) as it does not assume dispersal, vicariant or extinction events, without denying that they occur (Nelson 1982).

Application of Assumption 2 shows that all indivual area cladograms share two components (Fig. 3*A*). In addition, the area cladograms derived from Embothriinae, *Oreobolus* and *Cyttaria* support the extra component shown in Fig. 3*B*. However, the area cladograms obtained from *Nothofagus* support two other components involving New Zealand (Fig. 3*C*). These two alternative components are both in conflict with component 2 in Fig. 3*B*. The area cladogram for Eriococcidae is congruent with all three above area cladograms.

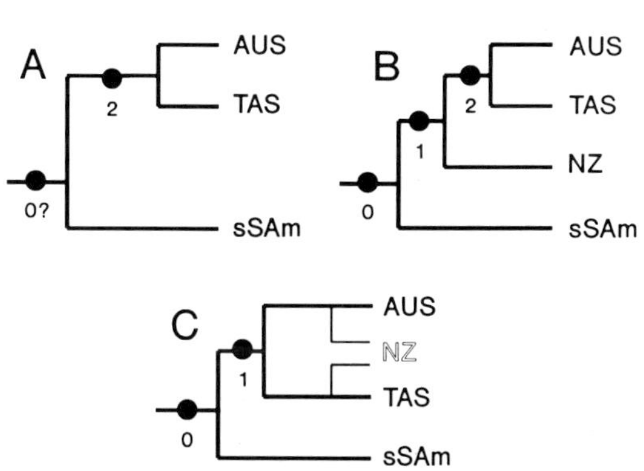

Fig. 3. (*A*) The general area cladogram supported by all the investigated taxa. (*B*) The area cladogram supported by all taxa except *Nothofagus*. (*C*) The two alternative positions of New Zealand supported by *Nothofagus*. AUS, south-eastern Australia; TAS, Tasmania, NZ, New Zealand; sSAm, southern South America. Component 0 (the defining component), AUS, TAS, NZ and sSAm; component 1, AUS, TAS and NZ; and component 2, AUS and TAS.

Apart from more obvious *ad hoc* explanations (dispersal, extinction) the incongruence between the area cladograms (Fig. 3*B, C*) could be due to different ages of the involved taxa.

The general area cladogram found by Humphries *et al.* (1986) has the same components as the one found in the present study (Fig. 3*B*).

The General Area Cladogram and Historical Geology

Several widely different hypotheses, attempting to explain evolution of continents by continental drift, have been proposed for the history of the earth. These hypotheses may be divided into three categories: those assuming an earth of fixed dimension (Smith *et al.* 1973); those assuming an earth of increasing dimensions (Carey 1976; Owen 1976, 1983; Shields 1979); those incorporating the hypothetical continent of Pacifica (Melville 1981, 1982; Nur and Ben-Avraham 1981).

These geological hypotheses have been brought into cladogram form by Humphries (1981: 195, fig. 11; 196, fig. 12; 197, fig. 13) and by Humphries and Parenti (1986: 85, fig. 4.14).

If the general biological area cladogram (Fig. 3*B*) is compared with the traditional, geological hypotheses for the break-up sequence of Gondwana published by Smith *et al.* (1973), Owen (1976) and Shields (1979), several interesting conclusions are evident. The biological area cladogram is congruent with the geological hypothesis of Smith *et al.* (1973) (Fig. 4*A*) and Shields (1979) (Fig. 4*B*), whereas reconciliation between the biological area cladogram and Owen's (1976) geological area cladogram requires at least eight items of error (Fig. 4*C*); eight extra terms and components (Nelson 1979). In other words, the biological area cladogram is congruent with both a fixed earth (Smith *et al.* 1973) and a fast expanding earth model (Shields 1979), but less congruent with a slow expanding earth model (Owen 1976).

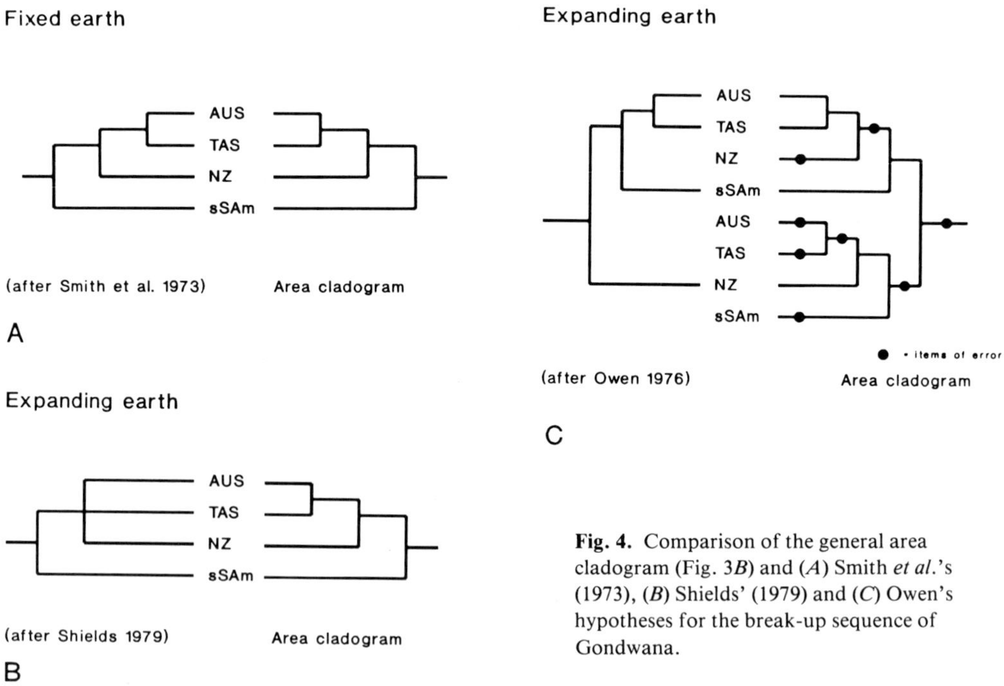

Fig. 4. Comparison of the general area cladogram (Fig. 3*B*) and (*A*) Smith *et al.*'s (1973), (*B*) Shields' (1979) and (*C*) Owen's hypotheses for the break-up sequence of Gondwana.

The general biological area cladogram is congruent with both geological hypotheses incorporating the hypothetical continent of Pacifica (Nur and Ben-Avraham 1978, 1981, and supplemented by information from Kamp 1980) and Melville (1981, 1982). The hypothesis of Nur and Ben-Avraham is founded on geological data only, whereas Melville's hypothesis is founded on a mixture of geological and biological data. Note that NZ is composed of three microplates: CI, Chatham I., LH, Lord Howe, and CP, Campbell Plateau (Craw 1982; Cooper 1989).

I must add that several of the biologically defined area relationships unconsidered in the present study have parallels on one or more of the geological area cladograms. This applies, for example, to the disjunction between taxa occurring in northern and southern South American (e.g. *O. obtusangulus* ssp. *unispicus* and *O. obtusangulus* ssp. *obtusangulus*). This disjunction may be explained by Melville's concepts of a Peruvian South America (PeSAm) and a Chilean South America (CSAm) (See also e.g. Humphries and Parenti 1986: 81–82).

Among the more important statements made by Croizat (1962: 707) is the idea 'that life and earth evolved together'. However, what is the relationship between such a statement and a hypothesis such as Pacifica? Or how does it affect one's choice of an expanding or fixed earth model as a causal explanation of an observed pattern?

For more than a century, historical geology has been a heavy burden on biogeographic thinking (see Darwin 1859; Thiselton-Dyer 1909; Darlington 1965; Raven and Axelrod 1974). However, the current state of affairs in historical geology is also also one of flux, with several competing hypotheses for the evolution of the earth. Widely different biologically defined area cladograms could find support in one or other of the historical geological hypotheses. The lost continent of Pacifica is a provocative and interesting hypothesis, but not the only one.

Thus geological data can no longer be a strait-jacket forcing interpretations onto biological data, but may be useful for inspiring new biological hypotheses (Michaux 1989).

A final conclusion drawn from this study is that, despite more than 10 years of progress, we still urgently need reliable phylogenetic hypotheses as a foundation for biogeographical inferences.

References

Balgooy, M. M. J. van (1975). 'Pacific Plant Areas 3.' (J. J. Groen en Zn.: Leiden.)

Balgooy, M. M. J. van (1984). 'Pacific Plant Areas 4.' (Rijksherbarium: Leiden.)

Ball, I. R. (1975). Nature and formulation of biogeographical hypotheses. *Systematic Zoology* **24**, 207–430.

Baumann, F. A. (1988). Geographische Verbreitung und Ökologie südamerikanischer Hochgebirgspflanzen. Beitrag zur Rekonstruktion der quartären Vegetationsgeschichte der Anden. (Lang Druck AG: Bern.)

Brundin, L. (1966). Transantartic relationships and their significance, as evidenced by Chironomid Midges. *Kungliga Svenska Vetenskapsakademiens Handlingar, 4. Serien* **11** (1): 1–472.

Carey, S. W. (1976), 'The Expanding Earth.' (Elsevier: Amsterdam.)

Carpenter, J. M. (1988). Choosing among multiple equally parsimonious cladograms. *Cladistics* 4, 291-6.

Cooper, R. A. (1989). New Zealand tectonostratigraphic terranes and panbiogeography. *New Zealand Journal of Zoology* **16**, 699–712.

Cracraft, J. (1973). Continental drift, paleoclimatology, and the evolution and biogeography of birds. *Journal of Zoology* **169**, 455–545.

Cracraft, J. (1974). Continental drift and vertebrate distribution. *Annual Review of Systematics and Ecology* **5**, 215–61.

Cracraft, F. (1975). Mesozoic dispersal of terrestrial faunas around the southern end of the world. *Memoires du Museum National d'Historie Naturelle, Nouvelle Series A, Zoologie* **88**, 29–52.

Craw, R. (1982). Phylogenetics, areas, geology and the biogeography of Croizat: a radical view. *Systematic Zoology* **31**, 304–16.

Crisci, J. V., Gamundí, I. J., and Cabello, M. N. (1988). A cladistic analysis of the genus *Cyttaria* (Fungi-Ascomycotina). *Cladistics* **4**, 279–90.

Croizat, L. (1962). 'Space, Time, Form: The Biological Synthesis.' (The author: Caracas.)

Croizat, L., Nelson, G. J. and Rosen, D. E. (1974). Centers of origin and related concepts. *Systematic Zoology* **23**, 265–87.

Darlington, J. P. Jr (1957). Zoogeography: 'The Geographical Distribution of Animals.' (Wiley & Sons: New York.)

Darlington, J. P. Jr (1965). 'Biogeography of the Southern End of the World.' (Harvard University Press: Cambridge, Mass.)

Darwin, C. (1859). 'On the Origin of Species by Means of Natural Selection or the Preservation of Favoured Races in the Struggle for Life.' (John Murray: London.)

Engel, J. J. (1990). Falkland Islands (Isla Malvinas) Hepaticae and Anthocerotophyta: A taxonomic and phytogeographic study. *Fieldiana, Botany, New Series* **25**, 1-209.

Farris, J. S. (1969). A successive approximations approach to character weighting. *Systematic Zoology* **18**, 374–85.

Farris, J. S. (1970). Methods for computing Wagner trees. *Systematic Zoology* **19**, 83–92.

Farris, J. S. (1988). 'Hennig86 Reference. Version 1.5.' (The author: New York.)

Felsenstein, J. (1982). Numerical methods for inferring evolutionary trees. *Quartenary Review of Biology* **57**, 379–404.

Fulford, M. (1951). Distribution patterns of the genera of leafy Hepatics of South America. *Evolution* **5**, 243–64.

Good, R. (1974). 'The Geography of the Flowering Plants.' 4th Edn. (Longman: London.)

Hill, R. S. (1983). *Nothofagus* macrofossils from the Tertiary of Tasmania. *Alcheringa* **7**, 169–83.

Hooker, J. D. (1853[-55]). Introductory essay. In 'The Botany of the Antarctic Voyage of H. M. Discovery Ships Erebus and Terror in the Years 1839–1843. II. Flora Novæ-Zelandiæ'. pp. i–xxxix. (L. Reeve: London.)

Humphries, C. J. (1981). Biogeographical methods and the Southern Beeches (Fagaceae: *Nothofagus*). In 'Advances in Cladistics'. (Eds V. A. Funk and D. R. Brooks.) pp. 177–207. (New York Botanical Garden: New York.)

Humphries, C. J. (1983). Biogeographical explanations and the Southern Beeches. In 'Evolution, Time and Space: The Emergence of the Biosphere'. (Eds R. W. Sims, J. H. Price and P. E. S. Whalley.) pp. 335–65. (Academic Press: London.)

Humphries, C. J., Cox, J. M., Nielsen, E. S. (1986). *Nothofagus* and its parasites: a cladistic approach to coevolution. In 'Coevolution and Systematics'. (Eds A. R. Stone and D. L. Hawksworth.) pp. 55–76. (Academic Press: London.)

Humphries, C. J., and Parenti, L. R. (1986). 'Cladistic Biogeography.' (Clarendon Press: Oxford.)

Kamp, P. J. J. (1980). Pacifica and New Zealand, proposed eastern elements in Gondwanaland's history. *Nature* **288**, 659–64.

Keast, A. (1973). Contemporary biotas and the separation sequence of the southern continents. In 'Implications of Continental Drift to the Earth Sciences. Vol. 1'. (Eds D. H. Tarling and S. K. Runcorn.) pp. 309–43. (Academic Press: London.)

Kluge, A. G., and Farris, J. S. (1969). Quantitative phyletics and the evolution of anurans. *Systematic Zoology* **18**, 1–32.

Meacham, C. A. (1980). Phylogeny of Berberidaceae with an evaluation of classifications. *Systematic Botany* **5**, 149–72.

Melville, R. (1981). Vicarious plant distributions and paleogeography of the Pacific region. In 'Vicariance Biogeography: a Critique'. (Eds G. Nelson and D. E. Rosen.) pp. 238–74. (Columbia University Press: New York.)

Melville, R. (1982). The biogeography of *Nothofagus* and *Trigonobalanus* and the origin of the Fagaceae. *Botanical Journal of the Linnean Society* **85**, 75–88.

Michaux, B. (1989). Generalized tracks and geology. *Systematic Zoology* **38**, 390–8.

Mora-Osejo, L. E. (1987). El género *Oreobolus* R. Br. en el Nuevo Mundo. In 'Estudios Morfológicos, Autoecológicos y Sistemáticos en Angiosperms'. (Ed L. E. Mora-Osejo.) pp. 147–95. (Academia Colombiana de Ciencias Exactas, Fisicas y Naturales. Serie Jorge Alvarez Lleras 1.) (Editorial Kimpres: Bogotá.)

Nelson, G. (1979). Cladistic analysis and synthesis: principles and definitions, with a historical note on Adanson's *Familles des Plantes* (1763–1764). *Systematic Zoology* **28**, 1–21.

Nelson, G. (1982). Cladistique et biogeographie. *Compte Rendus de la Société de Biogeographie* **58**, 75–94.

Nelson, G., and Platnick, N. I. (1981). 'Systematics and Biogeography. Cladistics and Vicariance.' (Columbia University Press: New York.)

Nur, A., and Ben-Avraham, Z. (1978). Speculations on mountain building and the lost Pacifica continent. *Journal of Physics of the Earth* **26**, 221–37.

Nur, A., and Ben-Avraham, Z. (1981). Lost Pacifica continent: a mobilistic speculation. In 'Vicariance Biogeography: A Critique'. (Eds G. Nelson and D. E. Rosen.) pp. 341–58. (Columbia University Press: New York.)

Owen, G. H. (1976). Continental displacement and expansion of the Earth during the Mesozoic and Cenozoic. *Philosophical Transactions of the Royal Society* **281**, 223–90.

Owen, G. H. (1983). 'Atlas of Continental Displacement, 200 Million Years to the Present.' (Cambridge University Press: Cambridge.)

Page, R. D. M. (1989). 'Component User's Manual. Release 1.5.' (The author: Auckland.)

Page, R. D. M. (1990). Component analysis: a valiant failure? *Cladistics* **6**, 119–36.

Philipson, W. R., and Philipson, M. N. (1988). A classification of the genus *Nothofagus* (Fagaceae). *Botanical Journal of the Linnean Society* **98**, 27–36.

Platnick, N. I., and Nelson, G. J. (1978). A method of analysis for historical biogeography. *Systematic Zoology* **27**, 1–16.

Raven, P. H., and Axelrod, D. I. (1974). Angiosperm biogeography and past continental movements. *Annals of the Missouri Botanical Garden* **61**, 539–673.

Schlinger, E. I. (1974). Continental Drift, *Nothofagus*, and some ecologically associated insects. *Annual Review of Entomology* **19**, 323–43.

Schuster, R. M. (1969). Problems of antipodal distribution in lower land plants. *Taxon* **18**, 46–91.

Schuster, R. M. (1972). Continental movements, 'Wallace's Line' and Indomalayan–Australasian dispersal and land plants: some eclectic concepts. *Botanical Review* **38**, 3–86.

Schuster, R. M. (1979). On the persistence and dispersal of transantarctic Hepaticae. *Canadian Journal of Botany* **57**, 2179–225.

Seberg, O. (1988). Taxonomy, phylogeny, and biogeography of the genus *Oreobolus* R. Br. (Cyperaceae), with comments on the biogeography of the South Pacific Continents. *Botanical Journal of the Linnean Society* **96**, 119–95.

Shields, O. (1979). Evidence for initial opening of the Pacific Ocean in the Jurassic. *Paleogeography, Palaeoclimatology, Paleoecology* **26**, 181–220.

Smith, A. G., Briden, J. C., and Drewry, G. E. (1973). Phanerozoic world maps. In 'Organisms and Continents through Time'. (Ed. N. F. Hughes.) pp. 1–42. (Palaeontological Association: London.)

Steenis, C. G. G. J. van (1963). Pacific plant areas. 1. *Monographs of the National Insititute of Science and Technology* **8** (1), 1–297.

Steenis, C. G. G. J. van (1971). *Nothofagus*, key genus of plant geography, in time and space, living and fossil, ecology and phylogeny. *Blumea* **19**, 65–98.

Steenis, C. G. G. J. van, and van Balgooy, M. M. J. (1966). Pacific plant areas 2. *Blumea (Supple)* **5**, 1–132.

Tanai, T. (1986). Phytogeographic and phylogenetic history of the genus *Nothofagus* Bl. (Fagaceae) in the Southern Hemisphere. *Journal of the Faculty of Sciences. Hokkaido University, Series IV*, **21**, 505–82.

Thiselton-Dyer, W. (1909). Geographical distribution of plants. In 'Darwin and Modern Science'. (Ed. A. C. Seward.) pp. 298–318. (Cambridge University Press: Cambridge.)

Thorne, R. F. (1972). Major disjunctions in the geographic ranges of seed plants. *Quartenary Review of Biology* **47**, 365–411.

Weston, P. H., and Crisp, M. D. (1987). Evolution and biogeography of the Waratahs. *Australian National Botanic Gardens, Occasional Publication* **9**, 17–34.

Wiley, E. O. (1988*a*). Vicariance biogeography. *Annual Review of Ecology and Systematics* **19**, 513–42.

Wiley, E. O. (1988*b*). Parsimony analysis and vicariance biogeography. *Systematic Zoology* **37**, 271–90.

Zandee, M. (1987). A computing environment for cladistic analyses. Preliminaries of a user manual for CAFCA/PC. In 'Systematics and Evolution: A Matter of Diversity'. (Ed. H. Hovenkamp.) pp. 11–140. (Institute of Systematic Botany: Utrecht.)

Zandee, M., and Ross, M. C. (1987). Component-compatibility in historical biogeography. *Cladistics* **3**, 305–32.

Appendix 1. The 10 equally parsimonious taxon cladograms for Embothriinae given in nested parentheses format

0: *Lomatia;* 1: *Embothrium coccineum;* 2: *Telopea speciosissima;* 3: *T.* 'Gibraltar Range' (a provisional name given to plants from the Gibraltar Range that may constitute a new species (Weston and Crisp 1987: 11–14)); 4: *T. truncata;* 5: *T. mongaensis;* 6: *T. oreades;* 7: *Oreocallis grandiflora;* 8: *O. mucronata;* 9: *O. pinnata;* and 10: *O. wickhamii*

1	(0, (7, 8, (9, 10)), (1, (4, ((2, 3), (5, 6)))));
2	(0, 1, (7, 8, (9, 10), ((2, 3), (4, (5, 6)))));
3	(0, 1, (7, (9, 10), (8, ((2, 3), (4, (5, 6))))));
4	(0, 1, (8, (9, 10), (7, ((2, 3), (4, (5, 6))))));
5	(0, 1, ((9, 10, (8, (7, ((2, 3), (4, (5, 6)))))));
6	(0, 1, ((9, 10), (7, (8, ((2, 3), (4, (5, 6)))))));
7	(0, 1, ((9, 10), (7, 8, ((2, 3), (4, (5, 6))))));
8	(0, 1, ((7, 8, (9, 10)), ((2, 3), (4, (5, 6)))));
9	(0, 7, 8, (9, 10), (1, ((2, 3), (4, (5, 6)))));
10	(0, (7, 8, (9, 10)), (1, ((2, 3), (4, (5, 6)))));

Appendix 2. The 10 equally parsimonious taxon cladograms for *Oreobolus* (and *Schoenoides*) given in nested parentheses format

1: *O. acutifolius;* 2: *O. ambiguus;* 3: *O. distichus;* 4: *O. ecuadorensis;* 5: *O. furcatus;* 6: *O. goeppingeri;* 7: *O. impar;* 8: *O. kuekenthalii;* 9: *O. obtusangulus* ssp. *obtusangulus;* 10: *O. obtusangulus* ssp. *unispicus;* 11: *O. oxycarpus* ssp. *oxycarpus;* 12: *O. oxycarpus* ssp. *brownii;* 13: *O. pectinatus;* 14: *O. pumilio* ssp. *clemensiae;* 15: *O. pumilio* ssp. *pumilio;* 16: *O. strictus;* 17: *O. venezuelensis;* and 18: *S. oligocephalus.*

1 (18, ((14, 15), ((7, (11, 12)), ((1, ((13, 16), (2, (5, 8)))), (4, (17, ((3, 6), (9, 10))))))));
2 (18, ((14, 15), ((7, (11, 12)), ((1, ((13, 16), (2, (5, 8)))), (4, ((9, 10), (17, (3, 6))))))));
3 (18, ((14, 15), ((2, 8), (5, ((7, (11, 12)), ((1, (13, 16)), (4, (17, ((3, 6), (9, 10))))))))));
4 (18, (14, 15, ((7, (11, 12)), ((1, ((13, 16), (2, (5, 8)))), (4, (17, ((3, 6), (9, 10))))))));
5 (18, (15, (14, ((7, (11, 12)), ((1, ((13, 16), (2, (5, 8)))), (4, (17, ((3, 6), (9, 10)))))))));
6 (18, (14, 15, ((7, ((11, 12)), ((1, ((13, 16), (2, (5, 8)))), (4, ((9, 10), (17, (3, 6))))))));
7 (18, (15, (14, ((7, (11, 12)), ((1, ((13, 16), (2, (5, 8)))), (4, ((9, 10), (17, (3, 6)))))))));
8 (18, ((14, 15), ((2, 8), (5, ((7, (11, 12)), (1, ((13, 16), (1, (4, (17, ((3, 6), (9, 10))))))))))));
9 (18, ((14, 15), ((2, 8), (5, ((7, (11, 12)), ((13, 16), (1, (4, (17, ((3, 6), (9, 10)))))))))));
10 (18, ((14, 15), ((2, 8), (5, ((7, (11, 12)), ((13, 16), (1, (4, ((9, 10), (17, (3, 6)))))))))));

Manuscript received 3 September 1990, accepted 4 March 1991

3 September 1991 *Aust. Syst. Bot.*, **4**, 137-49

Ocean Basins and the Biogeography of Freshwater Fishes

Lynne R. Parenti

Division of Fishes, National Museum of Natural History, Smithsonian Institution, Washington, D.C. 20560, U.S.A.

'The natural biogeographic regions for terrestrial and freshwater organisms are not present-day land areas but the world's major ocean basins.' (Craw and Page 1988: 180).

Abstract

Geological evolution of the Indian and Pacific Ocean basins, including the Tethys Sea, is relevant to the evolution and distribution of Indo-Pacific freshwater fishes. Area cladograms derived from phylogenetic analyses of atherinomorph fishes are compared, in part, with those for plant bugs, cicadas and bats. Geologically and biologically composite islands in the Indo-Australian archipelago such as New Guinea, Borneo and Sulawesi, are confirmed. If any lines (such as Wallace's, Weber's and so on) are to be drawn, they should pass through these complex islands, not between them, to mark the closing of the ancient Tethys Sea.

Sicydiines are a group of circumtropical, insular and coastal, amphidromous gobies: adults live and breed in freshwater, whereas larvae are transported to the sea where they undergo transformation. Sicydiine genera are restricted to ocean basins. Hypotheses of relationships among ocean basins, as indicated by phylogenetic relationships among sicydiine genera and by other, distantly related, aquatic vertebrates and plants, share components. Implications of oceanic biogeographic regions for continental biotas are outlined using South America as an example. A continent is part of the biogeographic regions of all the oceans it contacts.

Introduction

This paper addresses two points in historical biogeography: definition of global biogeographic realms or regions, and relationships among those regions. Terrestrial, including freshwater organisms are usually identified by the land mass, often a continent, in which they live. We are accustomed to hearing about 'South American freshwater fishes', 'Australian birds' (Cracraft 1986), 'African mammals' and so on. Similarly, we identify marine organisms by an ocean basin; 'Indo-Pacific fishes' is thought to mean marine fish generally (e.g. Springer 1982) and 'Indo-Pacific molluscs' means marine molluscs (e.g. Kohn 1990).

Botanist and biogeographer Leon Croizat (1964) provided extensive documentation to support his observation that major biotic tracks of both marine and terrestrial organisms cross ocean basins. Panbiogeographers Craw and Page (1988) followed Croizat's lead and proposed a new set of global biogeographic realms or regions that focus on ocean basins rather than continents. Springer (1982) identified the Pacific lithospheric plate as a biogeographic region for marine shorefishes and other organisms. Ocean biogeographic regions have been incorporated into biogeographic discussions of both terrestrial and marine organisms by members of the New Zealand school of biogeography (Page 1989) but rarely by others (e.g. Parenti 1989*a*; Platnick and Nelson 1989).

1030-1887/91/010137$05.00

All biogeographic regions will be altered during geologic history. My purpose is to demonstrate the relevance of ocean basin evolution to the distribution of freshwater fishes and other terrestrial organisms. I modify slightly the biogeographic regions of Craw and Page (1988) to include the Tethys Sea because of its importance in understanding the composite Indo-Australian biota (Fig. 1).

I support the goals and methods of vicariance biogeography as outlined by Nelson and Platnick (1981; see Page 1990). I use the area cladogram, a hierarchical representation of relationships among areas, to interpret relationships among ocean basins.

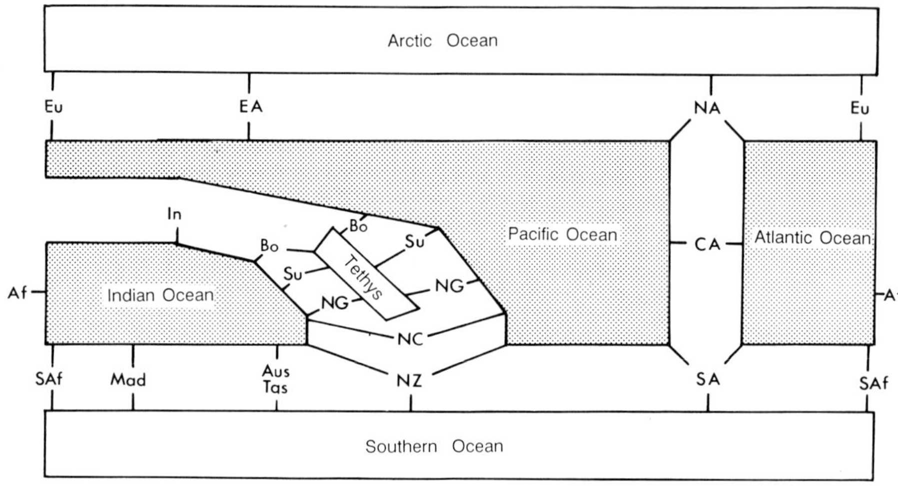

Fig. 1. Biogeographic realms modified from Craw and Page (1988: fig. 12) to show some composite areas of the Tethys Sea. Af, Africa; Aus, Australia; Bo, Borneo; CA, Central America; Eu, Europe; In, India; Mad, Madagascar; NA, North America; NC, New Caledonia; NG, New Guinea; NZ, New Zealand; SA, South America; SAf, South Africa; Su, Sulawesi; Tas, Tasmania.

The Indo-Australian Archipelago and the Tethys Sea

Schuh and Stonedahl (1986) presented a comprehensive cladistic summary of area relationships throughout much of the Indo-Pacific, including the Indo-Australian archipelago (Fig. 2). The scheme of area relationships was derived from their phylogenetic systematic studies of plant bugs (Heteroptera, Miridae) and incorporated the results of Duffels (1986) on cicadas and, to a limited degree, observations of Holloway (1984) on Lepidoptera. The sister-area of all those areas in Fig. 2 is Ghana (= West Africa or eastern Atlantic).

Many areas in Schuh and Stonedahl's analysis are islands or island groups (Fig. 2). The cladistic relationship of New Guinea is resolved (Fig. 2), even though Schuh and Stonedahl concluded that the island is a biological, as well as a geological, composite. The Philippines are divided into a northern and southern portion, with the southern portion showing a close relationship to northern Borneo. The relationships of southern India and Sulawesi are unresolved. Some of these area relationships were found by Cracraft (1988) for Indo-Australian birds: the Philippines do not form one area, but Palawan (the southern Philippines) is closely related to Sundaland (Java, Sumatra, Borneo and the Malay Peninsula).

Schuh and Stonedahl (1986) pointed out two limitations of their hypothesis that have been taken as a general criticism of cladistic biogeography (*sensu* Humphries and Parenti 1986): it does not adequately treat composite areas, and it does not incorporate trans-oceanic relationships (see Craw 1983). Panbiogeographers have long understood that these two perceived limitations really concern the same problem: major composite areas are trans-oceanic (Craw and Page 1988). Of course, some cladistic biogeographers have incorporated composite areas into their studies, and no aspect of the method precludes recognition of composites (Nelson and Platnick 1981; Schuh and Stonedahl 1986).

Smith and Hood (1981) analysed phylogenetic relationships and distribution of Indo-Pacific bats (Fig. 3). When New Guinea was treated as one area, two general patterns were found: (1) New Guinea is the sister-area to islands in the Banda Sea, the Bismark Archipelago and the Solomon Islands (Fig. 3*a*); (2) islands in the Banda Sea form the sister-area to New Guinea, the Bismark Archipelago and the Solomon Islands (Fig. 3*b*). A consensus tree between the area cladograms of Fig. 3*a* and 3*b* is trivial: the relationship among Banda Sea, New Guinea and the sister-areas Solomon Islands and the Bismark Archipelago is unresolved. Smith and Hood (1981) achieved resolution when they treated New Guinea as a composite, divided into two areas: nuclear New Guinea and tethyan New Guinea (Fig. 3c). What ocean does the composite New Guinea span? 'Tethyan New Guinea' implies some rendition of the ancient Tethys Sea.

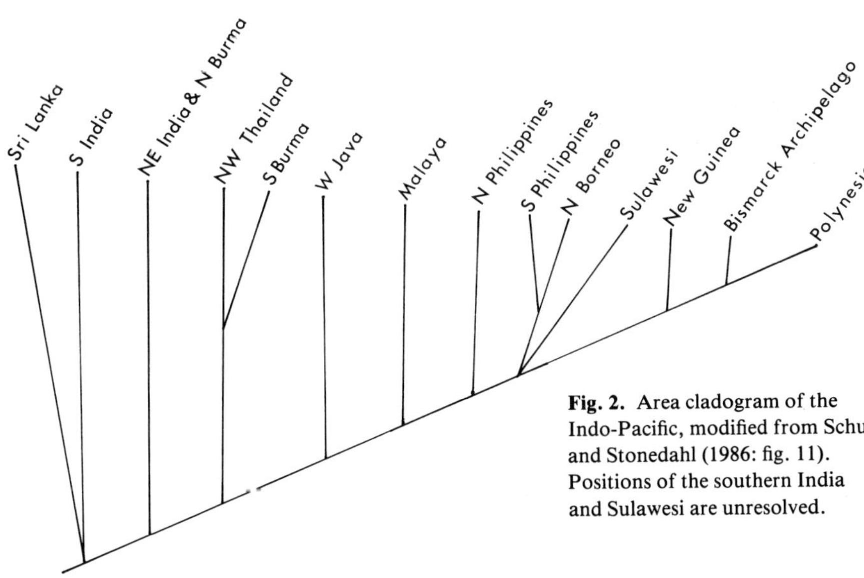

Fig. 2. Area cladogram of the Indo-Pacific, modified from Schuh and Stonedahl (1986: fig. 11). Positions of the southern India and Sulawesi are unresolved.

Plate tectonics has changed the way we look at the Indo-Australian archipelago. The area is complex. Modern large islands such as Sulawesi, New Guinea and Borneo are composites, as are political entities such as the Philippines. Historical geological details vary and may conflict (Hamilton 1979, 1988; Audley-Charles 1987; Cracraft 1988; Burrett *et al.* 1991). In one summary, by the Middle Mesozoic (Late Jurassic, 160 MYA), New Guinea rifted from Sumatra and was flanked by bits of Borneo and western Sulawesi; Timor rifted from Southern Tibet and Burma; and eastern Sulawesi rifted from Thailand (Audley-Charles 1987: fig. 2.1). The body of water that formed as these land masses separated has been referred to as the Tethys Sea (e.g. Audley-Charles 1987). This should not be confused with the larger, hypothetical Tethys, believed to have once separated the Indian subcontinent from Asia; that ocean is an artefact under theories of an expanding earth (Carey 1983).

Following the rift of widespread, ancestral biotas across an expanding Tethys Sea, sister-taxa on present-day Sumatra and nuclear New Guinea could be called trans-Tethyan. This is simple. What is less clear or less obvious is the distribution throughout present-day large islands such as Borneo, Sulawesi and New Guinea which, in late Jurassic times, were present as separate bits, rifted apart only to come back together again in combinations that give seemingly little indication of their former associations.

Are there trans-Tethyan relationships among freshwater fishes? Firstly, I consider the silversides or hardyheads of the bony fish family Phallostethidae, comprising 19 species (Parenti 1986, 1989*b*). Phallostethids are primarily coastal (freshwater and marine) fishes, distributed throughout the Malay Peninsula, north-western Borneo and the Philippines, entirely west of Wallace's Line (Parenti 1989*b*: fig. 1). Their hypothesised sister-group is the monotypic silverside *Dentatherina* (Parenti 1984), a widely distributed marine shorefish described from material collected from the Philippines to north-eastern Australia, and from the Moluccas to the Trobriand islands (Patten and Ivantsoff 1983). The distributions of phallostethids and *Dentatherina* are complementary, overlapping in Borneo and the southern Philippines.

I classified phallostethids in three tribes (Parenti 1989*b*), and discuss two here. The tribe Phallostethini contains four species with resolved phylogenetic relationships (Fig. 4). An area cladogram for the tribe is formed by replacing the name of the species with the area in which it occurs (Fig. 5). This cladogram adds little to our broad knowledge of area relationships but does show that Malayan phallostethids are not monophyletic (Fig. 5).

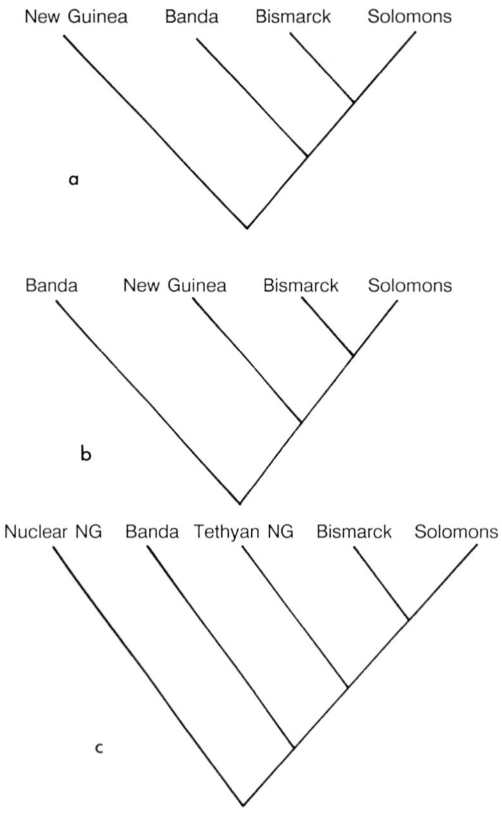

Fig. 3. Area cladogram of bats, from Smith and Hood (1981: fig. 11): (*a*) and (*b*) alternate area cladograms for different groups of bats; (*c*) resolution of alternate patterns in 3*a* and 3*b* when New Guinea (NG) is treated as a composite.

The tribe Neostethini comprises 10 species (Fig. 6). An area cladogram reveals that many taxa are widespread and therefore are also uninformative with respect to relationships among areas (Nelson and Platnick 1981; Fig. 7). However, general statements may be made: (1) Philippine species are not monophyletic; (2) Borneo species are not monophyletic; and (3) Malay Peninsula species are not monophyletic. There is no further resolution of Schuh and Stonedahl's hypothesis but, their composites are supported (Northern and Southern Philippines) and additional composites identified (Malaya).

The second group is ricefish family Adrianichthyidae (*sensu* Rosen and Parenti 1981). Ricefish are broadly distributed from the eastern U.S.S.R. and India through continental Asia, and along the Indo-Australian archipelago to Sulawesi and Timor (Yamamoto 1975: fig. 2–4; Uwa and Parenti 1988). Ricefish distribution spans Wallace's Line.

Ricefish are absent from Sumatra, Borneo and New Guinea. I agree that absences need not be explained in biogeography (e.g. Schuh and Stonedahl 1986). However, absence from Borneo/Sumatra/New Guinea makes sense if we consider that these land masses were adjacent prior to opening of the Tethys Sea; we may hypothesise that the ancestral distribution of ricefishes did not include these areas. This hypothesis is more parsimonious than creating a separate explanation for each absence on each large island.

Phylogenetic relationships among all ricefish species are under study by myself and others. Preliminary reports indicate that Sulawesi species are polyphyletic (Parenti 1987).

Lake Poso in north-eastern Sulawesi contains large, plesiomorphic ricefishes currently classified in the genera *Adrianichthys* and *Xenopoecilus*; a phylogenetically more derived *Oryzias* species lives in the south-western arm of Sulawesi (Parenti 1987; Kottelat 1990).

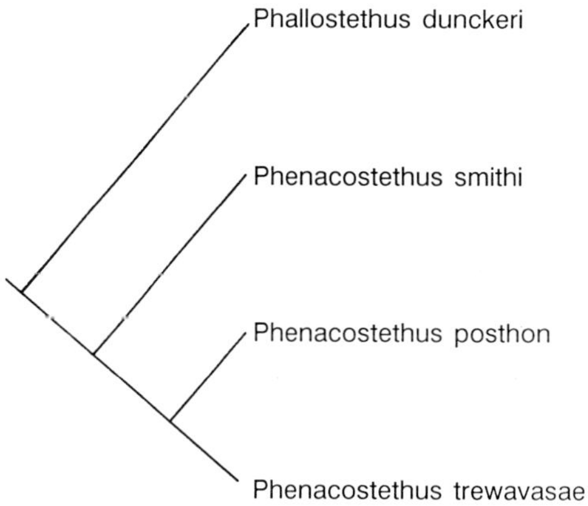

Fig. 4. Cladogram of relationships among species in the phallostethid tribe Phallostethini (from Parenti 1986: fig. 5).

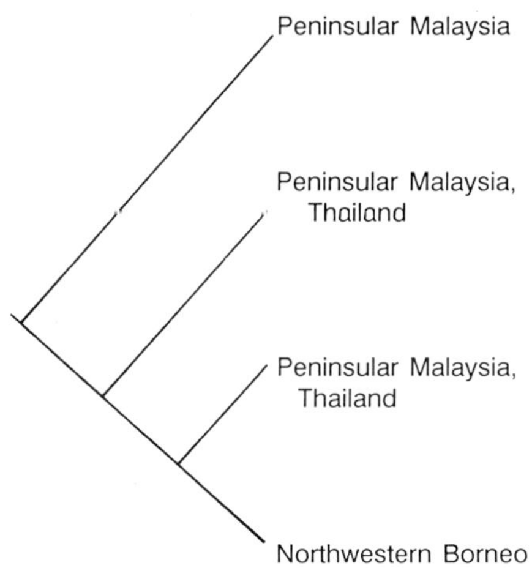

Fig. 5. Area cladogram for phallostethid species, following relationships in Fig. 4.

Distributions like those of ricefishes, phallostethids and their sister-taxon, plant bugs, cicadas and bats have been described in whole or part as Indo-Australian or Indo-Pacific. Relationships indicate that the taxa may also be described as trans-Tethyan; together they confirm the composite biotic relationships of geologically composite Sulawesi, New Guinea and Borneo. Some will argue that these taxa are too young to have been affected by geological events over 150 mya. However, I agree with Heads (1990: 224) that '. . . tectonics and biogeography suggest that the last main phase of modernisation for the world was in the upper Mesozoic . . .'. If modern species were not present their ancestral taxa were. Springer and Williams (1990) proposed Pliocene-Pleistocene extinction owing to lowered sea levels throughout the Indo-Australian archipelago as an explanation for some cases of Pacific plate endemism. Continuously distributed Indo-Pacific taxa went extinct in parts of their ranges with loss of marine habitats. If true, this explanation requires that ancestral populations were widely distributed throughout the Indo-Pacific prior to the Pliocene-Pleistocene.

Relationships among taxa help us identify or confirm composite areas such as nuclear and tethyan New Guinea which can be treated as two areas in further studies. This aspect of

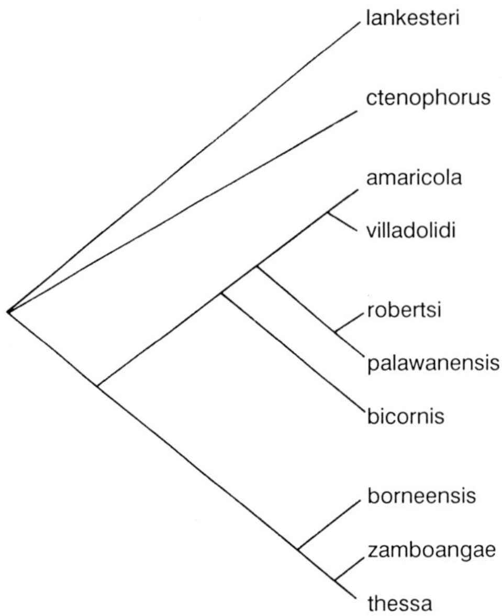

Fig. 6. Cladogram of relationships among species in the phallostethid tribe Neostethini (= genus *Neostethus*) (from Parenti 1989*b*: fig. 8).

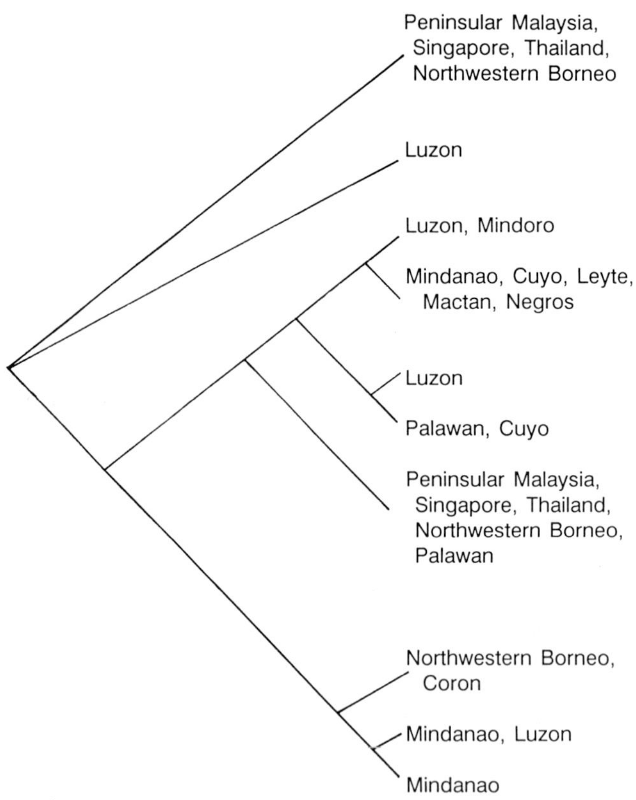

Fig. 7. Area cladogram for *Neostethus* species, following relationships in Fig. 6.

cladistic biogeography has been criticised as creating an endless regression: incongruent areas will demand finer and finer subdivision of areas (Patterson 1981). I claim that our experience tells us the regression will not be endless. We may stop dividing areas after we identify where phylogenetic relationships span an ocean basin, either past or present.

Recognition of composite areas and oceanic biogeographic realms is not meant to substitute for recognition of an area of endemism as that area occupied by a monophyletic group

(Nelson and Platnick 1981). Tethyan New Guinea may not be an area of endemism for all members of the tethyan New Guinea biota. However, knowing that it might be an area of endemism for much of the biota will prevent treating New Guinea as one area in biogeographic studies. New Guinea can never be identified as part of the Australian or the Asian realm. Continental realms have little reality when identifying historical relationships of continental biotas; the biota is trans-oceanic. Composite areas treated as one will always occupy unresolved positions on area cladograms, or result in incongruence among area cladograms (see also Cracraft 1989).

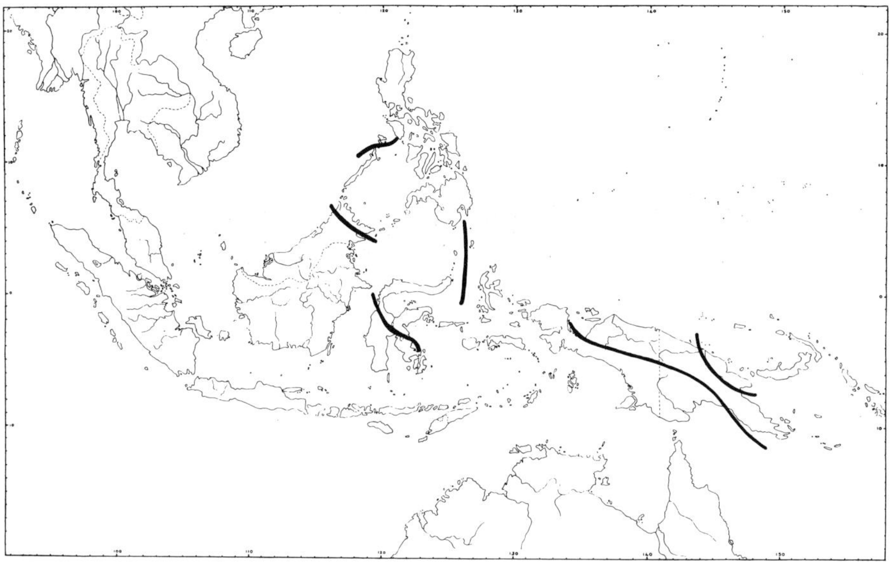

Fig. 8. Some major suture and fault lines through the Indo-Australian region (Hamilton 1988: fig. 2).

Wallace's and Other Lines

Simpson (1977) wrote that 'too many lines' had been proposed to separate the Asian from the Australian zoogeographic realm. He identified seven such lines: Müller's, Wallace's, Murray's, Huxley's, Sclater's, Lydekker's and Weber's. Wallace's Line has been given special status in biogeography of both terrestrial and marine organisms (e.g. Whitmore 1981; Woodland 1986). However, Humphries (1989) criticised Wallace's Line as an over-emphasised boundary that is no different from other major zones of biotic contact. I agree that too many vertical lines have been drawn to separate islands or island groups from one another. If we cannot decide whether to draw a line to the east or the west of modern Sulawesi, for example, then perhaps we should draw a line through it.

Were we to draw new lines to describe Indo-Australian distribution patterns, the lines would pass through and bisect the traditional lines, through New Guinea, through Sulawesi and through Borneo. Huxley's line separates the southern and northern Philippines (Simpson 1977). Suture and fault lines mark the closing of the Tethys Sea (Hamilton 1988; Fig. 8). These lines may be connected to create a new line, perhaps to be called Hamilton's Line. Like Wallace's and other lines, it too would separate areas but would remind us that there is an ancient ocean basin hidden inside the Indo-Australian archipelago, and that distribution of organisms throughout the archipelago may be better understood if we consider the Tethys Sea when trying to resolve relationships among areas.

The Indo-Australian Archipelago and Modern Oceans

How is the Indo-Pacific region related to the rest of the world? To address this question for freshwater fishes, I add a more widespread group to the study. The sicydiine gobies comprise a diverse, circumtropical group of about 60 species. They are amphidromous: adults live and breed in freshwater, whereas larvae are transported to the sea where they undergo transformation, returning to rocky bottom, fast-flowing streams where they live as adults (Erdman 1986, personal observation, New Guinea and Maui). Life history requirements of sicydiines may affect distribution patterns on a small scale (e.g. Leis 1986); however, sicydiine distribution is interpreted below as part of a larger, global pattern.

Sicydiines may be classified in five genera: *Sicyopus, Lentipes, Sicydium, Stiphodon* and *Sicyopterus* (see Hoese 1984). A preliminary hypothesis of phylogenetic relationships among genera is principally on a transition series of outer jaw teeth morphology (Fig. 9; see Sakai and Nakamura 1979: fig. 8). Characters are polarised using the pantropical gobiid *Awaous* as an outgroup (Harrison 1989). Phylogenetic relationships among the genera are under study and new species remain to be described; however, this preliminary cladogram is well founded.

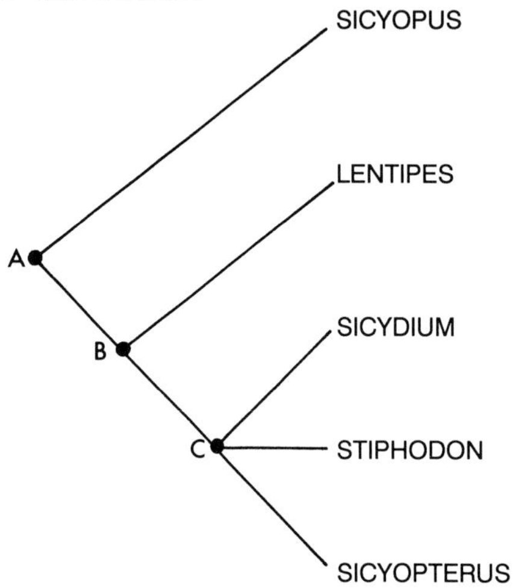

Fig. 9. Cladogram of relationships among sicydiine genera. Derived characters to support each node: (A) adnate tongue (skin covering basihyal and floor of mouth), fused, thickened and highly branched pelvic fin-rays, and fleshy pads at posterior tips of pelvic spines (Hoese 1984), (B) medial and anterolateral portion of upper and lower oral jaw with external row of thin, closely set teeth (Sakai and Nakamura 1979); (C) upper and lower oral jaw with external row of thin, closely set teeth (Akihito and Meguro 1979).

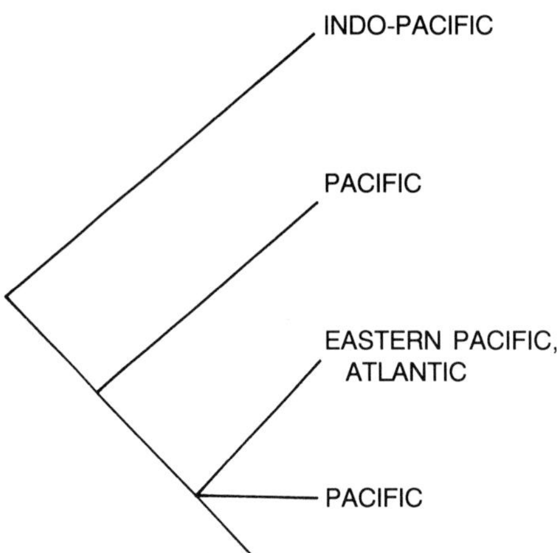

Fig. 10. Area (ocean) cladogram for sicydiine genera, with the name of a genus in Fig. 9 replaced by the name of the ocean basin(s) in which it occurs. See text for further discussion.

To construct an area cladogram for sicydiines, I could replace the name of each genus with a list of the areas in which it lives. However, the distribution of each genus can best and most easily be described in terms of an ocean basin (Fig. 10). *Sicyopus* lives in the Indo-Pacific (Klausewitz and Henrich 1986). *Lentipes* is known from Hawaii and the Ryukyus (Sakai and Nakamura 1979; although Springer (1982) questioned classification of the Japanese species). *Sicydium* is found in the eastern Pacific and the Atlantic; eastern Pacific includes freshwater coastal streams of Ecuador and western Central America, as well as insular streams of, for example, Cocos Island (Hertlein 1963). *Sicydium stimpsoni* Gill, common in Hawaii as well as throughout the central Pacific (Tinker 1978), has been reclassified in *Sicyopterus* (Akihito and Meguro 1979). *Sicyopterus* is Indo-Pacific, found from Madagascar, through the Indo-Australian archipelago to Rapa and Pitcairn Islands in the south-eastern Pacific (Koumans 1953; Randall *et al.* 1990).

The set of area (ocean) relationships (Fig. 10) is hierarchical. Were this a character state tree in phylogenetic systematic analysis, we could say that the Pacific, or the Indo-Pacific, is plesiomorphic. In cladistic biogeography, this means that vicariance between the Western (or Indo-) and eastern Pacific preceded vicariance between the Atlantic and the eastern Pacific. One contribution of cladistic biogeography that sets it apart from other biogeographic methods is that it allows the ordering of evolutionary events without relying on arbitrary measure of age, such as oldest known fossil, relative taxonomic rank, or conformation to a geologic hypothesis (Nelson and Platnick 1981; Humphries and Parenti 1986).

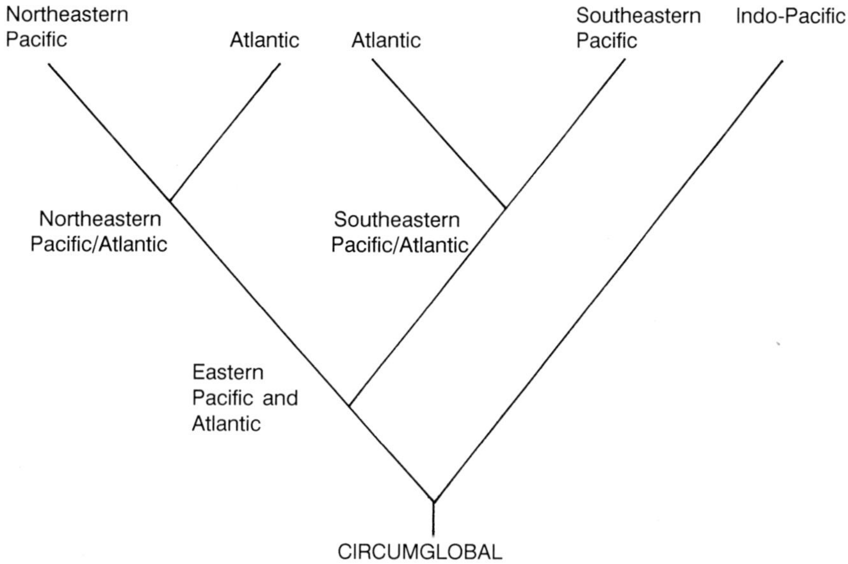

Fig. 11. Model of some relationships among modern ocean basins, modified in part from Brooks *et al.* (1981: fig. 16). Differentiation across the Indo-Pacific and the Atlantic is not considered here.

A model that incorporates some of the same ocean basin relationships was proposed by Brooks *et al.* (1981) for stingrays of the genus *Urolophus* and their helminth parasites. Their model may be rejected, in part, because monophyly of *Urolophus* may not be confirmed; South American species studied by Brooks *et al.* should be referred to *Urobatis* (John McEachran, personal communication). However, I treat the model like any hypothesis of area relationships, and ask if it has any general significance. The model was confirmed, in part, for cladistic area relationships of *Cetengraulis* anchovies by Nelson (1984: fig. 3): a South American freshwater species is plesiomorphic to a Eastern Pacific and an Atlantic sister-group pair. Valdebenito *et al.* (1990) documented the close relationship between species in the plant genus *Peperomia* from the Juan Fernandez Islands (Eastern Pacific) and the Tristan de Cunha archipelago (Southern Atlantic). Valdebenito *et al.*, (1990) proposed dispersal of plesiomorphic Juan Fernandez species to Tristan de Cunha. It is more parsimonious to suggest that stingrays and their parasites, anchovies, gobies and *Peperomia* are all part of the same general biogeographic pattern.

Brooks *et al.* (1981) considered a circumpacific distribution ancestral; the model works also with a circumglobal ancestral distribution (Fig. 11). They hypothesised vicariance first across the Pacific basin to create a western Pacific and an eastern Pacific biota, followed by a differentiation of the south-eastern Pacific into a South American freshwater and a Caribbean/Western Atlantic biota.

Brooks *et al.* (1981) proposed dispersal from the eastern Pacific to the Caribbean/ Western Atlantic to account for closely related taxa in both the eastern Pacific and the Atlantic. One could alternatively consider a circumglobal ancestral distribution disrupted by formation of the ocean basins with vicariance between the eastern Pacific and the Atlantic when the Panamanian isthmus was formed (Nelson 1984). Formation of antitropical distributions in conjunction with disruption of the Pacific basin is also implied by this model and has been discussed elsewhere (Nelson 1984; Humphries and Parenti 1986).

Implications for Continental Biotas: South America, an Example

If global biogeographic regions are ocean basins, then the regions cannot at the same time be continents. A continent is part of the biogeographic regions of all the oceans that it contacts. We can predict realistically that the South American biota will have relationships across three oceans: Atlantic, Pacific and the southern ocean (Fig. 1). Trans-Atlantic relationships are well known for many portions of the tropical, lowland South American biota. However, oceanic regions will be of little interest to those studying biogeographic patterns of taxonomic groups distributed solely within tropical lowland South America (e.g. Weitzman *et al.* 1988).

Brundin (1966) and Crisci *et al.* (1990) provide reasons for treating southern South America and the Andes as part of a southern oceans biota, rather than part of a continental biota. Southern South America and the Andes were not included in a discussion of areas of endemism of the widespread freshwater fish family Curimatidae, which has no representatives in those portions of South America (Vari 1988).

Trans-Pacific relationships for South America taxa are less obvious, principally because differentiation across the Pacific is proposed to have occurred before differentiation between the eastern Pacific and the Atlantic (Fig. 11). Sicyciine gobies of the genus *Sicydium* are found throughout the tropical eastern Pacific and the Atlantic. I consider the eastern Pacific distribution to include *Sicydium hildebrandi* from coastal streams of Ecuador (Barriga 1989) and the Atlantic distribution to include *Sicydium montanum* from coastal streams near Caracas, Venezuela (Hubbs 1920). Distributions of closely related taxa in, for example, coastal eastern Venezuela and coastal western Ecuador are often called trans-Andean. Andean orogeny is usually proposed as the vicariant event that disrupted a more widespread, hypothetical, ancestral biota to form trans-Andean sister-group pairs (e.g. Cracraft and Prum 1988). The model of Fig. 11 implies that such a distribution could also be called trans-eastern Pacific/Atlantic, and that formation of the Panamanian isthmus, along with expansion of ocean basins, were the vicariance events that disrupted the ancestral biota, separating eastern Pacific from western Atlantic taxa. Additional reinterpretations of distributions of South American freshwater (terrestrial) taxa as part of an oceanic biogeographic region, rather than a continental biogeographic region, are possible.

The history of continental biotas is undoubtedly complex (Cracraft 1989). Despite recent vigorous debate on the methods of cladistic biogeography (e.g. Brooks 1985; Humphries *et al.* 1988; Wiley 1988; Platnick and Nelson 1989; Page 1990), no consensus has been reached on how to choose among competing general area cladograms for continental taxa (e.g. Cracraft 1989). Recognition of composite areas and oceanic regions will help to eliminate incongruence (e.g. Fig. 3) from historical biogeographic studies.

Acknowledgments

I thank Pauline Ladiges for inviting me to participate in the symposium and contributed paper session on Austral Biogeography at the Ninth Willi Hennig Society meetings, Canberra, Australia, August 1990, and to submit a manuscript to be considered for publication in this issue. My attendance at the meetings was supported by the National Museum of Natural History Research Opportunities Fund.

This research was initiated at the California Academy of Sciences when I was funded by NSF grants BSR 87-00351 and BSR 89-06649; support from the Academy and NSF is gratefully acknowledged.

This paper was improved by the comments of A. C. Gill, V. G. Springer and S. H. Weitzman, National Museum of Natural History, Smithsonian Institution, and R. D. M. Page, American Museum of Natural History. C. J. Humphries, the Natural History Museum, London, provided some references and beneficial discussion. John McEachran, Texas A&M University, discussed stingray classification.

References

Akihito, (Prince) and Meguro, K. (1979). On the differences between the genera *Sicydium* and *Sicyopterus* (Gobiidae). *Japanese Journal of Ichthyology* **26**, 192–202.

Audley-Charles, M. G. (1987). Dispersal of Gondwanaland: relevance to evolution of the angiosperms. In 'Biogoegraphical Evolution of the Malay Archipelago'. (Ed. T. C. Whitmore.) pp. 5–25 (Clarendon Press: Oxford.)

Barriga, R. (1989). Peces de la Reserva Etnica y Forestal Awa, Ecuador Noroccidental. *Politecnica* **14**, 7–55.

Brooks, D. R. (1985). Historical ecology: a new approach to studying the evolution of ecological associations. *Annals of the Missouri Botantical Garden* **72**, 66–80.

Brooks, D. R., Thorson, T. B., and Mayer, M. A. (1981). Fresh-water stingrays (Potamotrygonidae) and their helminth parasites. In 'Advances in Cladistics'. (Eds V. A. Funk and D. R. Brooks.) pp. 147–75. (New York Botanical Garden: New York.)

Brundin, L. (1966). Transantarctic relationships and their significance, as evidenced by chironomid midges. *Kungliga Svenska Vetenskapsakademiens handlingar, Fjärde ser.* **11**, 1–472.

Burrett, C., Duhig, N., Berry, R., and Varne, R. (1991). Asian and south west Pacific continental terranes derived from Gondwana-biogeographic significance. *Australian Systematic Botany* **4**, 13–24.

Carey, S. W. (1983). 'The Expanding Earth: a Symposium.' (University of Tasmania: Hobart.)

Cracraft, J. (1986). Origin and evolution of continental biotas: speciation and historical congruence within the Australian avifauna. *Evolution* **40**, 977–96.

Cracraft, J. (1988). From Malaysia to New Guinea: evolutionary biogeography within a complex continent-island arc contact zone. In 'Acta XIX Congressus Internationalis Ornithologici, Volume II'. (Ed. H. Ouellet.) pp. 2582–93. (University of Ottawa Press: Ottawa.)

Cracraft, J. (1989). Deep-history biogeography: retrieving the historical pattern of evolving continental biotas. *Systematic Zoology* **37**, 221–36.

Cracraft, J., and Prum, R. O. (1988). Patterns and processes of diversification: speciation and historical congruence in some neotropical birds. *Evolution* **42**, 603–20.

Craw, R. (1983). Panbiogeography and vicariance cladistics: Are they truly different? *Systematic Zoology* **32**, 431–38.

Craw, R., and Page, R. (1988). Panbiogeography: method and metaphor in the new biogeography. In 'Evolutionary Processes and Metaphors'. (Eds M.-W. Ho and S. W. Fox.) pp. 163–89. (John Wiley and Sons: London.)

Crisci, J. V., Cigliano, M. M., Morrone, J. J., and Roi-Junent, S. (1990). 'Historical Biogeography of Southern South America: a Cladistic Approach.' p. 11. (Abstracts, Ninth meeting of the Willi Hennig Society, Canberra, Australia).

Croizat, L. (1964). 'Space, Time, Form: the Biological Synthesis.' (Published by the author: Caracas.)

Duffels, J. P. (1986). Biogeography of Indo-Pacific Cicadoidea: a tentative recognition of areas of endemism. *Cladistics* **2**, 318–36.

Erdman, D. S. (1986). The green stream goby, *Sicydium plumieri*, in Puerto Rico. *Tropical Fish Hobbyist* **34**, 70–4.

Hamilton, W. B. (1979). Tectonics of the Indonesian region. *Professional Papers of the United States Geological Survey* No. 1987, 1–338.

Hamilton, W. B. (1988). Plate tectonics and island arcs. *Geological Society of America Bulletin*, **100**, 1503–27.

Harrison, I. J. (1989). Specialization of the gobioid palatopterygoquadrate complex and its relevance to gobioid systematics. *Journal of Natural History* **27**, 325–53.

Heads, M. J. (1990). Mesozoic tectonics and the deconstruction of biogeography: a new model of Australasian biology. *Journal of Biogeography* **17**, 223–25.

Hertlein, L. G. (1963). Contribution to the biogeography of Cocos Island, including a bibliography. *Proceedings of the California Academy of Sciences* **32**, 219–89.

Hoese, D. F. (1984). Gobioidei: relationships. In 'Ontogeny and Systematics of Fishes'. (Eds H. G. Moser, W. J. Richards, D. M. Cohen, M. P. Fahay, A. W. Kendall, Jr and S. L. Richardson.) pp. 588–91. American Society of Ichthyologists and Herpetologists Special Publication No. 1.

Holloway, J. D. (1984). Lepidoptera of the Melanesian arcs. In 'Biogeography of the Tropical Pacific'. (Eds F. J. Radovsky, P. H. Raven and S. H. Sohmer.) pp. 353–73. (Bishop Museum Special Publication 72.)

Hubbs, C. (1920). *Sicydium montanum*, a new species of goby from Venezuela. *Proceedings of the Biological Society of Washington* **33**, 89–90.

Humphries, C. J. (1989). The importance of Wallacea to biogeographical thinking. In 'Rain Forest Insects of Wallacea'. pp. 7–18. (The Natural History Museum: London).

Humphries, C. J., Ladiges, P. Y., Roos, M., and Zandee, M. (1988). Cladistic biogeography. In 'Analytical Biogeography'. (Eds. A. A. Myers and P. S. Giller.) pp. 371–404. (Chapman and Hall: London.)

Humphries, C. J., and Parenti, L. R. (1986). 'Cladistic Biogeography.' (Clarendon Press: Oxford).

Klausewitz, W., and Henrich, S. (1986). *Sicyopus jonklaasi* n. sp., a new freshwater goby from Sri Lanka (Pisces, Perciformes, Gobioidei, Gobiidae, Sicydiaphiinae). *Revue Française, d'Aquariologie, Herpetologie* **12**, 117–22.

Kohn, A. (1990). Biogeography and evolution of Indo-Pacific marine Mollusca: patterns, progress, problems and prospect. *Bulletin of Marine Science* **47**, 2–9.

Kottelat, M. (1990). Synopsis of the endangered Buntingi (Osteichthyes: Adrianichthyidae and Oryziidae) of Lake Poso, Central Sulawesi, Indonesia, with a new reproductive guild and descriptions of three new species. *Ichthyological Exploration of Freshwaters* **1**, 49–67.

Koumans, F. P. (1953). 'Gobioidea. The Fishes of the Indo-Australian Archipelago. X.' (E. J. Brill: Leiden.)

Leis, J. M. (1986). Ecological requirements of Indo-Pacific larval fishes: a neglected zoogeographic factor. In 'Indo-Pacific Fish Biology: Proceedings of the Second International Conference on Indo-Pacific Fishes'. (Eds T. Uyeno, R. Arai, T. Taniuchi and K. Matsuura.) pp. 759–66 (Ichthyological Society of Japan: Tokyo.)

Nelson, G. (1984). Identity of the anchovy *Hildebrandichthys setiger* with notes on relationships and biogeography of the genera *Engraulis* and *Cetengraulis*. *Copeia* **1984**, 422–7.

Nelson, G., and Platnick, N. I. (1981). 'Systematics and Biogeography; Cladistics and Vicariance.' Columbia University Press: New York.)

Page, R. D. M. (1989) New Zealand and the new biogeography. *New Zealand Journal of Zoology* **16**, 471–83.

Page, R. D. M. (1990). Component analysis: a valiant failure? *Cladistics* **6**, 119–36.

Parenti, L. R. (1984). On the relationships of phallostethid fishes (Atherinomorpha), with notes on the anatomy of *Phallostethus dunckeri* Regan, 1913. *American Museum Novitates* No. 2779, 1–12.

Parenti, L. R. (1986). Bilateral asymmetry in phallostethid fishes (Atherinomorpha), with description of a new species from Sarawak. *Proceedings of the California Academy of Sciences* **44**, 225–36.

Parenti, L. R. (1987). Phylogenetic aspects of tooth and jaw structure of the medaka, *Oryzias latipes*, and other beloniform fishes. *Journal of Zoology, London* **211**, 561–72.

Parenti, L. R. (1989*a*). Fundamental biogeography. *Journal of Biogeography* **16**, 401–02.

Parenti, L. R. (1989*b*). A phylogenetic revision of the phallostethid fishes (Atherinomorpha, Phallostethidae). *Proceedings of the California Academy of Sciences* **46**, 243–77.

Patten, J. M. and Ivantsoff, W. (1983). A new genus and species of atherinid fish, *Dentatherina merceri* from the western Pacific. *Japanese Journal of Ichthyology* **29**, 329–39.

Patterson, C. (1981). [Response], In 'Vicariance Biogeography, a Critique'. (Eds G. Nelson and D. E. Rosen.) pp. 498–500. (Columbia University Press: New York.)

Platnick, N. I., and Nelson, G. (1989). Spanning-tree biogeography: shortcut, detour, or dead-end? *Systematic Zoology* **37**, 410–19.

Randall, J. E., Smith, C. L., and Feinberg, M. N. (1990). Report on fish collections from Rapa, French Polynesia. *American Museum Novitates* No. 2966, 1–42.

Rosen, D. E. and Parenti, L. R. (1981). Relationships of *Oryzias*, and the groups of atherinomorph fishes. *American Museum Novitates* No. 2719, 1–25.

Ryan, P. (1986). A new species of *Stiphodon* (Gobiidae, Sicydiaphiinae) from Vanuatu. In 'Indo-Pacific Fish Biology: Proceedings of the Second International Conference on Indo-Pacific Fishes'. (Eds T. Uyeno, R. Arai, T. Taniuchi and K. Matsuura.) pp. 655–62. (Ichthyological Society of Japan: Tokyo.)

Sakai, H., and Nakamura, M. (1979). Two new species of freshwater gobies (Gobiidae: Sicydiaphiinae) from Ishigaki Island, Japan. *Japanese Journal of Ichthyology* **26**, 43–54.

Schuh, R. T., and Stonedahl, G. M. (1986). Historical biogeography in the Indo-Pacific: a cladistic approach. *Cladistics* **2**, 337–55.

Simpson, G. G. (1977). Too many lines; the limits of the Oriental and Australian zoogeographic regions. *Proceedings of the American Philosophical Society* **121**, 107–20.

Smith, J. D., and Hood. C. S. (1981). Preliminary notes on bats from the Bismark Archipelago (Mammalia: Chiroptera). *Science in New Guinea* **8**, 81–121.

Springer, V. G. (1982). Pacific plate biogeography, with special reference to shorefishes. *Smithsonian Contributions to Zoology* **390**.

Springer, V. G., and Williams, J. T. (1990). Widely distributed Pacific plate endemics and lowered sea-level. *Bulletin of Marine Science* **47**, 631–40.

Tinker, S. W. (1978). 'Fishes of Hawaii. A Handbook of the Marine Fishes of Hawaii and the Central Pacific Ocean.' (Hawaiian Service Inc.: Honolulu.)

Uwa, H. and Parenti, L. R. (1988). Morphometric and meristic variation in ricefishes, genus *Oryzias*: a comparison with cytogenetic data. *Japanese Journal of Ichthyology* **35**, 159–66.

Valdebenito, H. A., Stuessey, T. F., and Crawford, D. J. (1990). A new biogeographic connection between islands in the Atlantic and Pacific oceans. *Nature* **347**, 549–50.

Vari, R. (1988). The Curimatidae, a lowland neotropical fish family (Pisces: Characiformes): distribution, endemism and phylogenetic biogeography. In 'Proceedings of a Workshop on Neotropical Distribution Patterns'. (Eds W. R. Heyer and P. E. Vanzolini.) pp. 343–77. (Academia Brasiliera de Ciências: Rio de Janeiro.)

Weitzman, S. H., Menzies, N. A., and Weitzman, M. J. (1988). Phylogenetic biogeography of the Glandulocaudini (Teleostei: Characiformes, Characidae) with comments on the distributions of other freshwater fishes in eastern and southeastern Brazil. In 'Proceedings of a Workshop on Neotropical Distribution Patterns'. (Eds W. R. Heyer and P. E. Vanzolini.) pp. 379–427. (Academia Brasiliera de Ciências: Rio de Janeiro.)

Whitmore, T. C., Ed. (1981). 'Wallace's Line and Plate Tectonics.' (Clarendon Press: Oxford.)

Wiley, E. O. (1988). Vicariance biogeography. *Annual Review of Ecology and Systematics* **19**, 513–42.

Woodland, D. J. (1986). Wallace's Line and the distribution of marine inshore fishes. In 'Indo-Pacific Fish Biology: Proceedings of the Second International Conference on Indo-Pacific Fishes'. (Eds T. Uyeno, R. Arai. T. Taniuchi and K. Matsuura.) pp. 453–60. (Ichthyological Society of Japan: Tokyo.)

Yamamoto, T. (1975). 'Medaka (Killifish) Biology and Strains. Series of Stock Culture in Biological Field.' (Keigaku Publishing Company: Tokyo.)

Manuscript received 9 November 1990, accepted 28 February 1991

3 September 1991 *Aust. Syst. Bot.*, **4**, 151–63

Cladistic Biogeography of Marine Water Striders (Insecta, Hemiptera) in the Indo-Pacific

N. Møller Andersen

Zoological Museum, University of Copenhagen, Universitetsparken 15, DK-2100 Copenhagen, Denmark.

Abstract

More than 120 species of marine water striders (Hemiptera, Gerromorpha), representing three families and eight genera, are distributed throughout the Indo-Pacific region. They live in marine habitats such as mangroves, intertidal coral reef flats and the sea surface near coral and rocky coasts. Five species of sea skaters, *Halobates* (Gerridae), have colonised the surface of the open ocean. Adult marine water striders are wingless but may disperse along coasts, chains of islands and possibly across wider stretches of open sea. Although some species of coral bugs, *Halovelia* (Veliidae) and *Halobates* are widespread, most species of marine water striders have rather restricted distributions. Cladistic hypotheses are now available for the genera *Halovelia*, *Xenobates* (Veliidae) and *Halobates*. Based upon distributional data for about 110 species, a number of areas of endemism can be delimited within the Indo-Pacific region. The results of component analyses of taxon-area cladograms for several monophyletic species-groups of marine water striders are presented.

The faunas of northern New Guinea, the Bismarck and Solomon Islands (Papuasia) are closely related and show much greater affinity with Maluku, Sulawesi and the Philippines than with the fauna of northern Australia. Relationships between the faunas of Papuasia + Sulawesi + the Philippines and those of Borneo + Jawa + Malaya are relatively weak. Marine water striders endemic to islands of the western Pacific show relationships among themselves and with Australia. Most marine water striders from the Indian Ocean (East Africa, Madagascar, Mauritius, Seychelles and Maldives) can be derived from the Indian–South-east Asian fauna. Composite faunas of marine water striders (either of different age or origin) are found in New Guinea, New Caledonia, Fiji Islands, the Philippines, tropical Australia and East Africa. The biogeography of marine water striders does not support the traditional division of the Indo-Pacific into the Ethiopian, Oriental and Australian regions. The distributional patterns are more compatible with a set of hierarchical relationships between more restricted areas of endemism.

Introduction

The Indo-Pacific has traditionally been divided into three major zoogeographical regions: the Ethiopian, Oriental and Australian regions. The boundary between the classic Oriental and Australian regions has been delimited by several lines, the most famous of which is Wallace's line. Based primarily on the distribution of the beetle families Chrysomelidae and Cerambycidae, Gressitt (1956, 1963) and Gressitt *et al.* (1961) suggested that some areas east of Wallace's line, including the Cape York Peninsula and the islands of the western Pacific, should be treated as parts of the Oriental rather than the Australian region. Although work by Gressitt and contemporary entomologists contained considerable detail, the hierarchical information on interrelationships of land areas (as areas of endemism) was very limited.

1030-1887/91/010151$05.00

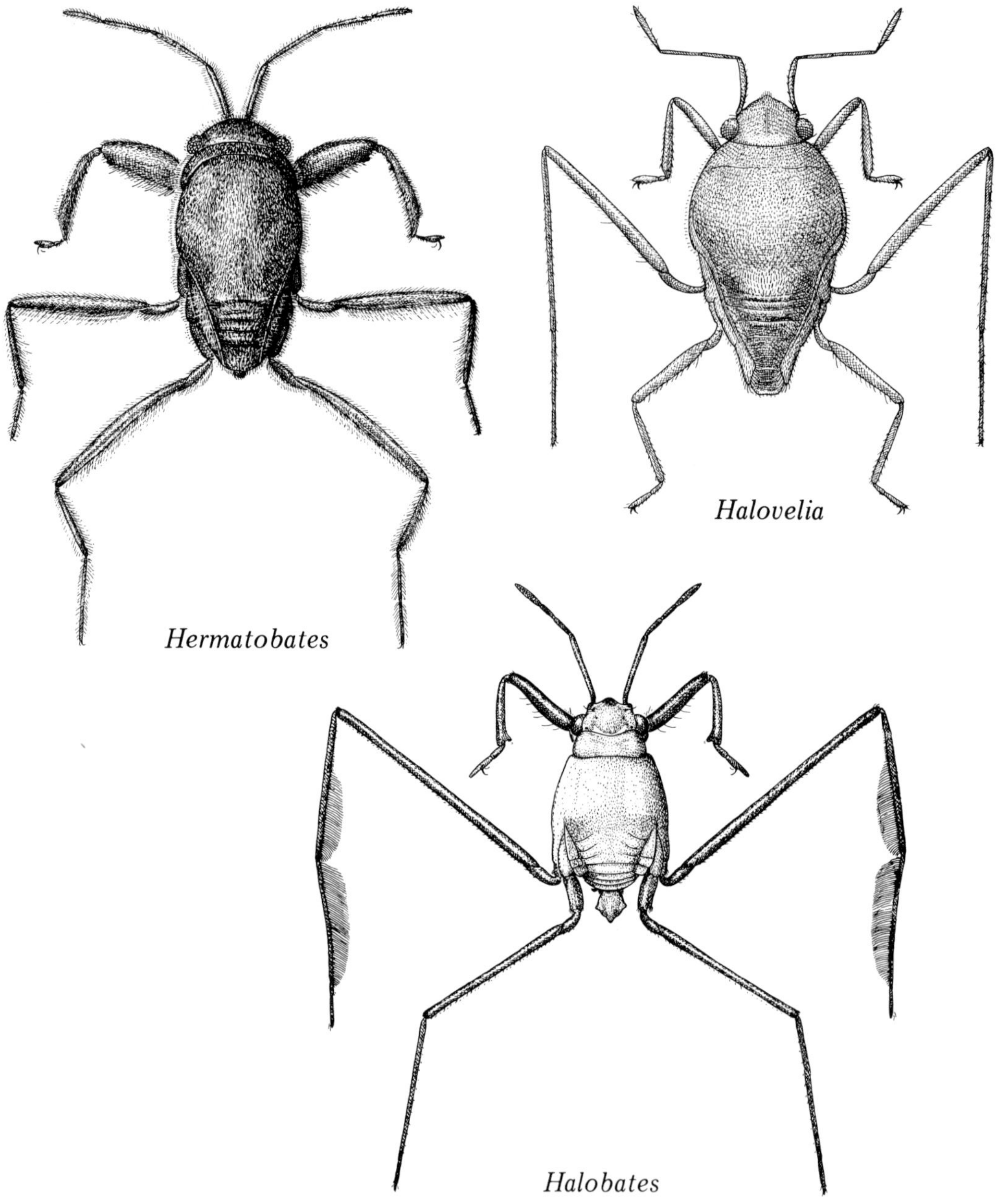

Fig.1. Marine water striders. *Hermatobates weddi* China (family Hermatobatidae), *Halovelia septentrionalis* Esaki (Veliidae, Haloveliinae) and *Halobates micans* Eschscholtz (Gerridae, Halobatinae).

The dramatic new discoveries in geology and palaeogeography during the early 1970s renewed an interest in historical biogeography of the Malay Archipelago and areas east of Wallace's line, collectively known as Malesia (Whitmore 1981, 1987). Although some progress has ensued in understanding the palaeogeography of this region, biologists have done little to integrate knowledge about the biotas of the Malay Archipelago, the Pacific Islands, Australia, continental Asia, Africa and the islands of the Indian Ocean into a comprehensive biogeographical scheme of the whole Indo-Pacific region (see e.g. reviews by Raven 1982 on Keast 1981 and Schuh 1982 on Gressitt 1982).

Most studies of Indo-Pacific biogeography have been based upon terrestrial animals and plants (see Schuh and Stonedahl 1986 for a review). In this paper, the results of biogeographical studies of a group of marine insects belonging to the hemipterous infraorder Gerromorpha are presented (Andersen 1979, 1982). These bugs show a remarkable diversity in habitat preferences, ranging from humid terrestrial habitats to the surface of stagnant and flowing waters, where water striders or pondskaters are familiar insects. Although the majority of the 1400 gerromorphan species are limnic, about 140 species belonging to five families occur in the marine environment (Andersen and Polhemus 1976; Andersen 1989*a*, 1989*b*, 1991*a*, 1991*b*).

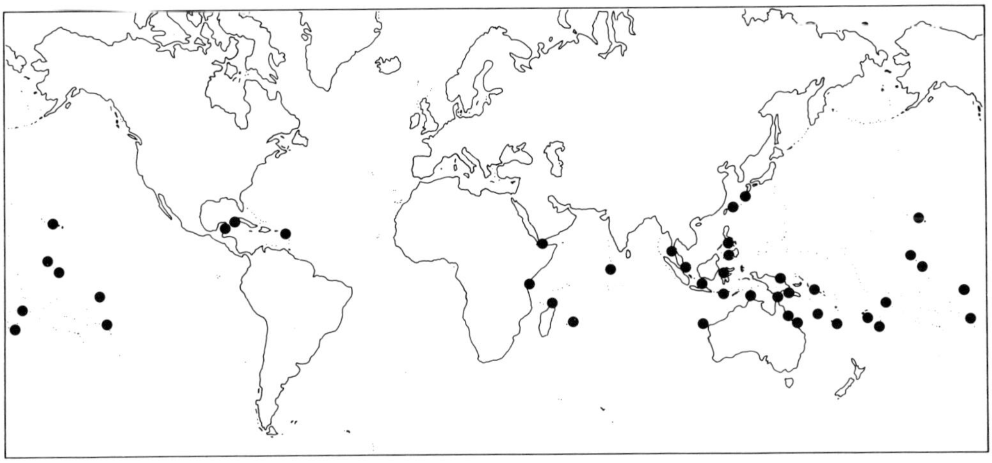

Fig. 2. Distribution of *Hermatobates* (Hermatobatidae). (Modified with permission from Andersen 1982.)

The Study Groups and their Distributions

Representatives of the major groups of marine water striders are illustrated in Fig. 1. These are a coral treader, *Hermatobates weddi* China (family Hermatobatidae), a coral bug, *Halovelia septentrionalis* Esaki (Veliidae, Haloveliinae), and a sea skater, *Halobates micans* Eschscholtz (Gerridae, Halobatinae). Common characteristics of these insects are dull coloration that usually is obscured by a thick, greyish hair-layer, strongly modified thorax with long and slender mid and hind legs, complete absence of wings and strongly reduced abdomen.

More than 120 species of marine water striders, representing three families and eight genera, are distributed throughout the Indo-Pacific region. They live in marine habitats such as mangroves, intertidal coral reef flats, and on the sea surface near coral reefs and rocky coasts. Five species of sea skaters, *Halobates*, have colonised the surface of the open ocean. Adult marine water striders are always wingless but may disperse along coasts, chains of islands and, for a few species of *Halobates*, possibly across wider stretches of open sea. Although some species belonging to the genera *Halovelia* and *Halobates* are widespread, most species of marine water striders have rather restricted areas of distribution.

Hermatobates

The genus *Hermatobates* contains eight described species, but a taxonomic revision is required. The biology of these odd marine insects was studied by Foster (1989), who named them coral threaders because they live on the tidal flats of coral reefs. Both adults and nymphs retreat to holes in porous blocks of dead coral during high tides. *Hermatobates* is unique (Andersen 1982), belonging to its own family, one of few exclusively marine insect families. It has no close freshwater relatives and it is difficult to trace its geographical origin.

Hermatobates species are found along continental coasts and islands throughout the Indo-Pacific (Fig. 2). The occurrence of one species in the Caribbean is interesting and it is possible that the genus was already widely distributed before the postulated Late Mesozoic island connection between North and South America. *Hermatobates* is probably the oldest group of marine water striders.

Halovelia

The subfamily Haloveliinae belongs to the Veliidae, one of the most speciose families of water striders, with about 600 species. Most haloveliines are marine. Freshwater relatives

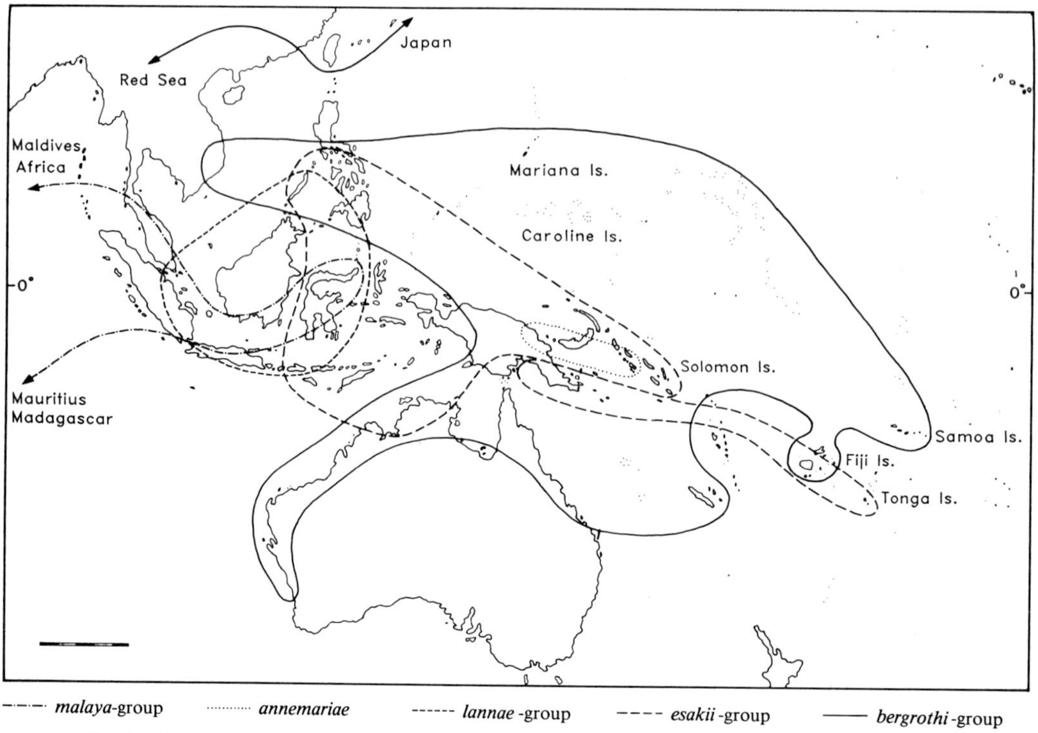

Fig. 3. Distribution of the species-groups of *Halovelia* (Veliidae, Haloveliinae). (Reproduced with permission from Andersen 1989*b*.)

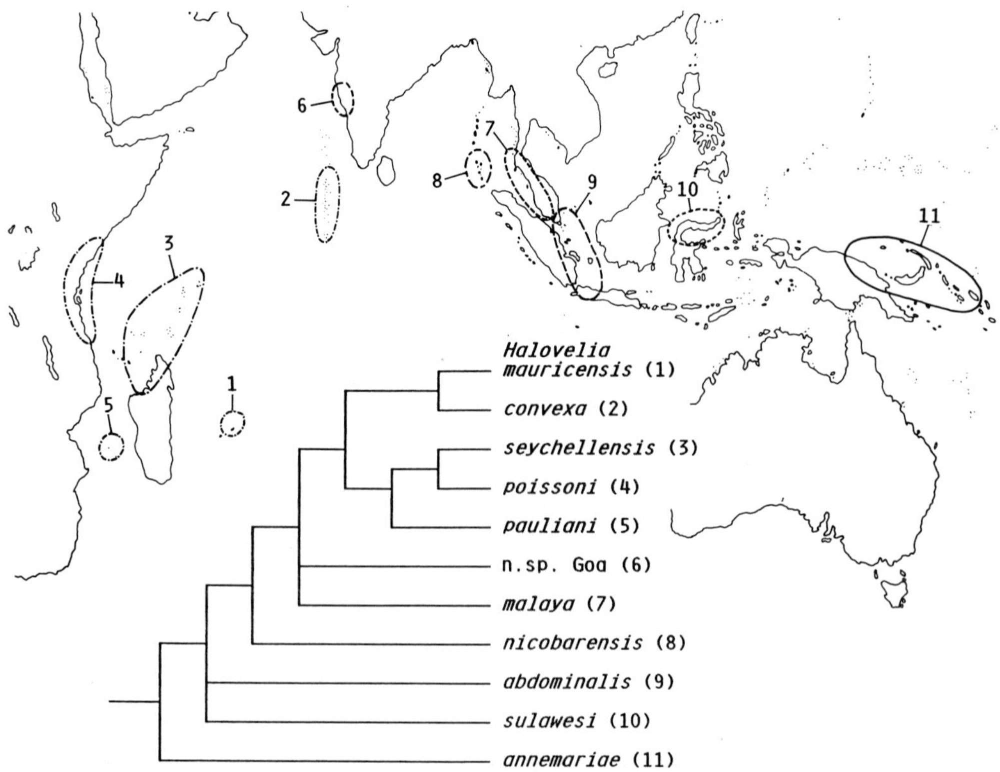

Fig. 4. Distribution and cladogram of the *Halovelia malaya* group (Veliidae, Haloveliinae).

belong to two genera: *Entomovelia* with one described and several undescribed species in Burma, Malaya and Borneo, and *Strongylovelia* with several, mostly undescribed species in the Indo-Australian region.

The marine Haloveliinae, currently under revision (Andersen 1989*a*, 1989*b*, 1991*b*), have been classified into two genera, *Halovelia* and *Xenobates*. So far, 38 species have been described, but during the past 10 years marine haloveliines have been collected in many new localities, chiefly through the efforts of Lanna Cheng of the Scripps Institution of Oceanography, California; William A. Foster, Cambridge; Dan A. Polhemus, Honolulu; and John T. Polhemus, Colorado. This has increased the number of known species from 13 to about 60.

Species of *Halovelia* or coral bugs inhabit the intertidal zone of rocky coasts or coral reefs where they can be found on the surface of tidal pools. At high tide they retreat to holes in blocks of coral or other porous rocks where they rest, surrounded by an air bubble, until the next low tide (Andersen 1989*b*).

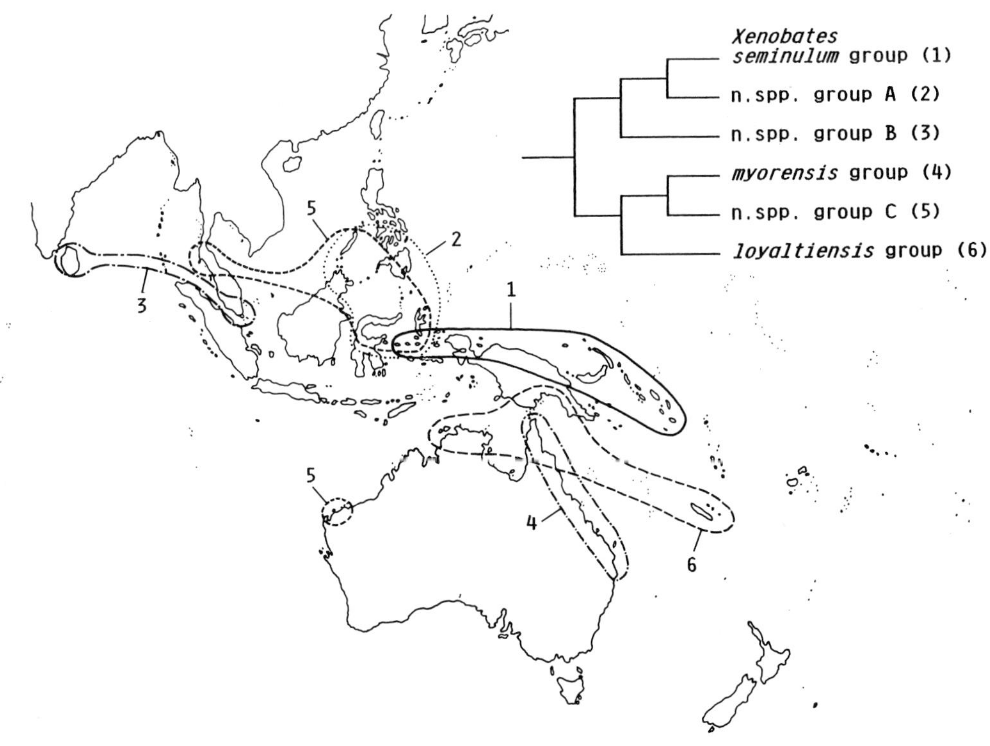

Fig. 5. Distribution and cladogram of the species groups of *Xenobates* (Veliidae, Haloveliinae).

The 30 described species of the genus *Halovelia* range from the Red Sea and East African coast to the West Pacific islands as far as Samoa (Andersen 1989*a*, 1989*b*). The distributional limits of monophyletic species groups of *Halovelia* are shown in Fig. 3. Species belonging to the *H. esakii* group occur along the coasts of the Philippines, Sulawesi, northern New Guinea, the Solomons, northern Australia, south-eastern New Guinea and the Fiji and Tonga Islands. Species of the *H. bergrothi* group have a similar distribution but two species are widespread in Australia and in the islands of the western Pacific respectively. Two closely related species, the *H. lannae* group, are found on both sides of Wallace's line.

The distribution and phylogeny of the *H. malaya* group (including *H. annemariae* Andersen), the largest species-group of *Halovelia*, is shown in Fig. 4. It is noteworthy that species found along both African and Asian coasts and islands bordering the Indian Ocean belong to the same species-group. Additionally, one species occurs in Sulawesi (*H. sulawesi* Andersen) and another one (*H. annemariae*) in northern New Guinea and the Solomons. The latter is the sister-group of all the others as shown by the cladogram.

Xenobates

Species of *Xenobates* typically live in mangroves where they can be found on the surface of tidal canals at low tide. Until recently this genus was thought to be monotypic, but now more than 25 species have been collected throughout the Indo-Australian region. Other species have been placed in *Halovelia* (China 1957; Polhemus 1982; Lansbury 1989).

A preliminary classification of *Xenobates* species has revealed at least seven monophyletic species-groups (Andersen, unpublished data). The distributions of six of these species-groups are shown in Fig. 5. The *X. seminulum* group (four species) occurs in Maluku, northern New Guinea, the Bismarck Archipelago and Solomon Islands. A related group (*Xenobates* spp. group A) of four species is found in Maluku, Sulawesi and the South Philippines while two species (*Xenobates* spp. group B) are known from Singapore and Sri Lanka. Another group of six species (*Xenobates* spp. group C) occurs in Maluku, Sulawesi, South Philippines, North Borneo and Malaya. Eight species of *Xenobates*, most of them undescribed, are found along the coast of tropical Australia. They belong to two separate species-groups. The *X. loyaltiensis* group includes also a new Caledonian species (*X. loyaltiensis* China, 1957). The *X. myorensis* group is confined to Queensland and south-eastern New Guinea.

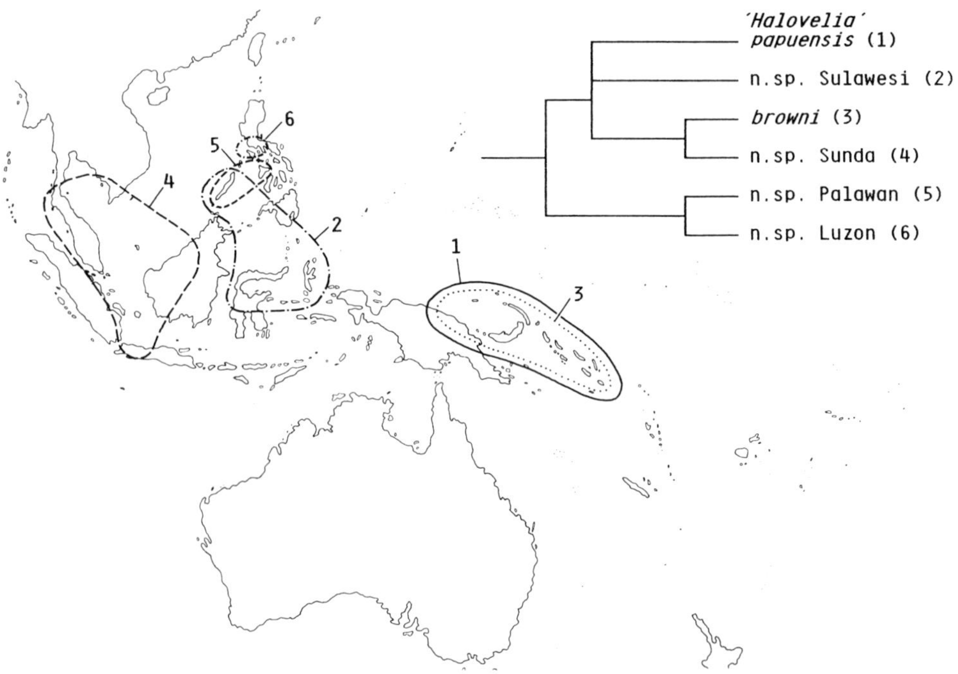

Fig. 6. Distribution and cladogram of the '*Halovelia*' *papuensis* group (Veliidae, Haloveliinae).

A monophyletic group composed of six species, the '*Halovelia*' *papuensis* group, deserves separate generic status (Andersen 1991*b*). It is closely related to *Xenobates* but most species have left the protecting mangroves and live on the sea surface at some distance from the coast. One species is found in areas of the Sunda Shelf (Fig. 6), another one in Sulawesi, Maluku and South Philippines (Palawan), two species in Palawan and Luzon, the Philippines, and another two species in northern New Guinea, the Bismarck Archipelago and Solomon Islands.

Stenobates and *Rheumatometroides*

Another large family of water striders, the Gerridae (500 species), contains several groups of marine species. The genera *Stenobates* and *Rheumatometroides* both have five mangrove-inhabiting species in the Indo-Australian area (Polhemus and Cheng 1982; Lansbury 1988; J. Polhemus, personal communication). They belong to the subfamily Trepobatinae and are related to the freshwater genus *Naboandelus* found in Africa, India and South-east Asia (Andersen 1982).

Halobates and *Asclepios*

Subfamily Halobatinae contains the well known sea skaters, genus *Halobates*, with 41 described species. Most species are confined to coastal, marine habitats such as estuaries, mangroves and coral reefs. Freshwater relatives belong to the tribe Metrocorini (Andersen 1982) and live in running freshwater throughout Africa, South and South-east Asia, and the Malay Archipelago, including Sulawesi (Polhemus 1990). The sister-group of *Halobates* is *Asclepios* with three species distributed along the coasts of South-east and East Asia. They are more primitive than *Halobates* in some characters but otherwise difficult to separate from the most primitive species, *H. mjobergi* Hale, found in tropical Australia. Based upon this information, it is most likely that *Halobates* originated somewhere in the Indo-Australian area.

The taxonomy of *Halobates* is well known, chiefly through monographic revision by Herring (1961). Herring delimited a number of species-groups of *Halobates* and discussed their relationships. Some of his species-groups, however, are certainly not monophyletic and his phylogenetic scheme is too crude to be of much use in a biogeographical analysis.

A cladistic analysis of *Halobates* based on new characters, mainly genital structures (Andersen 1991*a*), allows the definition of a number of monophyletic species-groups, each showing a characteristic distribution pattern.

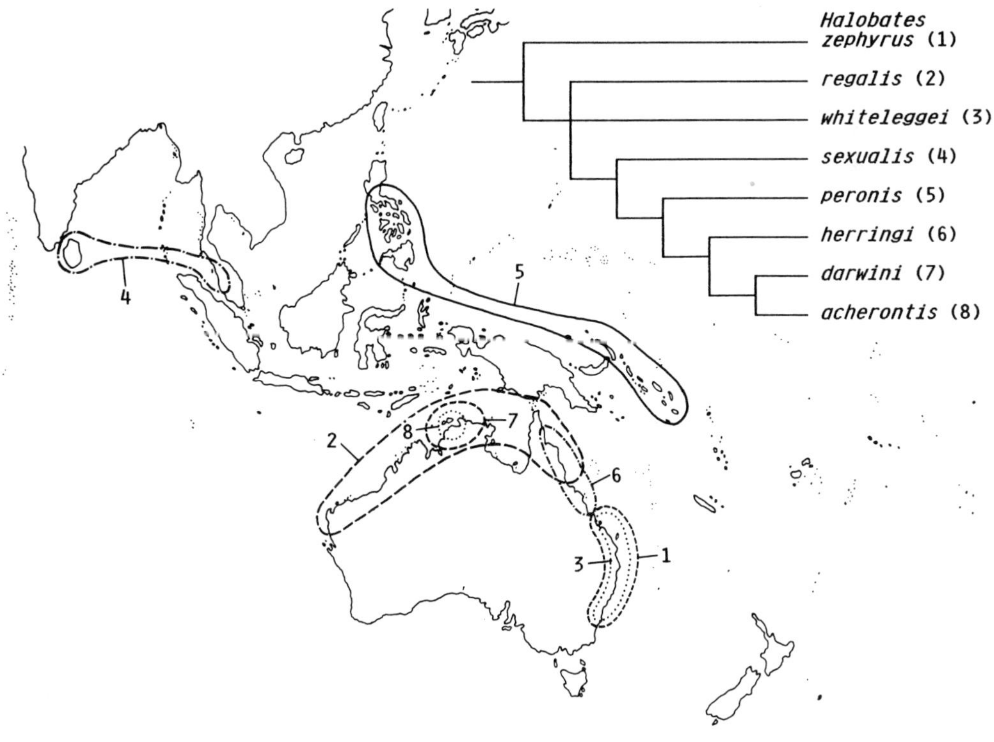

Fig. 7 Distribution and cladogram of the *Halobates regalis* group (Gerridae, Halobatinae).

The distribution of the eight species belonging to the *Halobates regalis* group is shown in Fig. 7. Most species occur in tropical Australia but two species (*H. zephyrus* Herring and *whiteleggei* Skuse) live along the coast of New South Wales as far south as Sydney. There are considerable distributional overlaps between species. *H. peronis* Herring, ranging from the Philippines to the Solomons, and *H. sexualis* Distant, from Sri Lanka and Malaya, are both closely related to species of tropical Australia.

Species belonging to the *H. princeps* group occur throughout the Indo-West Pacific region (Fig. 8). Most species are endemic to single islands or island groups. There is one subgroup of closely related species in the Indian Ocean (*H. alluaudi* group) including the west coast of India, and another subgroup (*H. mariannarum* group) on islands of the West Pacific. *H. princeps* White, the sister taxon of these subgroups, is widely distributed in Malesia.

Five species of *Halobates* spend their entire life on the surface of the open ocean, sometimes hundreds of kilometres from the nearest coast (Cheng 1985, 1989). Herring (1961) considered this group monophyletic but cladistic analysis of *Halobates* (Andersen 1991*a*) shows that two oceanic species are closer to some coastal, but widespread species, than to other oceanic species. Either the oceanic habit has evolved more than once, or there has been a reversal in preference, from far-shore to near-shore habitats. Undoubtedly, the oceanic species have originated somewhere in the Indo-Pacific and have subsequently dispersed to all tropical areas of the Atlantic, Indian, and Pacific Oceans.

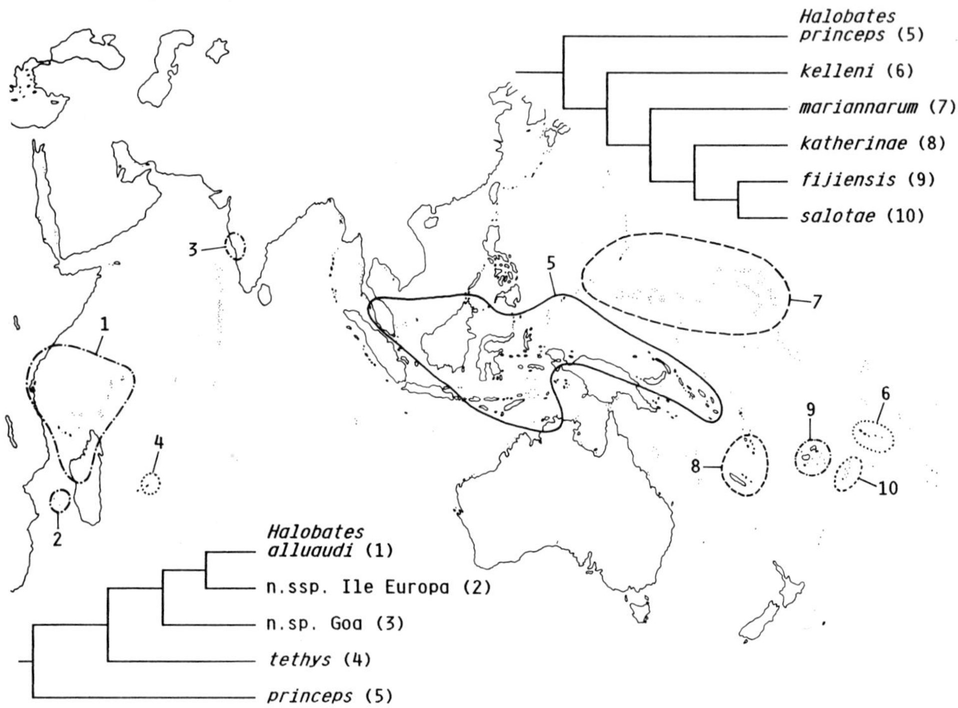

Fig. 8. Distribution and cladogram of the *Halobates princeps* group (Gerridae, Halobatinae).

Methods

The methods of cladistic biogeography as outlined by Nelson and Platnick (1981) and Humphries and Parenti (1986) are applied in studying the historical biogeography of the marine water striders of the Indo-Pacific, based upon distribution records for about 120 species. Firstly, about 40 areas of endemism were delimited in the Indo-Pacific, each characterised by having at least one endemic species of marine water striders. Secondly, information about the relationships between species was obtained from cladograms generated by analyses using Hennig86 (Farris 1988). In this way, cladistic hypotheses have been obtained for 25 monophyletic species-groups in the genera *Halovelia* (Andersen 1989*b*), *Halobates* (Andersen 1991*a*) and *Xenobates* (Andersen 1991*b*, and unpublished).

Taxon cladograms (showing cladistic relationships between species) were combined with distributional records (for the same species) in terms of areas of endemism to produce a number of taxon-area cladograms. These cladograms were subjected to component analysis using COMPONENT (Page 1989, 1990). The cladograms were analysed by the options *Build* and under Assumption 2 (Nelson and Platnick 1981; Humphries and Parenti 1986). The results were cladograms showing the relationships between geographical areas. The options *Compare*, *Consensus* and *Nelson* were used in cases which yielded more than a single area cladogram, obtaining a consensus cladogram for all possible area cladograms.

Results

Preliminary analyses using all 40 areas of endemism were unsuccessful due to insufficient memory space to store all possible area cladograms. With so many areas in the analysis, there

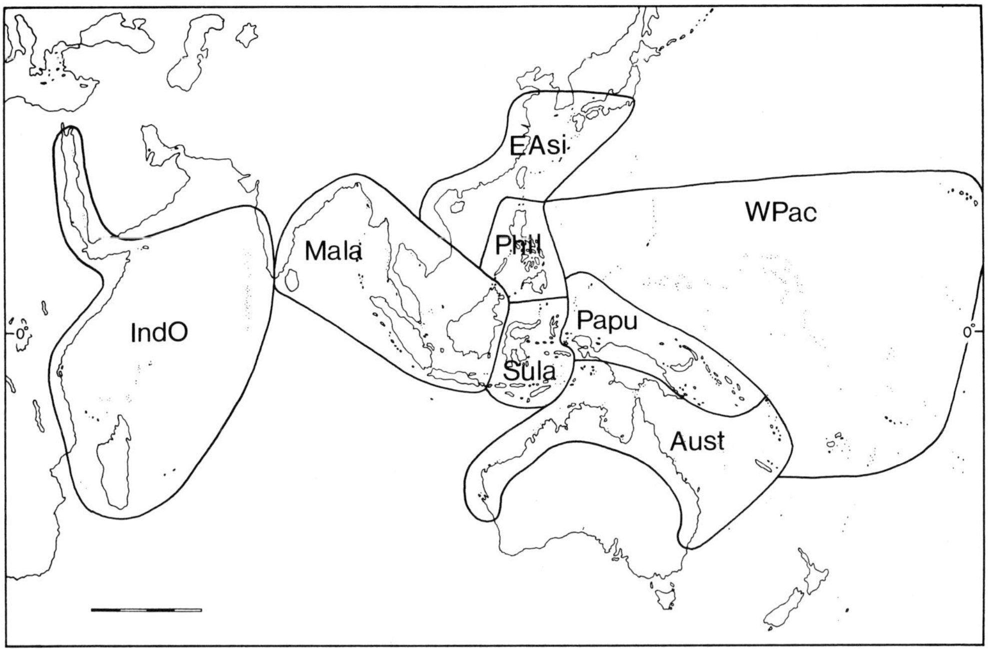

Fig. 9. Map of the Indo-Pacific showing areas of endemism used in the component analyses. IndO, Indian Ocean; Mala, Malayan; Phil, Philippine; Sula, Sulawesi; Papu, Papuasia; Aust, Australian; WPac, West Pacific; EAsi, East Asia.

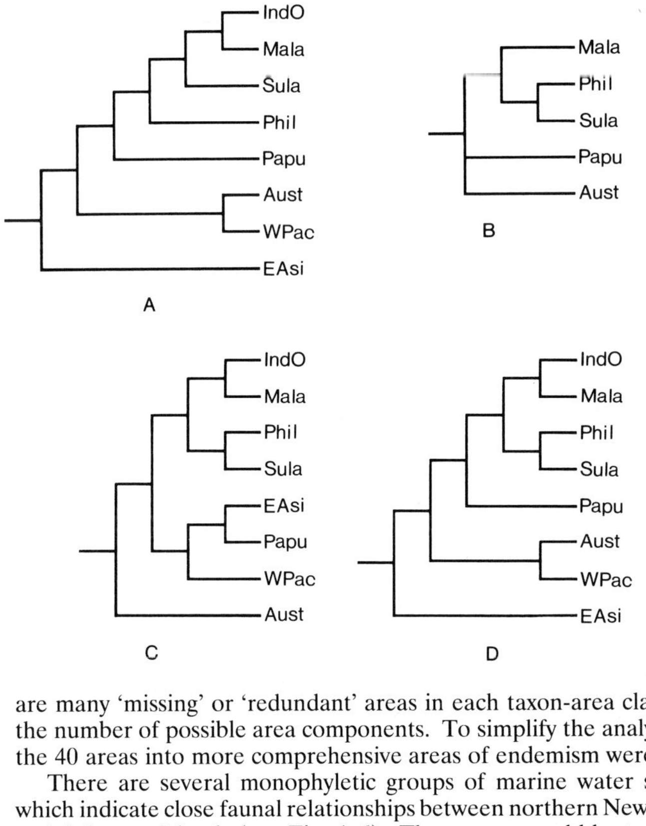

Fig. 10. Area cladograms for *Halovelia* (*A*), *Xenobates* and the '*Halovelia*' *papuensis* group (*B*), *Halobates*, (*C*) and summary area cladogram for all genera of marine water striders (*D*). Abbreviations of areas as in Fig. 9.

are many 'missing' or 'redundant' areas in each taxon-area cladogram which greatly increase the number of possible area components. To simplify the analyses, various ways of combining the 40 areas into more comprehensive areas of endemism were investigated.

There are several monophyletic groups of marine water striders that have distributions which indicate close faunal relationships between northern New Guinea, Bismarck Archipelago and Solomon Islands (e.g. Figs 4–6). These areas could be treated as one area of endemism

(Papuasia). Strong relationships exist also between species in Maluku and Sulawesi (Figs 5 and 6). Likewise, areas of Sunda shelf, Malayan Peninsula and Sri Lanka (Figs 5–7) were joined into one area of endemism. The same applies to islands of the western Pacific (Fig. 9). Finally, most marine water striders found along continents and islands of the Indian Ocean are closely related (Figs 4 and 8) and thus these areas (including the west coast of India) were combined into an Indian Ocean area of endemism.

Consequently, the original 40 areas of endemism were reduced to eight (Fig. 9): Indian Ocean (IndO), Malayan (Mala), Philippine (Phil), Sulawesi (Sula), Papuasia (Papu), Australia (Aust), West Pacific (WPac) and East Asia (EAsi). The loss of information due to this procedure is discussed below. For each genus, taxon-area cladograms were reduced to cladograms of eight areas of endemism.

The area cladogram for *Halovelia* is shown in Fig. 10*A*. The hierarchical structure of relationships between areas is noteworthy, particularly the Indian Ocean areas joined to the Malayan areas and Australia with the western Pacific islands.

The cladogram in Fig. 10*B* shows relationships based on the distribution of *Xenobates* and the '*Halovelia*' *papuensis* group. Records are missing for three of the eight areas. However, under Assumption 2 (Nelson and Platnick 1981) these areas can be placed anywhere on the cladogram. There is much congruence with the cladogram in Fig. 10*A*, except that the Philippines join with Sulawesi rather than with the Malayan areas.

Component analysis of the *Halobates* distribution is more complicated, mainly because there are some widespread species of sea skaters (including the open-ocean species) and several species with overlapping distributions and therefore many redundant areas. The cladogram shown in Fig. 10*C* is a consensus diagram for 34 area cladograms. Nevertheless, the area relationships are fully resolved. Again, there is much congruence with the area cladograms for the haloveliine genera. The most noteworthy difference is the relative position of East Asia.

The three area cladograms were then combined into a general cladogram of relationships between Indo Pacific areas (Fig. 10*D*). This summary area cladogram is fully resolved and all area components are compatible with components of at least two of the three area cladograms above (Fig. 10*A*–*C*).

Discussion and Conclusions

The high degree of congruence amongst the area cladograms of haloveliine and halobatine water striders is expected. Species belonging to different groups live together in similar habitats and areas and probably evolved together. They may have responded (speciated) in response to the same vicariance events and may have quite similar means of dispersal.

It could be argued that marine insects are less suited as study groups in biogeography because of better mechanisms for dispersal than terrestrial organisms. However, most marine water striders are confined to coastal habitats, especially mangroves, and are seemingly unable to cross wide stretches of open sea. The high level of endemism (about 60%) opposes the idea that all marine water striders disperse. Furthermore, they are wingless and unable to fly, actively or passively.

The reduction, for analytical reasons, of the original 40 areas of endemism to eight areas causes information loss for conflicting area interrelationships and composite areas. For example, New Caledonia is treated as part of the Australian area of endemism. However, *Halobates katherinae* Herring in New Caledonia clearly has western Pacific affinities. Thus, the fauna of New Calendonia seems to be composite. Composite faunas of marine water striders are seemingly found also in the islands of Fiji, Philippines, tropical Australia, East Africa and the Red Sea. The composite nature of New Guinea (as repeatedly stressed, e.g. by Duffels 1983, 1986; Schuh and Stonedahl 1986; Duffels and Boer 1990) is supported also by the distributions of marine water striders (Andersen 1989*b*).

There is some evidence of older, possibly relictual patterns of distribution among marine water striders. The most basal clades of the halobatine phylogeny have terminal species with very disjunct distributions (Andersen 1991*a*) in East Asia (*Asclepios* spp.), Australia (*Halobates mjobergi*), East Africa (*H. poseidon* Herring) and the Galapagos Islands (*H. robustus* Barber). A similar pattern was found in species of *Halovelia* (Andersen 1989*b*). Such older patterns of area interrelationships tend to obscure more recent patterns. Both composite relationships of areas and patterns of different age may be revealed by cladistic biogeography.

Marine water striders did not evolve in isolation but as part of more comprehensive biotas. How general is the pattern of interrelationships between areas of the Indo-Pacific derived from the present study of marine water striders? One would expect a high degree of congruence between the biogeography of marine insects and other organisms inhabiting coastal marine habitats. However, the work of Springer (1982) on the biogeography of Pacific plate shorefish and other marine animals is not based upon cladistic information and does not contain enough detail to allow comparisons with the present work. Winterbottom (1985), using cladistic methods, found several Indo-Pacific taxon-area relationships among congrogadine fish.

So far, very few biologists working on terrestrial animals or plants have recognised the Indo-Pacific as a biogeographic region. Although studies on the flora and fauna of the Indo-Pacific (or some parts of this region) are more numerous (e.g. Whitmore 1981, 1987), those using cladistic methods are still limited. Schuh and Stonedahl (1986) reviewed both methodological and empirical problems of Indo-Pacific biogeography. Using terrestrial insects, plant-bugs (Miridae) and Cicadas as examples and cladistic methods of biogeographical analysis, these authors found area relationships comparable to, although not fully congruent with, my results (Fig. 10*D*). Similar patterns to those for both marine water striders and terrestrial insects were found by Holloway (1987) in some groups of moths in the Malay Archipelago. Naturally, there are deviations from the pattern derived here. Duffels (1983, 1986, 1988) found only weak relationships between the Cicadas inhabiting the Philippines and Sulawesi, Maluku and New Guinea, while strong relationships exist between western Pacific (Fiji, Tonga) and Papuasia.

In conclusion, my study supports the view that the Indo-Pacific is a genuine biogeographic region. It does not support the traditional division of the Indo-Pacific into the Ethiopian, Oriental and Australian regions. The distributional patterns of marine water striders, supported by evidence from some terrestrial groups of insects, are more compatible with a set of hierarchical relationships between more restricted areas of endemism (Fig. 10*D*). These areas (Fig. 9) are very similar to those delimited by Gressitt (1956) and Gressitt *et al.* (1961), although the areas of the Indian Ocean were not recognised as part of the 'Pacific' region. Malesia (as defined by Whitmore 1981, 1987) is seemingly not a genuine, monophyletic area of endemism unless we want to include the Indian Ocean, tropical Australia, the Solomon Islands and the islands of the western Pacific. More studies on animals and plants with Indo-Pacific distributions are required in order to show the generality of the hierarchy of area interrelationships found in the present study (Fig. 10*D*). In my opinion, only cladistic biogeography will produce hierarchies of area interrelationships worth comparing.

Acknowledgments

Comments submitted by two anonymous referees greatly improved the content and style of the paper. I am indebted to Dr L. Cheng, La Jolla, Dr W. A. Foster, Cambridge, and my colleagues at the Zoological Museum, University of Copenhagen, for presenting material of marine water striders, and especially to Drs D. A. Polhemus, Honolulu, and J. T. Polhemus, Englewood, Colorado, for placing their large material of marine water striders (collected on expeditions sponsored by the National Geographic Society) at my disposal and for giving me access to their field notes. The present study would have been incomplete without the efforts of all these people. A grant from the Danish Natural Science Research Council enabled me to spend 3 weeks, January 1987, at the Phuket Marine Biological Center, Thailand, collecting and studying the biology of marine water striders. Another grant from the Danish Natural Science Research Council enabled me to attend the symposia 'Indo-Pacific Biogeography: at the Crossroads' (organised by the Australian Systematic Botany Society) and 'Systematics and Biogeography of the Austral Biota' (IX Willi Hennig Society Meeting), August 1990, Canberra.

References

Andersen, N. M. (1979). Phylogenetic inference as applied to the study of evolutionary diversification of semiaquatic bugs (Hemiptera: Gerromorpha). *Systematic Zoology* **28,** 554–78.

Andersen, N. M. (1982). The semiaquatic bugs (Hemiptera, Gerromorpha). Phylogeny, adaptations, biogeography, and classification. *Entomonograph* **3,** 1–455.

Andersen, N. M. (1989*a*). The coral bugs, genus *Halovelia* Bergroth (Hemiptera, Veliidae). I. History, classification, and taxonomy of species except the *H. malaya*-group. *Entomologica Scandinavica* **20,** 75–120.

Andersen, N. M. (1989*b*). The coral bugs, genus *Halovelia* Bergroth (Hemiptera, Veliidae). II. Taxonomy of the *H. malaya*-group, cladistics, ecology, biology, and biogeography. *Entomologica Scandinavica* **20,** 179–227.

Andersen, N. M. (1991*a*). Marine insects: genital morphology, phylogeny and evolution of sea skaters, genus *Halobates* (Hemiptera, Gerridae). *Zoological Journal of the Linnean Society*. (In press.)

Andersen, N. M. (1991*b*). A new genus of marine water striders (Hemiptera, Veliidae) with five new species from the Indo-Malayan subregion. *Entomologica Scandinavica*. (In press.)

Andersen, N. M., and Polhemus, J. T. (1976). Water-striders (Hemiptera: Gerridae, Veliidae, etc.). In 'Marine Insects'. (Ed. L. Cheng.) pp. 187–224. (North-Holland Publishing Company: Amsterdam.)

Cheng, L. (1985). Biology of *Halobates* (Heteroptera: Gerridae). *Annual Review of Entomology* **30,** 111–35.

Cheng, L. (1989). Factors limiting the distribution of *Halobates* species. In 'Reproduction Genetics and Distributions of Marine Organisms'. (Eds J. S. Ryland and P. A. Tyler.) pp. 357–62. (Olsen and Olsen: Fredensborg.)

China, W. E. (1957). The marine Hemiptera of the Monte Bello Islands, with descriptions of some allied species. *Journal of the Linnean Society, London, Zoology* **40**, 342–57.

Duffels, J. P. (1983). Taxonomy, phylogeny and biogeography of the genus *Cosmopsaltria*, with remarks on the historic biogeography of the subtribe Cosmopsaltraria (Homoptera: Cicadidae). *Pacific Insects Monograph* **39,** 1–127.

Duffels, J. P. (1986). Biogeography of Indo-Pacific Cicadoidea: A tentative recognition of areas of endemisms. *Cladistics* **2,** 318–36.

Duffels, J. P. (1988). The Cicadas of the Fiji, Samoa and Tonga Islands, their taxonomy and biogeography (Homoptera, Cicadoidea) with a chapter on the geological history of the area by A. Ewart. *Entomonograph* **10,** 1–108.

Duffels, J. P., and Boer, A. J. de (1990). Areas of endemism and composite areas in East Malesia. In 'The Plant Diversity of Malesia'. (Eds P. Baas *et al.*) pp. 249–72. (Kluwer Academic Publishers: Amsterdam.)

Farris, J. S. (1988). 'Hennig86 Reference, Version 1.5'. (41 Admiral Street, Port Jefferson Station, New York.)

Foster, W. A. (1989). Zonation, behaviour and morphology of the intertidal coral-treader *Hermatobates* (Hemiptera: Hermatobatidae) in the south-west Pacific. *Zoological Journal of the Linnean Society* **96,** 87–105.

Gressitt, J. L. (1956). Some distribution patterns of Pacific island faunae. *Systematic Zoology* **6,** 12–32.

Gressitt, J. L. (1963). 'Pacific Basin Biogeography.' (Bishop Museum Press: Honolulu.)

Gressitt, J. L. (Ed.) (1982). 'Biogeography and Ecology of New Guinea.' 2 Vols. (W. Junk: The Hague.)

Gressitt, J. L., Maa, T. C., Mackerras, I. M., Nakata, S., and Quate, L. W. (1961). Problems in the zoogeography of Pacific and Antarctic insects. *Pacific Insects Monograph* **2,** 1–127.

Herring, J. L. (1961). The genus *Halobates* (Hemiptera: Gerridae). *Pacific Insects* **3,** 223–305.

Holloway, J. D. (1979). A survey of the Lepidoptera, biogeography and ecology of New Caledonia. *Series Entomologica* **15,** i–xii, 1–588.

Holloway, J. D. (1987). Lepidoptera patterns involving Sulawesi: what do they indicate of past geography? In 'Biogeographical Evolution of the Malay Archipelago. Oxford Monographs on Biogeography'. (Ed. T. C. Whitmore.) pp. 103–11. (Clarendon Press: Oxford.)

Humphries, C. J., and Parenti, L. E. (1986). 'Cladistic Biogeography. Oxford Monographs on Biogeography.' (Clarendon Press: Oxford.)

Keast, A. (Ed.) (1981). 'Ecological Biogeography of Australia.' 3 Vols. W. Junk: The Hague.)

Lansbury I., (1988). Notes on marine *Rheumatometroides* from the Solomon Islands (Hem., Gerridae). *Entomologist's Monthly Magazine* **125,** 63–71.

Lansbury, I. (1989). Notes on the Haloveliinae of Australia and the Solomon Islands (Insecta, Hemiptera, Heteroptera: Veliidae). *Reichenbachia* **26,** 93–109.

Nelson, G., and Platnick, N. (1981). 'Systematics and Biogeography. Cladistics and Vicariance.' (Columbia University Press: New York.)

Page, R. D. M. (1989). 'Component User's Manual. Release 1.5.' (R. D. M. Page: Auckland.)

Page, R. D. M. (1990). Component analysis: a valiant failure? *Cladistics* **6,** 119–36.

Polhemus, D. A. (1990). A revision of the genus *Metrocoris* Mayr (Heteroptera: Gerridae) in the Malay Archipelago and the Philippines. *Entomologica Scandinavica* **21,** 1–28.

Polhemus, J. T. (1982). Marine Hemiptera of the Northern Territory, including the first fresh-water species of *Halobates* Eschscholtz (Gerridae, Veliidae, Hermatobatidae and Corixidae). *Journal of the Australian Entomological Society* **21,** 5–11.

Polhemus, J. T., and Cheng, L. (1982). Notes on marine water-striders with descriptions of new species. Part I. Gerridae (Hemiptera). *Pacific Insects* **24,** 219–27.

Raven, R. J. (1982). [Review of] Keast, A., (Ed.) (1981). Ecological biogeography of Australia. *Systematic Zoology* **31,** 218–22.

Schuh, R. T. (1982). [Review of] Gressitt, J. L. (Ed.) (1982). Biogeography and ecology of New Guinea. *Systematic Zoology* **31,** 222–5.

Schuh, R. T., and Stonedahl, G. (1986). Historical biogeography in the Indo-Pacific: A cladistic approach. *Cladistics* **2,** 337–55.

Springer, V. C. (1982). Pacific Plate biogeography, with special reference to shorefishes. *Smithsonian Contributions to Zoology* **367,** 1–182.

Whitmore, T. C. (Ed.) (1981). 'Wallace's Line and Plate Tectonics. Oxford Monographs on Biogeography.' (Clarendon Press: Oxford.)

Whitmore, T. C. (Ed.) (1987). 'Biogeographical Evolution of the Malay Archipelago. Oxford Monographs on Biogeography.' (Clarendon Press: Oxford.)

Winterbottom, R. (1985). Revision and vicariance biogeography of the subfamily Congrogadinae (Pisces: Perciformes: Pseudochromidae). *Indo-Pacific Fishes* **9,** 34.

Manuscript received 3 September 1990, accepted 11 February 1991

3 September 1991 *Aust. Syst. Bot.*, **4**, 165–82

The Eucnemidae of South-east Asia and the Western Pacific — a Biogeographical Study

J. Muona

Oulanka Biological Station, University of Oulu, SF-90570, Oulu, Finland.

Abstract

Eighty-nine eucnemid genera occur in the region from South-east Asia to the south-west Pacific. The phylogenies of 84 of these were used together with the present-day distributions of the species to analyse the biogeographical history of the area. Fifty-seven genera shared a pattern coinciding with the traditional model of Laurasia–Gondwana break-up. Six genera showed a pattern contradicting the model. The remaining 21 genera neither supported nor refuted the model. Twenty-five genera were observed to include an Indomalesian clade younger than the South America–Australia connection. This biogeographical unit consisted of present-day South-east Asia and the Sunda islands, but did not include the Philippine Islands and Sulawesi. In addition to this Indomalesian clade, three separate clades involving northern Australia or New Guinea were observed, New Guinea–Australia, New Guinea–Philippines–Sulawesi and New Guinea–Fiji. The possible presence of four separate areas in the general region of New Guinea–north Australia as the result of the Cretaceous geological events is suggested. Three of these, in the area of present-day New Guinea, originally sharing sister-groups with the north-eastern Australian isolate, are regarded as the sources of the New Guinea–Indomalesia, New Guinea–Philippines and New Guinea–Fiji faunas after northward drifting of the Australian continent. During the Oligocene–Miocene these source areas were flooded and their original fauna lost. When the present-day New Guinea emerged, it was invaded from the north-eastern Australian region. This invasion created new New Guinea–Australia connections and brought in the sister-groups of the old New Guinea source areas as well.

The eucnemids of Vanuatu, Samoa and Tonga are regarded as having originated in connection with dispersal from Fiji.

The New Zealand fauna has strong, old connections with that of south-eastern Australia, but other complex connections are indicated.

The Eocene Baltic Amber fauna agrees well with the results obtained from extant species. The species belonging to five fossil genera belong to Gondwanan groups that seem to have invaded the Holarctic via Central America. Four other fossil genera showing discordant patterns belong to the group of six genera exhibiting these aberrant patterns even today. The eucnemid fauna of the region is of Gondwanic origin. Only six Laurasian genera have invaded the area, all of them apparently quite recently.

Introduction

This paper discusses the biogeography of the beetle family Eucnemidae in the area from south-east continental Asia to the south-west Pacific. The emphasis is on the origin of the Australian fauna and its connections with the surrounding areas.

1030-1887/91/010165$10.00

The geological history of the region is complex. It has been discussed in several recent works (e.g. Audley-Charles 1987; Ewart 1988; Hamburger and Isacks 1988; McCabe and Cole 1989; Muruyama *et al.* 1989; Spörli and Ballance 1989), but conflicting theories still exist. The details on the origin of the Indomalesian archipelago are not settled and some areas, e.g. the Solomon Islands, have received little attention.

A competent review of the existing biological data is badly needed but such a proposition is beyond my capabilities. An attempt of this kind should be based on a large number of adequate phylogenies but it appears that the required information will not be available in the near future. This is unfortunate, as useful data sets are needed to test the considerable theoretical advances in biogeography during the last two decades. I present here a set of phylogenies that can be used for this purpose.

Table 1. Eucnemidae, number of genera and species, South-east Asia to western Pacific

	Genera	Species
Burma to China	46	140
Sunda + Philippines	51	290
New Guinea	54	260
Australia	51	150
Solomon Islands	27	170
Vanuatu	6	9
Fiji	10	41
Samoa	7	19
Tonga	4	5
New Caledonia	5	12
New Zealand	7	15

Materials and Methods

The eucnemids are polyphagan beetles belonging to the superfamily Elateroidea (e.g. Lawrence 1987). They are often regarded as having evolved from the click-beetles (family Elateridae) but this view should be rejected (Muona 1992*b*). It is clear that they are quite primitive and probably the oldest group among the Elateriformia families with specialised propleurocoxal mechanism (see e.g. Lawrence 1987). With the exception of the most primitive forms (Perothopinae, Phyllocerinae), the larvae develop in dying or dead wood — a few forms in conifers, the majority in angiosperms (Muona 1992*b*). The great majority of species and genera are found in the tropics or subtropics, but the number of temperate forms is low. Typical Elateroidea beetles are already known from Triassic fossils (e.g. Lawrence and Newton 1982) and there is reason to believe that the eucnemids were in existence before the break-up of the Pangaean supercontinent. Thus, one could expect their present-day distributions to reflect the fragmentation of the supercontinent, provided that subsequent events have not obscured the picture. The 'traditional' view of Pangaean break-up is understood here as in Nelson and Platnick (1980).

The number of eucnemid genera and species occurring in the area from South-east Asia to the south-west Pacific is given in Table 1. Current knowledge is unequal for all parts of the area: Australia, New Guinea, Solomon Islands, Fiji, Samoa and Tonga are quite well known, whereas New Caledonia and Vanuatu have been little studied. The remaining areas fall in a range between the two extremes.

Altogether 89 genera are known from the region, 38 additional ones from other geographical areas are involved in the phylogenies studied. The phylogenetic position of these 127 genera is given in Table 2. Five genera with numerous species occurring worldwide had to be omitted, as it was impossible to produce phylogenies for them (*Dromaeolus*, *Entomophthalmus*, *Fornax*, *Microhagus* and *Nematodes*). All these genera belong to the derived tribes Dirhagini, Macraulacini and Nematodini.

The phylogenetic relationships between genera could not be resolved satisfactorily in all cases within the huge tribes Dirhagini and Macraulacini. Consequently, most of the information had to be extracted from clades within genera. The phylogenetically old, usually small tribes, posed another problem. They were frequently the sister-groups of world-wide groups which themselves showed distinct archaic vicariant patterns.

Table 2. Systematic position of genera within Eucnemidae

The order of suprageneric taxa is based on phyletic sequencing. Total 130, in the study area, 89. Five genera are excluded from the analysis. *Taxa described in Muona (1992*b*)

Phyllocerinae, Anelastini
Anelastes Kirby
Pseudomeninae*, Pseudomenini*
Pseudomenes Fleutiaux
Pseudomeninae*, Schizophilini*
Schizophilus Bonvouloir
Melasinae, Ceballosmelasini*
Ceballosmelasis Cobos
Melasinae, Melasini
Isorhipis Boisduval & Lacordaire, Melasis Olivier
Melasinae, Calyptocerini*
Calyptocerus Guérin-Méneville, Otho Lacordaire
Melasinae, Xylobiini
Agalba Broun, Saproxylophilus Leiler, Paraxylophilus Cobos, Xylophilus Mannerheim
Melasinae, Neocharini*
Neocharis
Melasinae, Epiphanini*
Epiphanis Eschscholtz, Hylis des Gozis, New genus 1
Melasinae, Dirhagini
Arganus Bonvouloir, Arrhipis Bonvouloir, Balistica Motschulsky, Cafolus Bonvouloir, Entomophthalmus Bonvoulor (excluded from analysis), Farsus Du Val, Freyiola Cobos, Fryanus Fleutiaux, Melocarvalhosia Cobos, Microrhagus Dejean (excluded from analysis), New genus 2, New genus 3, New genus 4, Porraulacus Fleutiaux, Prodirhagus Fleutiaux, Rhagomicrus Fleutiaux, Sarfus Fleutiaux, Sarpedon Bonvouloir, Talerax Sharp
Eucneminae, Phaenocerini*
Phaenocerus Bonvouloir, Pinaroo*
Eucneminae, Proutianini*
Myall*, Onya*, Proutianus*
Eucneminae, Perrotini*
Perrotius Fleutiaux
Eucneminae, Dendrocharini
Dendrocharis Guérin-Méneville, Bossionus Fleutiaux, Buckia Cobos, Scopulifer Fleutiaux
Eucneminae, Yangini*
Yanga*
Eucneminae, Entomosatopini*
Entomosatopus Bonvouloir
Eucneminae, Dyscharachthini*
Dyscharachthis Blackburn
Eucneminae, Mesogenini
Arisus Bonvouloir, Baryus Fleutiaux, Chropoecilus Fleutiaux, Euryostus Bonvouloir, Feaia Fleutiaux, Mesogenus Bonvouloir, Stethon LeConte, Submesogenus Fleutiaux, Temnillus Bonvouloir, Temnus Fleutiaux, Vitellius Bonvouloir
Eucneminae, Eucnemini
Epipleurus Fleutiaux, Eucnemis Ahrens, Gastraulacus Guérin-Méneville, Idiotarsus Bonvouloir, Lamprotrichus Bonvouloir, Modius Fleutiaux, Poecilochrus Bonvouloir
Eucneminae, Galbitini
Achaica Muona, Agastocerus Bonvouloir, Galbimorpha Fleutiaux, Galbites Fleutiaux
Macraulacinae, Jenibuntorini*
Jenibuntor*
Macraulacinae, Echtrogasterini
Dictyeucnemis Lea, Echtrogaster Blackburn, Hemiopsida MacLeay, Henecocrus Bonvouloir
Macraulacinae, Euryptychini
Euryptychus LeConte
Macraulacinae, Orodotini*
Aubailius Fleutiaux, Eudorus Laporte, Langurioscython Heller, Namolius Bonvouloir, Orodotes Bonvouloir

Table 2 *(continued)*

Macraulacinae, Macraulacini

Acedax Bonvouloir, Aruanus Fleutiaux, Bermillus Bonvouloir, Bothrion Fleutiaux, Ceratus Bonvouloir, Chapianus Fleutiaux, Cladidus Fleutiaux, Curtocephalus Fleutiaux, Diapodius Bonvouloir, Diphytaxis Horn, Discaptothorax Blackburn, Dromaeoloides Cobos, Dromaeolus Kiesenwetter (excluded from analysis), Eucalodemas Bonvouloir, Eucalosoma Laporte, Eurachis Horn, Euryaulacus Bonvouloir, Fornax Laporte (excluded from analysis), Galbodema Laporte, Henecosoma Horn, Heterotaxis Bonvouloir, Hodocerus Bonvouloir, Hylotastella Fisher, Hylotastes Bonvouloir, Macraulacus Bonvouloir, Macroscython Fleutiaux, Maelodrus Fleutiaux, Melanocoleus Bonvouloir, Melanoscython Fleutiaux, Neodiapodius Fisher, Nodema Fleutiaux, Pachyfornax Fleutiaux, Piestocera Perty, Plesiofornax Coquerel, Pocoelus Fleutiaux, Procladidus Fleutiaux, Prosassanus Fleutiaux, Pseudochapianus Lucht, Pseudodiaeretus Fleutiaux, Pseudoscython Fleutiaux, Raapia Fleutiaux, Sakalavus Fleutiaux, Sassanus Fleutiaux, Scython Laporte, Semnodema Bonvouloir, Serrifornax Fleutiaux, Spinifornax Fleutiaux, Tacerus Fleutiaux

Macraulacinae, Nematodini

Nematodes Berthold (excluded from analysis), Neomathion Fleutiaux, Trigonopleurus Bonvouloir

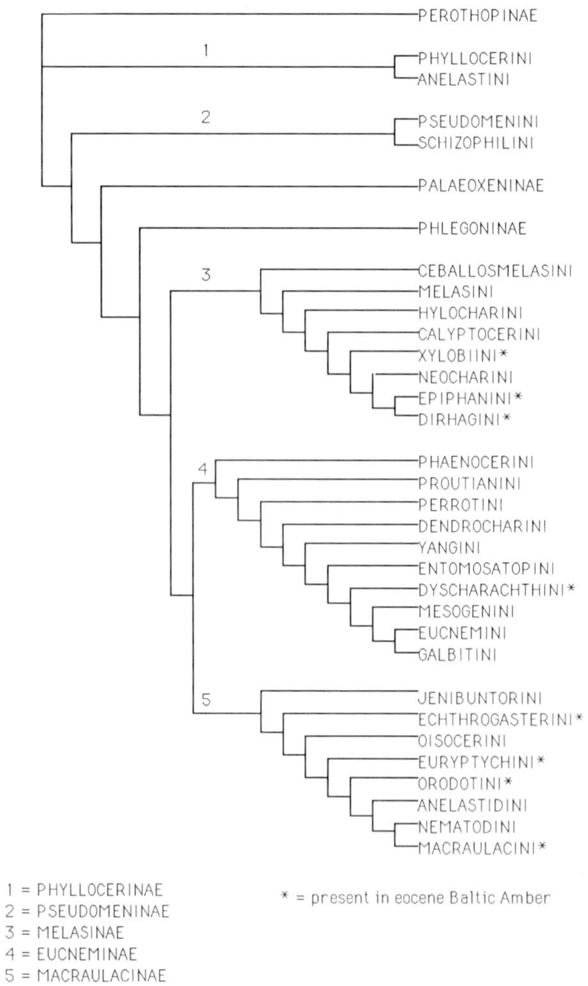

Fig. 1. Eucnemidae, phylogenetic relationships of higher taxa.

In addition to extant insects, the Baltic Amber fossil Eucnemidae were also utilised in the analysis (Muona 1992*a*). Altogether 14 Eocene fossil genera are known. Two of them appear to be extinct and three others (*Dromaeolus*, *Fornax* and *Microrhagus*) belong to a group of genera excluded from the present analysis. The remaining nine genera provided valuable information.

The biogeographical analysis is based on 55 phylogenies (Figs 1–3, 5). All the phylogenies were produced by parsimony analysis. They were constructed with Hennig86 (Farris 1988) using the successive weighting procedure (Farris 1970, 1988). The synapomorphies defining the monophyletic groups were evaluated in all cases. The cladogram depicting the phylogeny of the higher taxa (Fig. 1) includes some modifications to the result obtained with the program. The observed amount of homoplasy was about 50% and the value of several characters was clearly dubious. Consequently, some branches were ignored. It is not possible to present the reasoning behind the cladograms in this connection. They are treated in detail in (Muona 1991, 1992*b*) and in forthcoming revisions of the Australian, Melanesian and Pacific Eucnemidae (Muona, unpublished data).

The individual phylogenies include a variable number of 'genera'. Usually when more than one genus is named the other names actually refer to superfluous taxonomic splits. However, in order to make it possible for the user to determine the included species, I have retained established names.

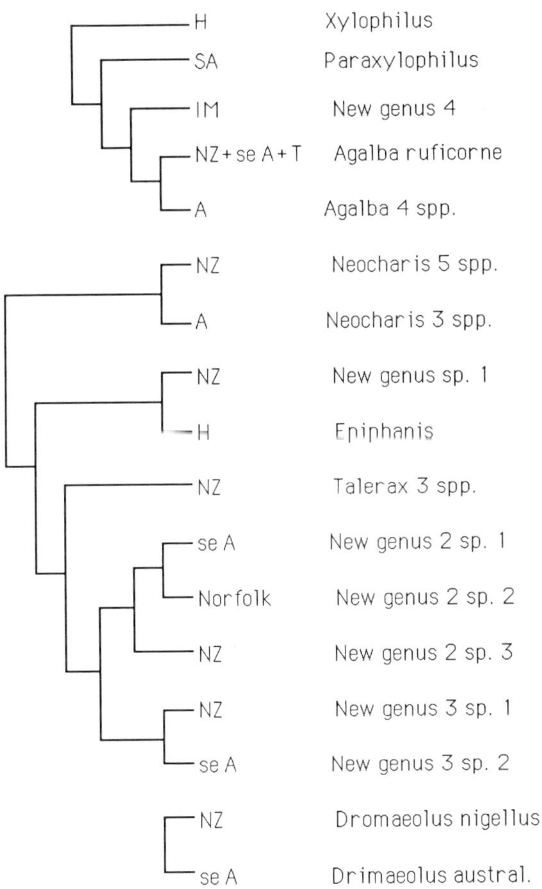

Fig. 2. New Zealand Eucnemidae, phylogenetic relationships. Abbreviations as in Fig. 5.

In some cases identical distributional patterns were observed between species and higher taxa in different clades. Obviously there are several possible reasons for this. There is no way of knowing whether certain clades have diverged more rapidly than others and this has traditionally resulted in recognition of superfluous higher taxa. On the other hand, even in a strictly phylogenetic system the choice of categories is a subjective one with respect to the most apical branches. In some cases it appears that fairly recent events have created a situation in which both recently diverged forms and much older ones are forced to occupy identical small areas. The recognition of multiple patterns masking older events of different ages is difficult unless there is further information to indicate that several different patterns involving outside areas are present. I have speculated on such possibilities while fully acknowledging the difficulties involved. In systematics the words primitive and derived are used in many different

Table 3. Distribution types of Eucnemidae occurring in SE. Asia to the SW. Pacific

*primitive genera; **bold type,** Eocene fossils known

A. Laurasia — Gondwana break-up pattern (51 of 84 genera)
  1. Vicariant Laurasian and Gondwanan groups (5)
    Agalba-**c**, Calyptocerus-c, Farsus-c, Proutianus-c, Pseudomenes-c
  2. Gondwanan groups (46)
    Africa (20): Cladidius-c, Entomosatopus, Eudorus-c, Feaia, Heterotaxis-c, Macroscython, Mesogenus-c, Phaenocerus, Plesiofornax, Prodirhagus-c, **Rhagomicrus**, Scopulifer, Scython, Serrifornax, Trigonopleurus
    India (3): Arganus-c, Porraulacus, **Spinifornax**
    Africa + *India* (2): Arrhipis, **Poecilochrus**-c
    C. & S. America (17): Aruanus-c, Bothrion-c, Dendrocharis, **Dyscharachthis**, Epipleurus, Eucalodemas-c, Fryanus, Galbodema-c, Hylotastes-c, Melanocoleus, Neomathion, Scopulifer, Semnodema-c, Temnus, Vitellius
    Indomalesian clade + Australiaw Guinea clade (25): Agalba-**c**, Arrhipis, Bothrion-c, Cladidius-c, Dendrocharis, Epipleurus, Fryanus, Galbodema-c, Heterotaxis-c, Hylotastes, Mesogenus, Plesiofornax, Poecilochrus, Porraulacus, Rhagomicrus, Scopulifer, Scython, Serrifornax, Semnodema-c, Spinifornax, Temnus, Vitellius

B. Endemic genera (41 of 84)
  1. Australian (12)
    Dictyeucnemis*, Discaptothorax, Echthrogaster*, Galbodema, Jenibuntor*, Myall*, Onya*, Orodotes, Pinroo*, New genus 5, Pseudomenes*, Yanga*
  2. Indomalesian (18)
    Achaica, Agastocerus, Galbimorpha, Galbites
    Ceballosmelasis*, Compsocnemis*, Euryostus
    Bermillus, Cafolus, Chapianus, Curtocephalus, Diapodius, Hodocerus, New genus 4, Nodema, Perrotius, Procladidus, Pseudochapianus
  3. SW. Pacific (11)
    Dromaeoloides, Freyiola, Maelodrus, New genus 4, Tacerus
    Agalba*, Neocharis*, New genus 2, New genus 3
    Talerax*, New genus 1*

C. Laurasian invaders (6 of 84)
  Anelastes*, Isorhipis*, Hylis*, Melasis*, Otho*, Pseudoscython

D. Discordant elements (6 of 84)
  Arisus, **Ceratus**, **Euryptychus***, **Hemiopsida***, Langurioscython, Sarpedon

ways. Here they refer to the relative phylogenetic positions of certain taxa. The term 'old taxon' refers to taxa which are involved in Laurasia–Gondwana patterns, either themselves or through their sister-group. Thus, forms descending from 'young taxa' never show this distributional pattern.

Distributional Patterns

Laurasia–Gondwana Pattern

The largest single block of genera showed distributional patterns coinciding with the traditional Pangaea break-up model, including 51 from the 84 genera studied (Table 3).

Five genera are involved in phylogenies with a basal split between Laurasian and Gondwanian clades: *Agalba*, *Calyptocerus*, *Farsus*, *Myall* and *Pseudomenes*. The African and Indian *Farsus* groups are very closely related, but all other geographically isolated clades within these genera have diverged considerably and are placed in separate genera. It is of interest to note that *Farsus*, which belongs to the fairly derived tribe Dirhagini, appears to have existed long enough to share this pattern.

Forty-six genera showed patterns coinciding with the traditional Gondwana break-up sequence, but did not have known Laurasian elements. The African clade was the most basal one in 20 cases, the Indian and the African + Indian in two cases. Seventeen phylogenies shared a disjunction between tropical America and the study area. These genera are listed in Table 3, their phylogenies are given in Fig. 5.

The *Poecilochrus*-group of genera, *Dyscharachthis* and *Rhagomicrus*, includes extant Holarctic species. These clades are known from Eocene fossils as is *Spinifornax*. The species in question do not, however, contradict the Gondwanan pattern observed. Both the fossil *Poecilochrus* species and the Holarctic 'genus' *Eucnemis* belong to a monophyletic group, the sister-group of which is found in southern North America and Central America. The sister-group of this combined clade is a South American group. This pattern [{[(Eocene E + A)] (NA)} (CA)] (SA) can scarcely be explained with the traditional Pangaean break-up model. These superficially discordant elements seem to have evolved in connection with an invasion of the Northern areas from South America via Central America after the initial split between Laurasia and Gondwana. The important point is that the Holarctic and Central American *Poecilochrus* clades form a monophyletic group that shares an ancestor with a typical Gondwanan clade found in South America. The situation is identical in the other three cases, except that no extant North American *Dyscharachthis* species is known.

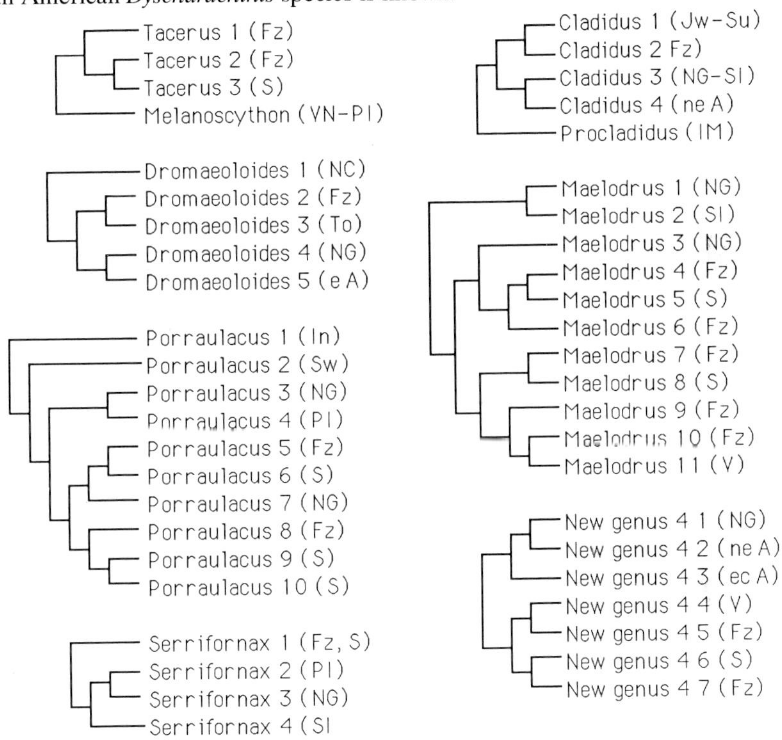

Fig. 3. South-west Pacific Eucnemidae, phylogenetic relationships. Abbreviations as in Fig. 4.

Perhaps the most interesting feature of the group sharing the Gondwana break-up pattern is the common presence of an anomalous Indomalesian clade between South America and Australia–New Guinea clades. Twenty-four phylogenies share this pattern, five of them including genera endemic to the Indomalesian region: *Chapianus*, *Diapodius*, *Hodocerus*, *New genus 4*, *Nodema*, *Procladidus* and *Pseudochapianus* (Table 3, Fig. 5). The genera belonging to this Indomalesian clade have radiated extensively in the area and derived species are found in a wide region from Burma-Vietnam to New Guinea and the Solomon Islands. However, the forms belonging to this Indomalesian clade can always be identified by a sister-group relationship with a separate Australian–New Guinea clade.

The taxa occurring in the Philippine Islands and Sulawesi do not usually belong to the Indomalesian clade. Their sister-groups belong mostly to the Australia–New Guinea clade, the sister-group of this combined clade being the Indomalesian one.

Endemic Forms

Altogether 41 genera are endemic to regions within the area, 12 to Australia, 18 to Indomalesia and 11 to the south-west Pacific (Table 3).

Australia

The Australian endemics are for the most part archaic. Their sister-groups frequently consist of other old Australian endemics and phylogenetic connections with other parts of the world are younger than the Australia–Australia relationship. Eight Australian endemic genera belong to old tribes of three subfamilies, Pseudomeninae, Eucneminae and Macraulacinae. To demonstrate the nature of these relationships it is best to discuss them in detail.

Pseudomeninae. *Pseudomenes* together with its Laurasian sister-group *Schizophilus* forms the most primitive known wood-boring eucnemid subfamily, Pseudomeninae. *Schizophilus* is associated with *Fagus* (Muona 1992*b*) and the distribution of *Pseudomenes*, south-eastern Australia and Tasmania, suggests the intriguing possibility that it may develop in *Nothofagus*.

Eucneminae: Phaenocerini. The sister-group of *Pinaroo*, *Phaenocerus*, contains one species in Australia and one in Madagascar.

Eucneminae: Proutianini. The sister-group of *Myall* is *Proutianus* in north-western North America. The sister-group of this combined clade is *Onya* with species in Australia and New Guinea. Proutianini is the only Eucneminae group showing the Laurasia–Gondwana pattern. All the more derived tribes of this subfamily are purely Gondwanic and appear to have evolved in Gondwanaland.

Eucneminae: Yangini. *Yanga* has no close extant relatives. It represents a clade whose sister-group includes the rest of the derived Eucneminae.

Macraulacinae: Jenibuntorini. *Jenibuntor* is the most primitive extant macraulacine genus with no known close relatives.

Macraulacinae: Echtrogasterini. The sister-group of the clade *Dictyeucnemis* + *Echtrogaster* contains the genera *Hemiopsida* and *Henecocerus*. *Hemiopsida* species are predominantly Australian and show aberrant sister-group relationships involving both hemispheres.

The remaining four Australian endemics, *Discaptothorax*, *Galbodema*, *Orodotes* and *New genus 5*, belong to typical Gondwanan groups.

Indomalesia

Only 2 of the 18 Indomalesian endemic genera are definite old forms. *Ceballosmelasis* is the most primitive extant Melasinae genus with no known close relatives. *Compsocnemis* belongs to the primitive Melasinae tribe Melasini, its sister-group formed by the Holarctic genera *Melasis* and *Isoriphis*.

The remaining 16 endemic Indomalesian genera fall into three groups. The first group includes distinct clades with widespread Gondwanan sister-groups. *Perrotius* is a Eucneminae genus with no known close relatives. The Eucneminae genus *Euryostes* forms the sister-group of the rest of the Mesogenini. *Achaica*, *Agastocerus*, *Galbimorpha* and *Galbites* form the tribe Galbitini, whose sister-group is the tribe Eucnemini. The sister-group of this combined clade is the tribe Mesogenini. Both the Mesogenini and the Eucnemini contain many genera showing Gondwana break-up patterns but neither includes vicariant Laurasian elements. These six genera belong to three distinct, well defined clades, none of which is involved in the traditional Gondwana break-up pattern. This is even more remarkable as they belong to the derived Eucneminae, a group that seems to have evolved on the southern continent before its break-up.

The second group consists of genera that are members of the anomalous Indomalesian clades within typical Gondwanan patterns: *Chapianus*, *Diapodius*, *Hodocerus*, *New genus 4*, *Nodema*, *Procladidus* and *Pseudochapianus*.

The third group includes distinct endemic genera whose relationships to other genera are not known: *Cafolus* (Dirhagini) and *Bermillus* and *Curtocephalus* (Macraulacini).

South-west Pacific

All the 11 endemic south-west Pacific genera are phylogenetically connected either with Australia–New Guinea or with New Zealand. The relationships of the endemic genera clearly show that the biogeographical history of the area is complex. I will discuss the New Zealand connections and the Melanesian–Polynesian connections separately.

All four old endemic south-west Pacific genera, *Agalba*, *New genus 1*, *Neocharis* and *Talerax*, are involved with patterns including New Zealand (Fig. 2).

Agalba ruficorne (Broun) occurs in New Zealand, south-eastern Australia and Tasmania. Its sister-group is a clade with four species in eastern Australia. As previously stated, the *Agalba* group of genera also includes distinct clades in the Holarctic, South America and

Indomalesia. *Neocharis* with several species both in east Australia and New Zealand is the sister-group of the combined clade Epiphanini + Dirhagini. Epiphanini includes two Holarctic genera, *Hylis* and *Epiphanis* and the *New genus 1* from New Zealand, which is the sister-group of *Epiphanis*. The Dirhagini is a huge world-wide group with three genera also in New Zealand. *Talerax* is an old form; its phylogenetic relationships with other genera are not known in detail. *New genera 2* and *3*, which are derived Dirhagini taxa not closely related to each other, form the sister-group of *Talerax* when only New Zealand forms are considered.

The only New Zealand eucnemid not belonging to a genus endemic to the south-west Pacific region is *Dromaeolus nigellus* (Broun). It is a derived species of Macraulacinae with a sister-species, *D. australasiae* Bonvouloir, in south-western Australia and Tasmania.

None of the old endemic Pacific genera occurs in the area Melanesia east of New Guinea and Polynesia north of New Zealand.

The Fijian species have their sister-groups in Indomalesia west of New Guinea (*Tacerus–Melanoscython*), in Australia–New Guinea (*Dromaeoloides*), both in Vanuatu and Samoa (*New genus 4*) and in Vanuatu, Samoa, New Guinea and Fiji (*Maelodrus*).

The sister-groups of Samoan species invariably occur in Fiji. The only endemic Pacific genus with more than one Samoan species is *Maelodrus*. Both the Samoan species belong to separate Fijian clades.

The Tongan fauna is very poor, but one species belonging to the endemic Pacific genus *Dromaeoloides* exists. Its sister-group occurs in Fiji.

The very poorly known fauna of New Caledonia includes one species belonging to *Dromaeoloides*. It forms the sister-group of all the other species of this genus.

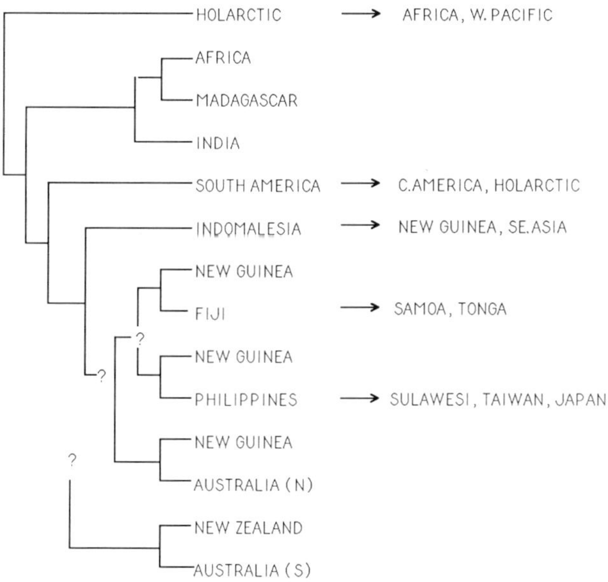

Fig. 4. The distributional pattern of the Eucnemidae found in the region from South-east Asia to the south-west Pacific. Arrows indicate secondary invasion.

Laurasian Invaders

This small group includes 6 out of the 84 genera (Table 3, Fig. 5). The species are either widespread in the Palaearctic (*Anelastes*, *Pseudoscython*) or have limited distributions in the study area (*Hylis*, *Isoriphis*, *Melasis*, *Otho*). In both cases the species occurring in the study area belong to derived clades within these otherwise Holarctic genera. Most of the species in question are restricted to the Malay Peninsula or the Greater Sunda Islands; Only *Hylis* extends all the way to the Solomon Islands and Australia. With the exception of *Pseudoscython*, all the genera are old forms.

A species of *Hylis*, whose relationships fully agree with the supposed Laurasian origin of the genus, is known from the eocene Baltic Amber.

Discordant Elements

Six of the 84 genera show distributional patterns that cannot be safely explained by the traditional Laurasia–Gondwana break-up model. Two of the genera are young, their relative position being high in the phylogenetic tree. *Arisus* species show extensive and complex sister-group relationships over the area from south-eastern continental Asia to the Solomon Islands.

The only known Indian species belongs to one such clade. *Ceratus* shows a similar pattern with one species belonging to an Indomalesian clade occurring in the Seychelles.

Four genera are old, their relative position being low in the phylogenetic tree. *Sarpedon* includes two clades, one with Nearctic and Japanese species, the other with species in Vietnam and New Guinea. *Languriosсythоn* has one species in India and another in the Philippine Islands. Both *Euryptychus* and *Hemiopsida* show internested north–south patterns. Both the *Sarpedon* and *Languriosсythоn* distributions could be explained by the traditional model with subsequent extinctions in many areas. On the other hand, the *Sarpedon* pattern could also be explained by invasions from the Holarctic.

Four of the six genera observed to be incongruent with the traditional Pangaean break-up pattern are present in the eocene Baltic Amber material, *Arisus*, *Ceratus*, *Euryptychus* and *Hemiopsida*. The phylogenetic position of the fossil forms strengthens the view that the distribution of these genera cannot be explained by large-scale historic events.

Discussion

Gondwana Pattern

The majority of genera occurring in the region from South-east Asia to the south-west Pacific, 51 of 84, share a distributional history coinciding with the traditional break-up model of Pangaea. The Laurasian and Gondwanan elements have diverged considerably. The species-groups within clades restricted to fragments of Gondwana tend to be fairly similar and are frequently placed in the same taxon. The phylogenies of the species belonging to six additional genera indicate that they are Laurasian elements that may have invaded the region from the north. Thus, there are 57 genera (68%) supporting the traditional model. Five of these genera are known as fossils from the Eocene Baltic Amber. Four of them belong to monophyletic groups, which all have ancestors in common with South American Gondwanan clades. Their phylogenetic position suggests that they may have invaded the Holarctic region via Central America.

Do the distributional patterns of the remaining 27 genera contradict the seemingly good agreement with the traditional Pangaean break-up model? Six genera are in conflict with the model. Two of them may be typical Gondwanan groups with unusual relict distributions today, but the evolutionary history of the remaining four genera definitely differs from that of the majority. They are all known as fossils from the Eocene Baltic Amber. The sister-group relationships of the fossil species are as aberrant as the ones between extant members of these genera.

The sister-groups of the remaining 21 genera are either unknown or large world-wide clades. Six are old forms endemic to Australia (3), Australia + New Zealand (1), New Zealand (1) and Indomalesia (1). Fifteen are young forms endemic to Australia (1), Indomalesia (9), south-west Pacific (3) and New Zealand (2). The biogeographical patterns shown by these forms neither contradict nor support the traditional model.

Indomalesian Clade and New Guinea Connections

Twenty-five groups showed an anomalous feature within the Gondwanan pattern having an Indomalesian clade as the sister-group of the Australia–New Guinea clade. These Indomalesian clades frequently include species that have reached as far east as New Guinea, Australia and the Solomon Islands, and as far north-west as northern Vietnam and China. All the eastern species are widespread ones, whereas the western ones are mostly endemic to restricted areas. It is tempting to speculate that several of the Indomalesian endemics with widespread or unknown sister-groups, 11 genera in all, actually belong to the anomalous Indomalesian clade as well. Candidates include the tribe Galbitini and the genera *Ceballosmelasis*, *Compsocnemis*, *Euryostes* and *Perrotius*.

The presence of this large biogeographical unit belonging to the Gondwanan break-up sequence is of considerable interest. How can this pattern be explained? Another feature, which has to be explained in connection with this pattern, is the presence of separate connections between Australia–New Guinea, New Guinea–Philippines and New Guinea–Fiji. The distinctly separate New Guinea–Indomalesia and New Guinea–Philippines–Sulawesi clades agree well with the pattern observed by Holloway (1987). In this connection, one might note that the few Japanese species included in the analysis show two distinct patterns. They are either derived from the Gondwanan New Guinea–Philippine clade or are strictly Laurasian.

With respect to the anomalous Indomalesian clade, two obvious types of explanations can be given. These clades may have evolved on a Gondwanan fragment that separated from the

AGALBA

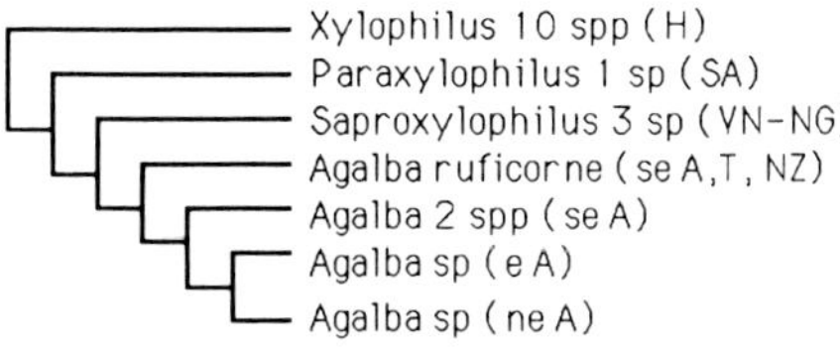

ARUANUS

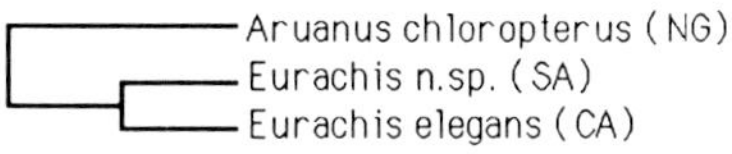

ARISUS

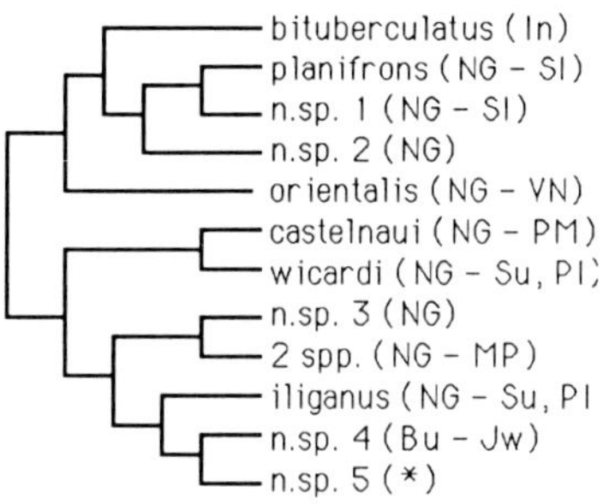

ARRHIPIS

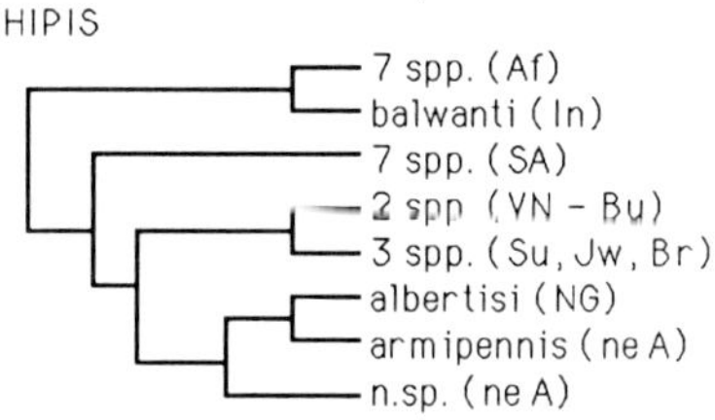

BOTHRION

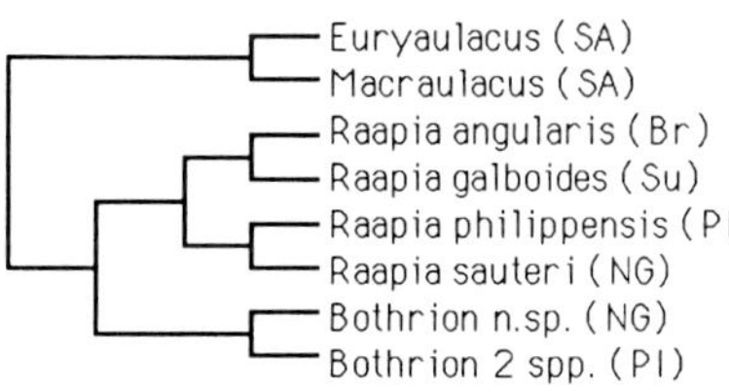

CAFOLUS

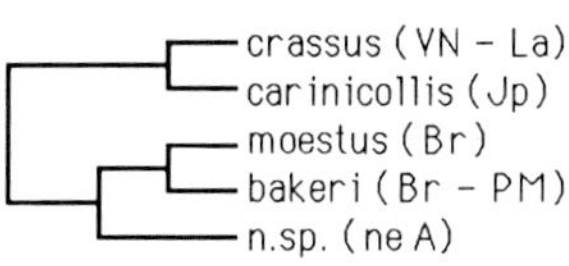

CALYPTOCERUS

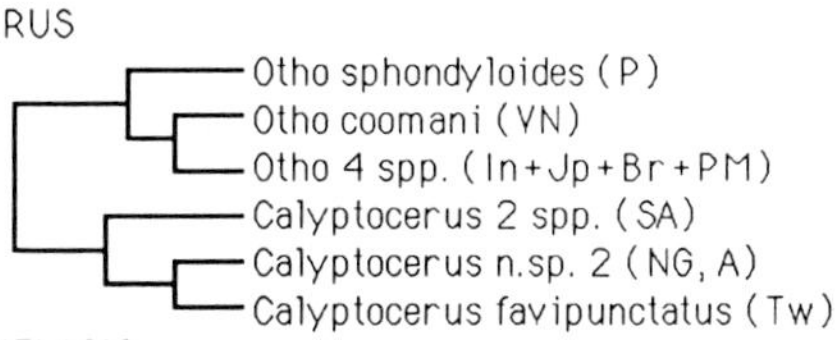

CEBALLOSMELASIS

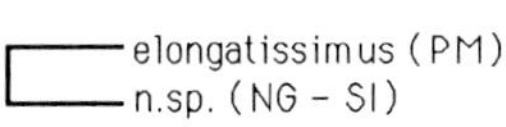

CERATUS

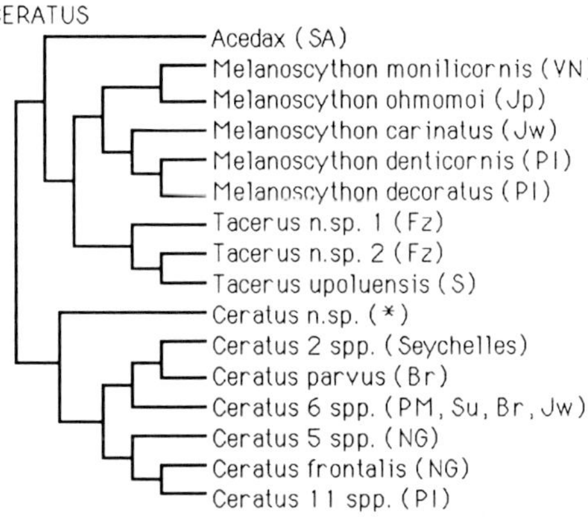

CLADIDUS

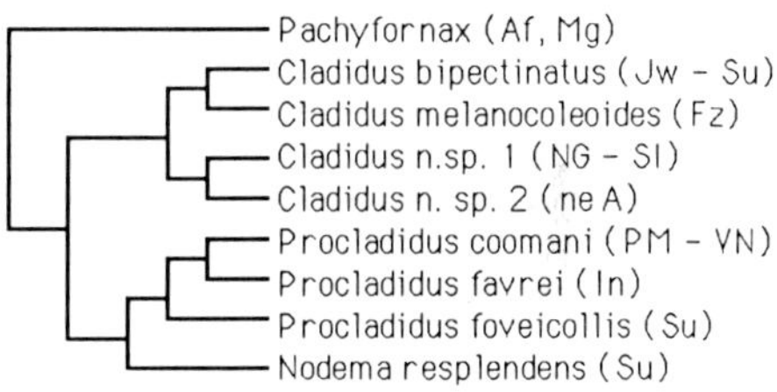

COMPSOCNEMIS

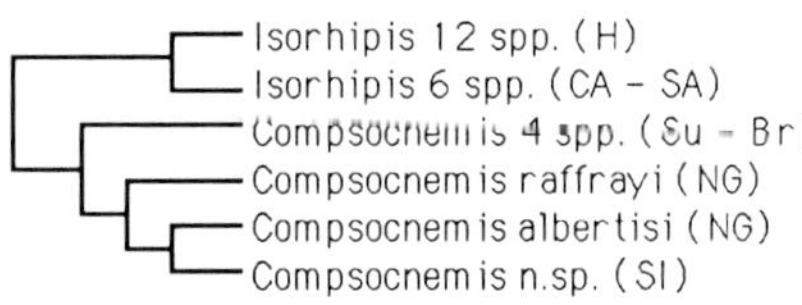

DENDROCHARIS

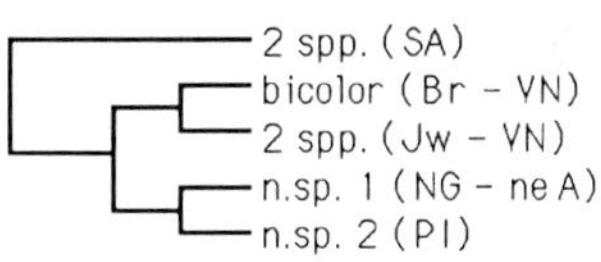

DROMAEOLOIDES

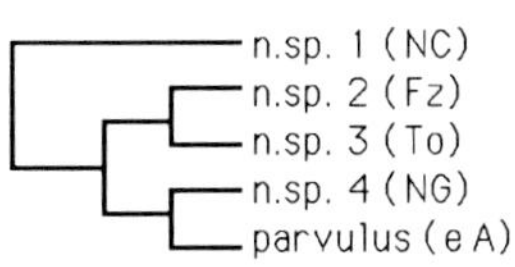

DYSCHARACHTHIS

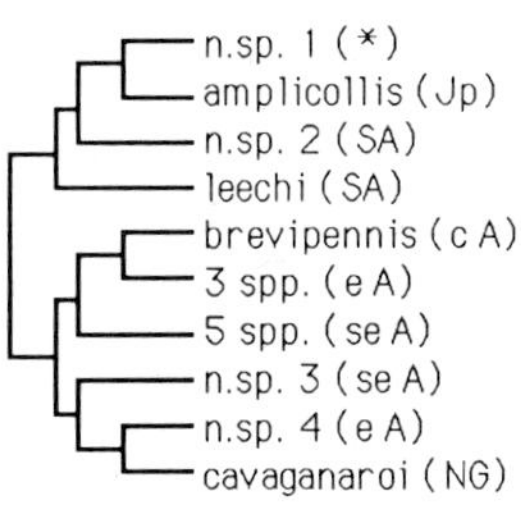

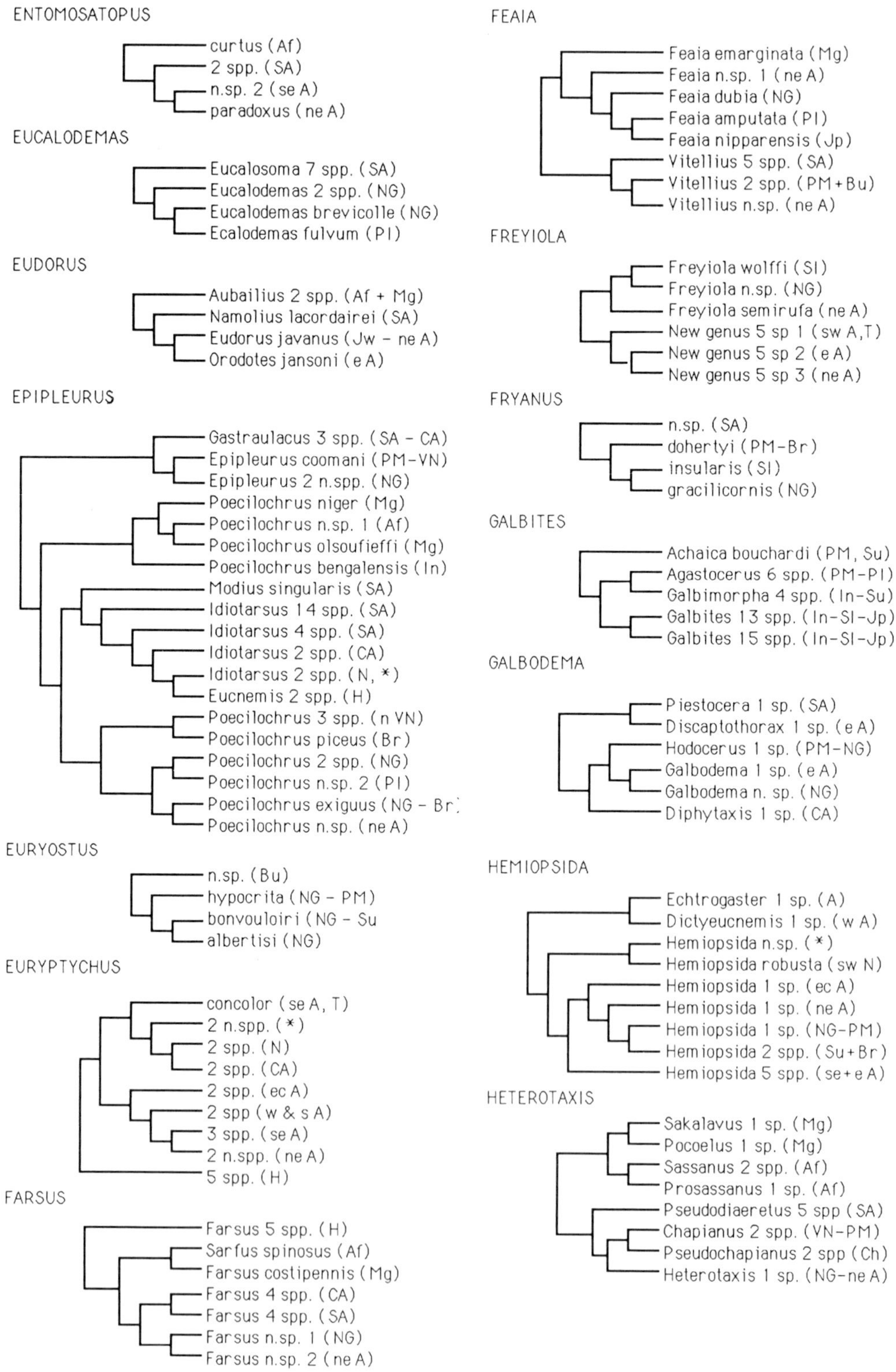
ENTOMOSATOPUS
curtus (Af)
2 spp. (SA)
n.sp. 2 (se A)
paradoxus (ne A)
EUCALODEMAS
Eucalosoma 7 spp. (SA)
Eucalodemas 2 spp. (NG)
Eucalodemas brevicolle (NG)
Ecalodemas fulvum (PI)
EUDORUS
Aubailius 2 spp. (Af + Mg)
Namolius lacordairei (SA)
Eudorus javanus (Jw - ne A)
Orodotes jansoni (e A)
EPIPLEURUS
Gastraulacus 3 spp. (SA - CA)
Epipleurus coomani (PM-VN)
Epipleurus 2 n.spp. (NG)
Poecilochrus niger (Mg)
Poecilochrus n.sp. 1 (Af)
Poecilochrus olsoufieffi (Mg)
Poecilochrus bengalensis (In)
Modius singularis (SA)
Idiotarsus 14 spp. (SA)
Idiotarsus 4 spp. (SA)
Idiotarsus 2 spp. (CA)
Idiotarsus 2 spp. (N, *)
Eucnemis 2 spp. (H)
Poecilochrus 3 spp. (n VN)
Poecilochrus piceus (Br)
Poecilochrus 2 spp. (NG)
Poecilochrus n.sp. 2 (PI)
Poecilochrus exiguus (NG - Br)
Poecilochrus n.sp. (ne A)
EURYOSTUS
n.sp. (Bu)
hypocrita (NG - PM)
bonvouloiri (NG - Su
albertisi (NG)
EURYPTYCHUS
concolor (se A, T)
2 n.spp. (*)
2 spp. (N)
2 spp. (CA)
2 spp. (ec A)
2 spp (w & s A)
3 spp. (se A)
2 n.spp. (ne A)
5 spp. (H)
FARSUS
Farsus 5 spp. (H)
Sarfus spinosus (Af)
Farsus costipennis (Mg)
Farsus 4 spp. (CA)
Farsus 4 spp. (SA)
Farsus n.sp. 1 (NG)
Farsus n.sp. 2 (ne A)
FEAIA
Feaia emarginata (Mg)
Feaia n.sp. 1 (ne A)
Feaia dubia (NG)
Feaia amputata (PI)
Feaia nipparensis (Jp)
Vitellius 5 spp. (SA)
Vitellius 2 spp. (PM+Bu)
Vitellius n.sp. (ne A)
FREYIOLA
Freyiola wolffi (SI)
Freyiola n.sp. (NG)
Freyiola semirufa (ne A)
New genus 5 sp 1 (sw A,T)
New genus 5 sp 2 (e A)
New genus 5 sp 3 (ne A)
FRYANUS
n.sp. (SA)
dohertyi (PM-Br)
insularis (SI)
gracilicornis (NG)
GALBITES
Achaica bouchardi (PM, Su)
Agastocerus 6 spp. (PM-PI)
Galbimorpha 4 spp. (In-Su)
Galbites 13 spp. (In-SI-Jp)
Galbites 15 spp. (In-SI-Jp)
GALBODEMA
Piestocera 1 sp. (SA)
Discaptothorax 1 sp. (e A)
Hodocerus 1 sp. (PM-NG)
Galbodema 1 sp. (e A)
Galbodema n. sp. (NG)
Diphytaxis 1 sp. (CA)
HEMIOPSIDA
Echtrogaster 1 sp. (A)
Dictyeucnemis 1 sp. (w A)
Hemiopsida n.sp. (*)
Hemiopsida robusta (sw N)
Hemiopsida 1 sp. (ec A)
Hemiopsida 1 sp. (ne A)
Hemiopsida 1 sp. (NG-PM)
Hemiopsida 2 spp. (Su+Br)
Hemiopsida 5 spp. (se+e A)
HETEROTAXIS
Sakalavus 1 sp. (Mg)
Pocoelus 1 sp. (Mg)
Sassanus 2 spp. (Af)
Prosassanus 1 sp. (Af)
Pseudodiaeretus 5 spp (SA)
Chapianus 2 spp. (VN-PM)
Pseudochapianus 2 spp (Ch)
Heterotaxis 1 sp. (NG-ne A)

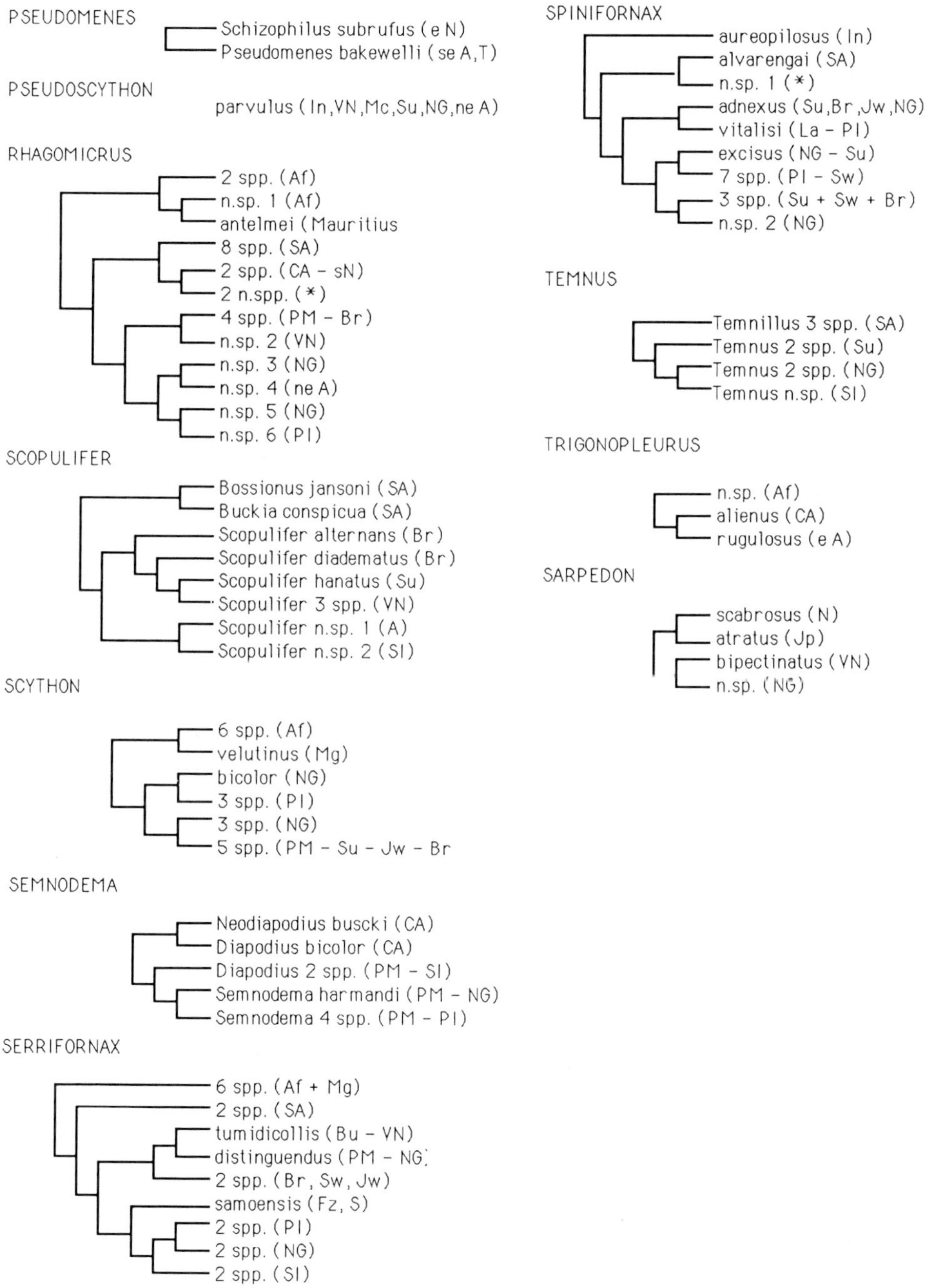
PSEUDOMENES
Schizophilus subrufus (e N)
Pseudomenes bakewelli (se A,T)
PSEUDOSCYTHON
parvulus (In,VN,Mc,Su,NG,ne A)
RHAGOMICRUS
2 spp. (Af)
n.sp. 1 (Af)
antelmei (Mauritius
8 spp. (SA)
2 spp. (CA - sN)
2 n.spp. (*)
4 spp. (PM - Br)
n.sp. 2 (VN)
n.sp. 3 (NG)
n.sp. 4 (ne A)
n.sp. 5 (NG)
n.sp. 6 (PI)
SCOPULIFER
Bossionus jansoni (SA)
Buckia conspicua (SA)
Scopulifer alternans (Br)
Scopulifer diadematus (Br)
Scopulifer hanatus (Su)
Scopulifer 3 spp. (VN)
Scopulifer n.sp. 1 (A)
Scopulifer n.sp. 2 (SI)
SCYTHON
6 spp. (Af)
velutinus (Mg)
bicolor (NG)
3 spp. (PI)
3 spp. (NG)
5 spp. (PM - Su - Jw - Br
SEMNODEMA
Neodiapodius buscki (CA)
Diapodius bicolor (CA)
Diapodius 2 spp. (PM - SI)
Semnodema harmandi (PM - NG)
Semnodema 4 spp. (PM - PI)
SERRIFORNAX
6 spp. (Af + Mg)
2 spp. (SA)
tumidicollis (Bu - VN)
distinguendus (PM - NG)
2 spp. (Br, Sw, Jw)
samoensis (Fz, S)
2 spp. (PI)
2 spp. (NG)
2 spp. (SI)
SPINIFORNAX
aureopilosus (In)
alvarengai (SA)
n.sp. 1 (*)
adnexus (Su,Br,Jw,NG)
vitalisi (La - PI)
excisus (NG - Su)
7 spp. (PI - Sw)
3 spp. (Su + Sw + Br)
n.sp. 2 (NG)
TEMNUS
Temnillus 3 spp. (SA)
Temnus 2 spp. (Su)
Temnus 2 spp. (NG)
Temnus n.sp. (SI)
TRIGONOPLEURUS
n.sp. (Af)
alienus (CA)
rugulosus (e A)
SARPEDON
scabrosus (N)
atratus (Jp)
bipectinatus (VN)
n.sp. (NG)

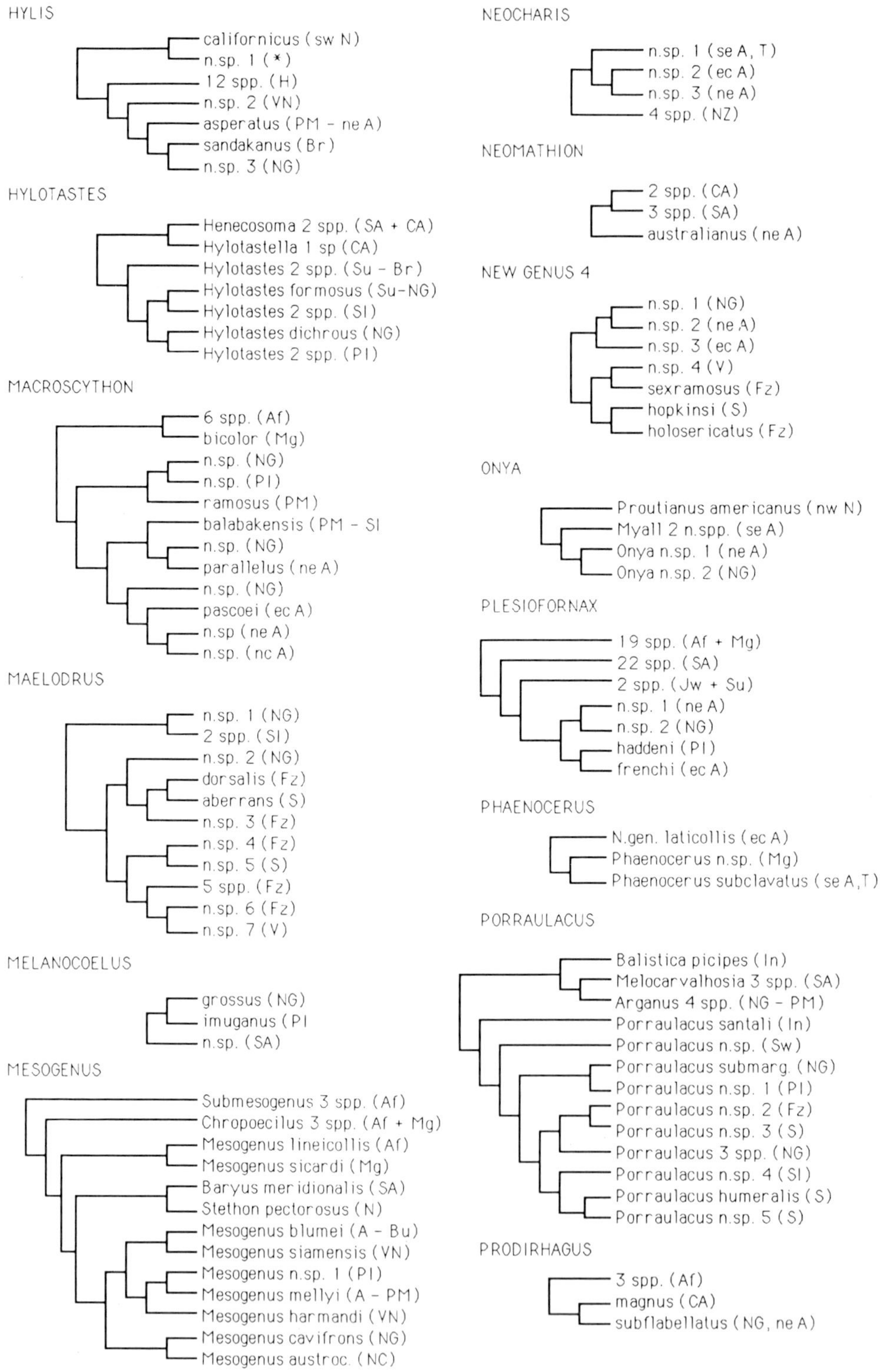

Fig. 5. The distributions and phylogenies of the taxa. The cladograms are identified by the name of one genus only. A list of the other genera involved in the phylogenies is given on the facing page: Abbreviations, capital letter: (*) = Eocene Baltic Amber, A = Australia, Af = Africa, Bu = Burma, Br = Borneo, CA = Central America, Fz = Fiji, H = Holarctic, IM = Indomalesia, excluding Philippines and Sulawesi, In = India, Jp = Japan, Jw = Jawa, La = Laos, Mg = Madagascar, N = Nearctic, NC = New Caledonia, NG = New Guinea, NZ = New Zealand, P = Palearctic, PI = Philippine Islands, PM = Peninsular Malaya, S = Samoa, SA = South America, SI = Solomon Islands, Su = Sumatra, Sw = Sulawesi, T = Tasmania, To = Tonga, Tw = Taiwan, V = Vanuatu, VN = Vietnam.

north shore of Australia–New Guinea. This would be in agreement with the observed speciation pattern as well. The forms found in the western areas, i.e. present-day continental South-east Asia and the Sundas, have diverged considerably, whereas the eastern ones are mostly widespread recent immigrants. Among others, Audley-Charles (1987) has suggested that such an event involving parts of present day continental South-east Asia and the Greater Sundas has taken place. However, the dating of this event to mid- or late-Jurassic is a definite problem. The Indomalesian clade is invariably more recent than the South American and the Indian + African one. Consequently, it cannot be a reflection of Jurassic events. One could speculate that the Indomalesian clade is a result of the South America–Australia connection via Antarctica, which is quite recent having existed at least in the Cretaceous (e.g. Raven and Axelrod 1974; Audley-Charles 1987). The climate of the area appears to have been at least temperate, although a definite deterioration took place from the late Cretaceous on (Frakes 1979). It seems clear that climatic barriers would not have prevented the eucnemids from utilising the Antarctic route for dispersal. However, there is no evidence supporting such a scenario.

Another, to me more plausible explanation, is offered by more recent events suggested by Audley-Charles (1987, e.g. Figs 2.5–2.9). During the early Cretaceous much of Australia was flooded, the southern region being connected with Antarctica, with parts of New Guinea and north-eastern Australia being isolated. Distinct faunas could have developed during this isolation at least in north-eastern and southern Australia and New Guinea. The fact that four different New Guinea patterns were observed in the material can be explained by postulating two or three separate isolated regions in the general area of the old New Guinea. The northward drift of Australia placed two of these areas in proximity with the Sundas and the Philippines making invasion easier and creating the observed separate Indomalesian and Philippine–Sulawesi connections. The New Guinea–Fiji groups may have evolved at this time as well. During the late Oligocene–Miocene period, the source areas in New Guinea were flooded and its old fauna destroyed. The fourth New Guinea connection, with Australia, evolved quite late when the north-eastern Australian fauna was able to invade the present-day New Guinea. The sister-groups of the old, already extinct, New Guinea fauna had survived in north-eastern Australia and are to be found among the recent New Guinea immigrants as well.

This hypothetical sequence is in agreement with the history of the area and explains the complex connections between Australia–New Guinea and the surrounding areas as well as the presence of the anomalous Indomalesian clade.

Australia

The five genera excluded from the analysis contain 65 of some 150 Australian eucnemids. It would be premature to draw definite conclusions about the relationships within the continent when such a major number of forms have been left out. The present-day Australian eucnemids are mainly restricted to the eastern parts of the continent. The general pattern observed was clear, the south-western clades, including Tasmanian ones, form the sister-groups of the northern New South Wales to southern Queensland + north-eastern Queensland clades. Both old endemic genera with no known relatives and young genera involved in major patterns

---

Abbreviations, lower case letter: c = central, e = eastern, ec = central-eastern, n = northern, s = southern, se = south-eastern, w = western.

Achaica see Galbites
Agastocerus see Galbites
Aubailius see Eudorus
Balistica see Porraulacus
Baryus see Mesogenus
Bossionus see Scopulifer
Buckia see Scopulifer
Chapianus see Heterotaxis
Chropoecilus see Mesogenus
Diapodius see Semnodema
Dictyeucnemis see Hemiopsida
Diphytaxis see Galbodema
Discaptothorax see Galbodema
Echthrogaster see Hemiopsida
Eucalosoma see Eucalodemas
Eurachis see Aruanus
Euryaulacus see Bothrion
Galbimorpha see Galbites
Gastraulacus see Epipleurus
Henecosoma see Hylotastes
Hodocerus see Galbodema
Hylotastella see Hylotastes
Isoriphis see Compsocnemis
Macraulacus see Bothrion
Melocarvalhosia see Porraulacus
Myall see Onya
Namolius see Eudorus
Neodiapodius see Semnodema
New genus 5 see Freyiola
Nodema see Cladidus
Orodotes see Eudorus
Otho see Calyptocerus
Pachyfornax see Cladidus
Paraxylophilus see Agalba
Piestocera see Galbodema
Pocoelus see Heterotaxis
Procladidus see Cladidus
Prosassanus see Heterotaxis
Proutianus see Onya
Pseudochapianus see Heterotaxis
Pseudodiaeretus see Heterotaxis
Raapia see Bothrion
Sakalavus see Heterotaxis
Sarfus see Farsus
Saproxylophilus see Agalba
Sassanus see Heterotaxis
Schizophilus see Pseudomenes
Submesogenus see Mesogenus
Stethon see Epipleurus
Temnillus see Temnus
Vitellius see Feaia
Xylophilus see Agalba

including outside regions share this basic Australian pattern. Evidently relatively recent events are responsible for this. This is further supported by the few west–east disjunctions known. *Dictyeucnemis* is an old endemic genus with one species restricted to south-western Australia. Its sister-group, another Australian endemic, *Echthrogaster*, includes one species with a disjunct distribution. It is sympatric with *Dictyeucnemis* in south-western Australia and in addition, is found in South Australia and northern Queensland. Both *Euryptychus porosus* (Lea) and its sister-species, *Euryptychus*, sp. nov., occur in the same area in south-western Australia as *Dictyeucnemis* and *Echthrogaster*. In addition to this, *E. porosus* is generally distributed from eastern South Australia to New South Wales. This pattern suggests that an old west–east disjunction has existed and produced divergent taxa. The eastern elements have later been able to invade much of Australia, but because of more recent events they have vanished from most regions.

Seven of the nine old Australian endemics occur in the south-west and/or in the New South Wales–Queensland border region. These are the areas where the relicts of the old eastern Gondwanan eucnemids are to be found at present. The western *Dictyeucnemis* and the north-eastern *Onya* appear to be the only representatives of old *in situ* divergence. Another species of *Onya* in New Guinea as well as two *Jenibuntor* species in Queensland have probably evolved in connection with recent events.

South-west Pacific, New Zealand Excluded

The phylogenies and distributions of the south-west Pacific Eucnemidae are given in Fig. 3.

All the six genera show a pattern supporting the postulated north-eastern Australia–New Guinea history. *Cladidus*, *Dromaeoloides*, *New genus 4* and *Porraulacus* include the recent New Guinea–Australia clade and parts of the old New Guinea–Fiji and/or New Guinea–Indomalesia clade. The other two genera lack the Australian element. The New Guinea–Philippines connection is present in *Serrifornax* and *Porraulacus*, the New Guinea–Indomalesia connection in *Tacerus*.

The Samoan species have their sister-groups in Fiji. In addition to the Pacific endemic *Maelodrus*, *Porraulacus* includes more than one Fiji–Samoa sister-group. In both genera the Samoan species are related to separate Fijian clades. It is clear that the Samoa–Fiji connection cannot be explained with only one major event.

The only Tongan species included, *Dromaeoloides*, sp. nov., has its sister-species in Fiji. The Tongan *Dromaeolus* and *Fornax* species, which belong to genera excluded from the analysis, all have their sister-groups in Fiji as well but in separate Fijian clades. This indicates that the Tongan eucnemids have arrived separately from Fiji.

As only one New Caledonian species is included, *Dromaeoloides*, sp. nov., little can be said of this area. However, this species forms the sister-group of the rest of the species in the genus. This suggests an old Australian–New Caledonia connection. A preliminary analysis of the New Caledonian *Fornax* species indicates that this is a general pattern.

The few Vanuatu species have all their sister-groups in Fiji. Almost invariably the endemic Solomon Islands forms have their sister-groups in New Guinea, many other species being widespread. In one instance, *Serrifornax*, the Solomon Islands species, has as its sister-group the New Guinea–Philippines clade. This indicates that the region may also be connected with the New Guinea–Fiji unit.

New Zealand

The phylogenies of all the known New Zealand eucnemids are given in Fig. 2. The New Zealand eucnemids clearly share a history in common with eastern Australia. However, as the relationships of most of the genera are not known beyond the Australian component, it is difficult to estimate the age of the connection. In the case of the *Agalba* group of genera this is partly possible, the New Zealand–Australia component being the sister-group of the Indomalesian one. This would place the New Zealand connection anywhere from cretaceous to very recent, not a particularly impressive piece of information. The Neocharini form the sister-group of the Epiphanini + Dirhagini clade. The Dirhagini are a world-wide group with derived species in New Zealand as well. The Epiphanini are a Holarctic group with two genera, one of which, *Epiphanis*, is the sister-group of New Zealand *New genus 1*. It is worthwhile to note that *Epiphanis* has a species in Hawaii as well. The relationships between the three tribes suggest that both *Neocharis* and *New genus 1* represent very old elements, but it is not possible to date the connections observed. The relationships between *Talerax* and *New genera 2* and *3* exhibit a similar complex pattern involving New Zealand and Australia.

The only definitely general pattern between New Zealand and Australia shows that the New Zealand clades form the sister-groups of Australian clades that have south-eastern species. A general Australian pattern shows that the south-eastern species form the sister-group of clades with species in both central-eastern and north-eastern Australia. Both in *Agalba* and *Neocharis* the Australian clades agree with this pattern. However, it is not possible to say whether the Australian speciation has taken place during the Cretaceous flooding of the continent (Audley-Charles 1987) or in connection with more recent events. In my opinion, the most plausible explanation is that the separation of the New Zealand–Australia connection took place in the cretaceous and the speciation within the Australian clades during a later time. It seems clear that the Australia–New Zealand relationship evolved in connection with one major event. This event does not, however, explain the relationships of all the groups.

Summary

The general pattern of relationships observed is given in Fig. 4. The biogeographical history of the Eucnemidae in the area from South-east Asia to the south-west Pacific cannot be understood without knowledge of the phylogeny of the family on a world-wide basis. The history of the eucnemids follows the separation of Laurasia–Gondwana as well as the fragmentation of the latter as postulated in the traditional model. The fauna has evolved predominantly in connection with the break-up of the southern continent, Gondwanaland. After the Antarctica connection between Australia and South America was lost, the Cretaceous flooding of most of Australia created discrete faunas in northern and southern Australia. According to the eucnemid phylogenies, three separate isolates in the general region of present day New Guinea may have existed, all of these showing sister-group relationships with taxa in the north-eastern Australian isolate. When the Australian plate drifted further north, these separate areas came into closer proximity with the archipelagos of the north. One of the areas produced a biogeographical unit involving present-day South-east continental Asia and the Sunda Islands, another the Philippine Islands–Sulawesi unit and the third the Melanesian unit. During subsequent flooding the old New Guinea fauna was eradicated. When the island emerged again, the eucnemids from the north-eastern Australian region migrated to the new New Guinea. This invasion brought in both species belonging to the old sister-groups of the early radiation as well as created new sister-group connections between Australia and the present-day New Guinea. The New Guinea–Philippine unit appears to be the main source of tropical eucnemids found in Sulawesi and Japan, whereas the South-east Asian region has obtained its eucnemid fauna from the Indomalesian unit.

The Solomon Islands fauna has been derived from present-day New Guinea, although some forms may be part of the old New Guinea–Fiji clade. Vanuatu, Samoa and Tonga have a eucnemid fauna that cannot be explained with any single event. The ancestors of these species appear to have arrived during several occasions, mainly from Fiji.

The New Zealand fauna is mostly old and shows strong sister-group relationships with south-eastern Australia. A Cretaceous separation of the areas explain many of the connections, but other complex ones remain to be solved.

The New Caledonian fauna is so poorly known that definite conclusions cannot be made. However, all the groups studied indicate that the New Caledonia–Australia connection is older than the complex old New Guinea–Australia connections.

The results suggest that a tropical eucnemid component of Laurasian origin is weak or may not even exist. The known cases of tropical clades occurring in the Holarctic belong to the Gondwanan groups and appear to have invaded the area from South America. The proximity of Africa and Europe in the late Cretaceous suggests that such forms might be found in Africa, but this does not seem to be the case either. The only Laurasian eucnemids to invade Africa seem to be the members of the tribe Phyllocerini, whose larvae are soil-dwelling forms not known from tropical or subtropical areas. If a major Laurasian tropical eucnemid fauna has existed, there is no evidence of it left. Few present-day tropical forests in areas of Laurasian origin exist. The climatic history of these regions may have been the main reason for the total destruction of this component.

Acknowledgments

I am grateful for the valuable help given by many curators and private collectors, who have provided the material needed for this study. I especially thank the following indivduals: Miss C. M. F. von Hayek (British Museum), Dr C. Girard and Mr J. Menier (Museum d'Histoire Naturelle, Paris), Drs J. F. Lawrence and A. Calder (ANIC, Canberra), Dr A. A. Samuelson (Bishop Museum, Honolulu) and Mr J. Sedlacek (Brookfield, Australia).

References

Audley-Charles, M. G. (1987). Dispersal of Gondwanaland: relevance to evolution of the angiosperms. In 'Biogeographical Evolution of the Malay Archipelago'. (Ed. T. C. Whitmore.) pp. 5–25. (Oxford University Press: Oxford.)

Ewart, A. (1988). Geological history of the Fiji–Tonga-Samoan region of the S.W.Pacific, and some palaeogeographic and biogeographic implications. In 'The Cicadas of the Fiji, Samoa and Tonga Islands, Their Taxonomy and Biogeography (Homoptera, Cicadoidea)'. (Ed. F. P. Duffels.) *Entomonograph* **10,** 5–25. (Brill: Leiden.)

Farris, J. S. (1970). Methods for computing Wagner trees. *Systematic Zoology* **19,** 83–92.

Farris, J. S. (1988). 'Hennig86 Reference. Version 1.5. Mimeograph.'

Frakes, L. A. (1979). 'Climates Throughout Geologic Time.' (Elsevier: New York.)

Hamburger, M. W., and Isacks, B. L. (1988). Duffuse back-arc deformation in the southwestern Pacific. *Nature* **322,** 599–604.

Holloway, J. D. (1987). Lepidoptera patterns involving Sulawesi: What do they indicate of past geography? In 'Biogeographical Evolution of the Malay Archipelago'. (Ed. T. C. Whitmore.) pp. 103–19. (Oxford University Press: Oxford.)

Lawrence, J. F. (1987). Rhinorhipidae, a new beetle family from Australia, with comments on the phylogeny of the Elateriformia. *Invertebrate Taxonomy* **2,** 1–53.

Lawrence J. F., and Newton A. F. (1982). Evolution and classification of beetles. *Annual Review of Ecology and Systematics* **13,** 261–90.

McCabe, R., and Cole, J. (1989). Speculations on the late Mesozoic and Cenozoic evolution of the Southeast Asia margin. In 'The Evolution of the Pacific Ocean Margins'. (Ed. Z. Ben-Avraham.) pp. 143–60. (Oxford University Press: Oxford.)

Muona, J. (1991). The phylogeny and classification of the Indomalesian tribe Galbitini new tribe (Coleoptera, Eucnemidae). *Entomologica Scandinavica Supplement* (in press).

Muona, J. (1992*a*). Baltic Amber Throscidae and Eucnemidae (Coleoptera). *Deutsche Entomologische Zeitschrift* (in press).

Muona, J. (1992*b*). The phylogeny, classification and biology of the family Eucnemidae (Coleoptera). *Entomologica Scandinavica Supplement* (submitted).

Muruyama, S., Liou, J. G., and Seni, T. (1989). Mesozoic and Cenozoic evolution of Asia. In 'The Evolution of the Pacific Ocean Margins'. (Ed. Z. Ben-Avraham.) pp. 75–99. (Oxford University Press: Oxford.)

Nelson, G. J., and Platnick, N. I. (1980). A vicariance approach to historical biogeography. *Bioscience* **30,** 339–43.

Raven, P. H., and Axelrod, D. I. (1974). Angiosperm biogeography and past continental movements. *Annals of the Missouri Botanical Garden* **61,** 539–673.

Spörli, K. B., and Ballance, P. F. (1989). Mesozoic–Cenozoic ocean floor/continental interaction and terrane configuration, southwest Pacific area around New Zealand. In 'Evolution of the Pacific Ocean Margins'. (Ed. Z. Ben-Avraham.) pp. 176–92. (Oxford University Press: Oxford.)

Manuscript received 3 September 1990, accepted 25 February 1991

3 September 1991 *Aust. Syst. Bot.*, **4**, 183–97

Transcending the Wallace Line: do the Western Edges of the Australian Region and the Australian Plate Coincide?

R. I. Vane-Wright

Biodiversity Programme, Department of Entomology, The Natural History Museum, Cromwell Road, London SW7 5BD, U.K.

Abstract

The island of Sulawesi (Celebes), which lies at the heart of the Malay Archipelago, occurs in a region of exceptional tectonic complexity. Since Wallace first drew attention to the anomalous fauna of the island, debate has continued regarding the biogeography and geology of the area. Through an analysis of the distribution of the 183 genera and 470 species of butterflies known from Sulawesi (of which more than 200 species are regional endemics), two classes of biotic patterns linking the island to surrounding regions can be demonstrated. All, or virtually all of the genera on Sulawesi are Asian, but with no special link to Borneo. A set of younger patterns, derived from analysing species' distributions, links Sulawesi to the Moluccas, Philippines and the Lesser Sunda Islands, in addition to Asia. Of these younger patterns, the link between Sulawesi and the Moluccas is most pronounced . This is interpreted to suggest that current geological models, in which Sulawesi consists of at least two terranes, one Asian and one Australian in origin, are consistent with butterfly biogeography only if certain assumptions or constraints are imposed. Firstly, it must be assumed that Sulawesi has had a long independent history from Borneo; it seems most unlikely that Sulawesi and Borneo could have been contiguous 2 mya, as one geological theory has suggested. Secondly, before collision of the Asian and Australian plates about 15 mya, the advancing edge of the Australian plate must have been submerged during most if not all of the approach phase. If the collision has created new land by uplift in the eastern Sulawesi, Banggai and Sula region, then the strong species-level link between Sulawesi and the Moluccas is explicable by local dispersion over the last 15 mya. It is concluded that there is no sharp distinction, at least within Wallacea, between the Asian and Australian biota, as Wallace originally tried to demonstrate and as geological theories might predict: the western edges of the Australian biogeographic area and the Australian tectonic plate do not coincide.

Introduction

According to Hutchison (1989), the collision of the Australian and Asian plates, which suture in or adjacent to eastern Sulawesi, offers a satisfying geological explanation for Wallace's Line. Wallace (1867) conceived such lines to represent 'ancient and permanent barriers to the diffusion of species'. Modern biogeographers, however, using different philosophies and more systematic methods than those employed by Wallace, and accepting the great significance of plate tectonics for our understanding of plant and animal geography, discount ideas of fixed biogeographical boundaries between fixed geographical regions. Instead, they seek to elucidate how the Earth's surface and its biota have evolved together (Humphries 1990; Michaux 1990, 1991).

1030-1887/91/010183$05.00

Sulawesi: Endemism and Uncertainty

The island of Sulawesi occurs on the western edge of the Pacific margin, at the heart of the great Malay Archipelago, a region of outstanding geological complexity (Michaux 1990). Hamilton (1979: fig. 2) shows how Sulawesi lies beside the junction of no less than five different plates: the Pacific, Philippine Sea, Caroline, Asian (= Eurasian or Indo-Chinese) and Australian (= Indian-Australian or Indo-Australian). Sulawesi can also be seen as the hub and spokes of a huge geographical 'wheel' (Fig. 1). The rim comprises a full circle of islands: Palawan, Mindanao, the Moluccas, the Lesser Sundas, Java and Borneo. Each island or island group (Fig. 1: 1–7) has its own rich and varied biota. Some species are endemic to particular islands. Others are widespread throughout all these islands, and often beyond. Still others are relatively restricted, within this system being confined to particular subsets of islands.

Fig. 1. Areas used in analysis of butterfly genera and species native to Sulawesi. 1, Palawan region; 2, Mindanao region; 3, northern Moluccas; 4, southern Moluccas (overlap: Obi); 5, Lesser Sunda Islands; 6, Java region (overlap: Bali); 7, Borneo; SR, Sulawesi Region; D indicates disjunct distribution of *Pseudergolis*, the only non-endemic Sulawesi butterfly taxon not found in at least one of the seven numbered regions; the main island of Sulawesi is hatched.

Occurring in the middle of so many potential sources, on the equilibrium theory of biogeography (MacArthur and Wilson 1967), Sulawesi might be expected to have a very rich but poorly endemic fauna, conceivably like that of the Malay Peninsula which, despite its remarkable total of 1000 butterfly species (Corbet and Pendlebury 1978; J. N. Eliot, personal communication), has a very small number of endemics. In fact, Sulawesi has a modest total of 470 butterfly species, of which more than 200 (43%) are not known to occur outside a narrowly defined Sulawesi Region (Vane-Wright *et al.*, in press; Fig. 1). Even though Sulawesi appears to be a geological composite (see below) there are many endemic animal and plant species that are widespread within it, and the whole island must be regarded as an area of endemism (van Balgooy 1987; Holloway 1987: 114; 1991; Duffels and de Boer 1990). We are therefore challenged to offer an explanation for its relatively low level of species richness coupled with high species endemism.

This paper considers the biogeography of this part of the Pacific margin, and in particular the Asian and Australian plates, through an analysis of the entire butterfly fauna of the island of Sulawesi in relation to the islands that surround it. In attempting to do this, we are faced immediately with two prior problems: a paucity of well resolved cladograms, and a conflict of ideas about both the biogeography and geology of the land between the Sunda and Sahul shelves (the region aptly named Wallacea by Dickerson 1928).

Regarding biogeography, uncertainty has been evident from the start: Wallace drew his 'line' to delimit the Asian and Australian faunal regions first to the east, then to the west, and finally again to the east to Sulawesi, while Huxley re-drew the second version to run west of the Philippines instead of south as Wallace had arranged it (George 1981). Weber, Lydekker and others proposed further modifications. Simpson (1977) eventually reviewed the growing confusion under an appropriate rubric, 'too many lines'.

Turning to the geological debate, Katili (1978) has suggested that, as recently as 2 mya, Sulawesi was attached to eastern Borneo, and that the Makassar Strait has only opened subsequently. Other geologists do not agree. In Hamilton's scheme (1979: fig. 77), virtually the whole of present-day Sulawesi is seen as a long separate part of the Asian plate, with the Australian/Asian collision bringing Australian continental fragments into contact with eastern Sulawesi about 15 mya, to form the offshore Banggai and Butung island groups. In Audley-Charles' scheme the whole Sula Peninsula, including Banggai and a fragment called eastern Sulawesi, converges northwards, with the result that Sulawesi is seen to consist of at least two terranes, one an eastern, Australian part, the other a western, Asian part (Audley-Charles 1981: figs 4·5, 4·6; 1987).

There is uncertainty as to whether the postulated Banggai/Butung/eastern Sulawesi terrane existed as dry land or not, during the approach phase (Holloway 1987). It is also notable that the south-western tip of Sulawesi (including Mt Lompobatang) could represent a third terrane, a southern 'micro-continental fragment' (Michaux 1991). If these theories of a complex origin for Sulawesi are true, we might expect to find evidence of this in an analysis of the butterfly data, challenging the notion that the island represents a single area of endemism.

An Analysis of Sulawesi Biogeography: Butterfly Genera and Species

Despite the difficulty posed by a lack of cladograms, I believe important insights can still be gained from an analysis of the distributions of all butterflies native to Sulawesi, by considering them at two conventional taxonomic levels, genera and species. Central to this approach is Croizat's (1952) view, recently re-expressed by Duffels and de Boer (1990) and Holloway (1991), that our first priority should not be the search for a single area-cladogram but, instead, the recognition of repeated biogeographic patterns amongst areas of endemism. If more than one pattern is found it is likely that we are dealing with historical events of differing ages. On average, 'genera' (being more widely distributed) will reflect older distribution patterns, while 'species' (being on average more narrowly distributed) will tend to reflect more recent distribution patterns (cf. Willis 1922). There is no reason to expect any consistent geographical bias in the recognition of false groups, or in the accordance of incorrect relative ranking. Consequently, when (native) genera and species are analysed in bulk, the existence of unrecognised polyphyletic and paraphyletic taxa among them will act as 'noise', tending only to obscure rather than misdirect our search for significant patterns. This noise will be maximal when a high proportion of genera and species are not reliably monophyletic. With better systematic data (i.e. samples rich in reliably monophyletic groups), the more patterns we can expect to see, and/or the more frequently particular patterns will be represented.

Predictions from the Geological and Biogeographical Theories

The various biogeographical and geological theories can be tested by using them to make simple predictions about the older and younger biogeographic patterns that we might expect to see among the Sulawesi biota. If, for example, the Wallace line were to represent a virtually 'impenetrable' barrier to the west of Sulawesi, we would expect both older (generic) and younger (specific) patterns to be overwhelmingly Papuan or Australian. If, for example, Borneo and Sulawesi were in contact as little as 2 mya (Katili's hypothesis), we might expect to see a number of species uniquely shared by the two islands (young pattern). Eight theories are identified, and the 'older versus younger' predictions they can be used to make are listed in Table 1.

Methods

Historical biogeography involves a search for general patterns of relationship between areas of endemism. Humphries and Parenti (1986) define an area of endemism as a 'region occupied by a monophyletic group of organisms or a species'. In this paper the genera and species of the entire native Sulawesi butterfly fauna are analysed in terms of their areas of endemism, or partial components of those areas, with reference to all the encircling islands or island groups of the Malay Archipelago. The analysis thus relates solely to taxa found on Sulawesi, except insofar as the surrounding unit areas have been recognised as areas of endemism in their own right (i.e. defined by possession of their own endemic genera or species, as above). Sulawesi is thus treated as a 'focal point' (cf. analysis of Sulawesi bats by Musser 1987). The exclusion of non-Sulawesi butterflies from the analysis is necessary because comparable data for the whole of the Malay Archipelago have yet to be assembled.

Table 1. Four hypotheses about biogeography and four theories about geology of the Sulawesi Region

All eight make different predictions about biogeographic patterns the terrestrial biota of Sulawesi can be expected to show. For the butterflies, none of the theories offers a fully satisfactory explanation of the patterns found without one or more modifying hypotheses, such as the idea that Australian crustal material in collision within or adjacent to eastern Sulawesi about 15 mya must have been submerged during most, if not all of the approach phase (see also text)

Theory	Expected patterns for Sulawesi terrestrial biota	
	Older	Younger
Wallace line west of Sulawesi	Papuan	Papuan
Wallace line east of Sulawesi	Asian	Asian
Wallace line south of Philippines	No Philippine link	No Philippine link
Wallace line west of Philippines	Philippine link	Philippine link
Katili: Borneo + Sulawesi 2 mya	Asian	Borneo link
Hamilton: suture Banggai 15 mya	Asian	No Papuan link
Audley-Charles: suture within Sulawesi	Asian + Papuan	Asian + Papuan
Michaux: three terranes	Asian + Papuan + Lesser Sundas	Asian + Papuan + Lesser Sundas

Sulawesi Region and Surrounding Areas

The native butterfly fauna of the Sulawesi Region is currently known to include 186 genera and 534 species (Vane-Wright *et al.*, in press). The Sulawesi Region (Fig. 1: SR) is defined as the main island of Sulawesi (Celebes), together with satellite islands to the north, east, south-east and south (see below). Butung and the Sula Islands are included, but not Buru, which is treated here as part of the southern Moluccas. The island areas of the surrounding 'wheel rim' are divided into seven numbered groups, 1–7, in a clockwise sequence starting (arbitrarily) with Palawan (Fig. 1). These groupings, enumerated below, have been chosen in light of recent attempts to recognise areas of endemism amongst South-East Asian butterflies (see Ackery and Vane-Wright 1984; Vane-Wright 1990; Vane-Wright *et al.*, in press).

Sulawesi. Defined here by the geographical shoreline boundary of present-day Sulawesi. Endemic butterflies found virtually throughout the island but not beyond include *Aoa* de Nicéville (monobasic: Pieridae), *Amathusia* subgenus *Pseudamathusia* Honrath (monobasic: Morphinae), and *Bletogona* Felder & Felder (dibasic: Satyrinae).

Sulawesi Region. The Region, as defined here, consists of Sulawesi plus Kep. Talaud and Kep. Sangihe to the north, Kep. Banggai and Kep. Sula to the east, the Butung Group and Kep. Tukangbesi to the south-east, and Salayar, Tanahjampea, Kalao, Nusa Penida and other islands to the south. Regional endemics amongst the butterflies include genus *Lohora* Moore (14 species: Satyrinae), *Papilio gigon* Felder & Felder (Papilionidae), and *Neptis ida* Moore (Limenitinae).

(1) *Palawan.* This comprises Palawan and a few satellites. These Philippine islands are, biogeographically, a peripheral part of Sundaland. Endemics include: *Troides (Trogonoptera) trojana* Staudinger (Papilionidae), *Elymnias parce* Staudinger (Satyrinae), and *Athyma speciosa* Staudinger (Limenitinae).

(2) *Mindanao.* This is the Mindanao Region of the Philippines, including Mindanao, Basilan, Bohol, Leyte and Samar. Endemics include: *Papilio antonio* Hewitson (Papilionidae), *Acrophtalmia leto* Semper (Satyrinae), and *Discophora philippina* Moore (Morphinae).

(3) *Northern Moluccas.* Include the Indonesian islands of Halmahera, Morotai, Ternate, Batjan and Obi. Endemics include: *Troides (Troides) croesus* Wallace (Papilionidae), *Delias chrysomelaena* Vollenhoven (Pieridae), and *Hypochrysops siren* Grose-Smith (Lycaenidae). Only one of these three, *H. siren*, occurs on Obi (in addition to Halmahera).

(4) *Southern Moluccas.* Include Buru, Seram, Ambon and Obi. In this scheme, the last named island thus holds an ambiguous or overlap position between the northern and southern Moluccas. Endemics include: *Delias dorimene* Stoll (Pieridae), *Hypochrysops anacletus* Felder (Lycaenidae), and *Hypolimnas pandarus* Linnaeus (Nymphalinae). None of these three is reliably recorded from Obi, but a number of (more widespread) butterfly species occurring in the southern Moluccas, including Obi, do not occur to the north (e.g. *Euploea sylvester* Fabricius and *E. alcathoe* Godart; Ackery and Vane-Wright 1984).

(5) *Lesser Sunda Islands.* Comprise the great chain of islands from Lombok east through Sumbawa, Sumba, Flores and Timor to the Tanimbar group, plus (in the west) Bali. (The Timor, Tanimbar and Kai island groups have a number of taxa in common with Australia, and may well prove to be incorrectly placed here.) Endemics include: *Melanitis belinda* Grose-Smith (Satyrinae), *Cyrestis nais* Wallace (Limenitinae), *Euploea caespes* Ackery & Vane-Wright (Danainae), and *Delias sambawana* Rothschild (Pieridae). The first three do not occur on Bali, but *D. sambawana* does.

(6) *Java.* This consists of just Java and Bali. The latter thus holds an overlap position between the Lesser Sunda Islands and Java. Endemics include: *Ypthima nigricans* Snellen (Satyrinae), *Prioneris autothisbe* Hübner (Pieridae), and *Cyrestis lutea* Zinken (Limenitinae). Of these three species, *Y. nigricans* and *C. lutea* are recorded from Bali, but *P. autothisbe* is not.

(7) *Borneo.* This comprises the geographical unit of Borneo island. Although Borneo is almost certainly complex both geologically and biogeographically (see Michaux 1991, and below), I have adopted the single region as a practical unit for the present exercise (cf. Rosen 1988: 446). Endemics include: *Troides (Troides) andromache* Staudinger (Papilionidae), *Lethe dora* Staudinger (Satyrinae), and *Chersonesia excellens* Martin (Limenitinae).

A butterfly genus or species found only on Sulawesi and Palawan is coded as distribution type (1); the same applies whether or not the taxon also occurs further beyond, outside the seven numbered areas (thus the coded groupings reflect a mixture of areas of endemism and partial areas). If, for example, the taxon occurred on Taiwan as well as Palawan, it would still be coded as a type (1) distribution. But if it occurred on, say, Borneo and Seram as well as Taiwan, Sulawesi and Palawan, it would then be coded by simple *combination* with respect to the specified surrounding areas, thus: (1+4+7). Butterflies found on Bali are grouped as follows: if they occur in the Lesser Sundas and Bali but not Java, they are coded as distribution type or component (5); if they occur in Java and Bali but not Lombok or further east in the Lesser Sundas, they are coded as (6). A corresponding system is used for Obi. This ambiguous coding system can be criticised as arbitrary. However, as there is no butterfly taxon endemic only to Obi plus Sulawesi, or Bali plus Sulawesi, separate coding for these two islands would only complicate the system.

All of the 470 Sulawesi butterfly species are either endemic to the Sulawesi Region or to Sulawesi itself, or can be described in terms of the seven surrounding island groups, singly or in various combinations. For the 183 genera this is also true, with just one exception: the nymphalid genus *Pseudergolis* Felder & Felder is disjunct, occurring only on Sulawesi and the Asian mainland (Fig. 1: D). The system of seven areas gives well over 100 potential categories, but only half of these are realised. Considering the total of 652 coded generic and specific distributions, this reflects many repeated patterns. Amongst these, those components neither wholly endemic to the Sulawesi Region nor widespread throughout all the surrounding islands are potentially informative, with respect to the present analysis. (Note: widespread taxa can never be informative, whereas the endemic taxa are likely to become highly informative once their cladistic relationships have been elucidated.)

Table 2. Sulawesi has 183 native butterfly genera, of which 65 have repeated distribution patterns describable as subsets of seven surrounding regions (the other 118 genera have endemic, totally widespread or unique distribution patterns)

The 65 genera represent just 10 combinatorial patterns, five of which account for more than 80% of this total. The coded areas are 1: Palawan; 2: Mindanao region; 3: northern Moluccas; 4: southern Moluccas; 5: Lesser Sunda Islands; 6: Java; 7: Borneo (see also text)

Informative generic components		
1+2 + 6+7:	33·8%	(22)
1+2 + 5+6+7:	23·1%	(15)
2 + 6+7:	9·2%	(6)
1+2+3+4+6+7:	9·2%	(6)
1 + 6+7:	6·2%	(4)
Five others:	18·5%	(12)
	100·0%	(65)

Table 3. The six most strongly supported biogeographic patterns shown by native Sulawesi butterfly species, identified by describing their distributions as subsets of seven surrounding regions (endemic or totally widespread species have been disregarded)

The coded areas are 1: Palawan; 2: Mindanao region; 3: northern Moluccas; 4: southern Moluccas; 5: Lesser Sunda Islands; 6: Java; 7: Borneo (see also text)

Informative specific components		
2	15·4%	(12)
3+4	16·7%	(13)
5+6	12·8%	(10)
6+7	12·8%	(10)
1+2 + 6+7	19·2%	(15)
1+2 + 5+6+7	23·1%	(18)
	100.0%	(78)

Results

Genera

Of the 183 butterfly genera found on Sulawesi, 164 exhibit repeated distribution patterns, representing just 12 different distribution components. The remaining 19 genera all have unique distributions with respect to Sulawesi, and are not considered further. Of the 164 repeated generic patterns, 93 are widespread and 6 are endemics, a total of 99 which are thus uninformative for Sulawesi in this analysis. This leaves 65 genera representing 10 repeated informative components. These 65 genera are taken as the baseline for calculating generic percentages discussed (i.e. 65 = 100%).

The strongest generic pattern is 1+2+6+7, with 22 examples, or 33.8%. This is followed by the related grouping, 1+2+5+6+7 (15 examples, 23.1%). Together they account for 56.9% of the informative components. Beyond that, only three components have more than three examples: 2+6+7 (6 examples, 9.2%), 1+2+3+4+6+7 (also 6 examples), and 1+6+7 (4 examples, 6.2%). The remaining five components account for 12 genera (18.5%). The first five components therefore make up more than 80% of the informative generic distributions, as listed in Table 2.

From Table 2 it is apparent that 47 (72%) of the 65 informative genera have distributions of the type 1+2+6+7 (Sulawesi plus Palawan, Mindanao, Java and Borneo) complete, or plus or minus one element. Note that only four of these 47 genera do not occur to the north, in the Mindanao Region of the Philippines (region 2). The only other strong generic pattern, accounting for a further 9.2%, is also very similar: 1+2+3+4+6+7, which is equivalent to

1+2+6+7 plus the Moluccas, or widespread minus the Lesser Sunda Islands (a pattern often noted by lepidopterists, and usually explained by the relatively dry and seasonal conditions found in the Lesser Sundas).

Thus the 183 butterfly genera found on Sulawesi point overwhelmingly to a South-East Asian facies for the fauna, but South-East Asia can be taken to include the Philippines as well as the Greater Sunda Islands. At this level there is no significant evidence for strong links to individual adjacent islands or island groups, including Borneo, the Moluccas or the Lesser Sunda Islands. The only unique link is one nymphalid genus (*Zethera* Felder) shared with the Philippines. Another special generic link of great potential interest is that shown by *Acrophtalmia* Felder & Felder, a nymphalid found in the Philippines (not Palawan), Sulawesi and northern Moluccas (Vane-Wright 1990).

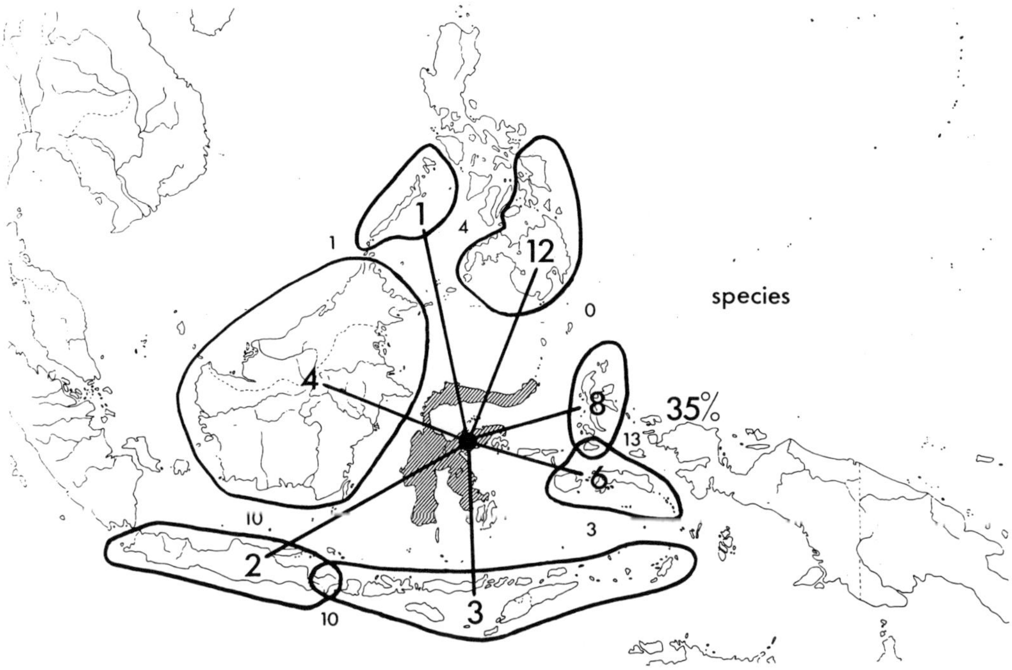

Fig. 2. Numbers of butterfly species present only on Sulawesi and any one of the seven surrounding numbered regions (large figures at ends of 'spokes'), and numbers of species present only on Sulawesi plus two adjacent surrounding areas (smaller figures between area 'envelopes'). See text for explanation.

Species

Turning to butterfly species, there is a total of 470 recorded from Sulawesi, with 534 known from the Sulawesi Region. Taking out all endemic, peripheral or widespread taxa as uninformative with respect to Sulawesi (a total of 345), this leaves 189 species distributed in 52 patterns, an average of 3.63 species per potentially informative component. The six strongest of these are supported by 10 or more examples: components 2 (12 species), 3+4 (13 spp.), 5+6 (10 spp.), 6+7 (10 spp.), 1+2+6+7 (15 spp.) and 1+2+5+6+7 (18 spp.) (Table 3). From this it is apparent that two of the strongest species-level components coincide with the strongest generic components, but the other four do not.

In Fig. 2 the species data are presented to demonstrate the number of species restricted to Sulawesi and single islands or island groups (large figures within area envelopes, at the ends of the 'spokes'), or to pairs of adjacent island groups (small figures between area envelopes). These total 77, of which 27 (or 35%) link Sulawesi with either the northern or southern Moluccas, or both. The strongest *single* link, however, is to the Mindanao Region, which claims over 15% of these 77 narrow links. Note also that the only two adjacent areas without any links are Mindanao and the northern Moluccas, suggesting that the strong Philippine and Moluccan linkages, or tracks, are sharply separate (this despite the outstanding distribution of the genus *Acrophtalmia* noted above).

Thus special links with the Philippines and Moluccas are evident, together with links to the greater area defined by the Sunda Shelf, Philippines and Lesser Sundas, alone or in combination. The question then arises: are the 27 Moluccan links evidence of a Papuan or even truly Australian element in the fauna? Analyses of the generic data and the few cladograms available do not suggest that this is the case, although such unusual genera as *Elodina* Felder & Felder, *Psychonotis* Toxopeus, *Hypochrysops* Felder & Felder and *Cephrenes* Waterhouse & Lyell might eventually prove to contain or represent such links.

With regard to eastern linkages, I need here to make an aside concerning the special links between Sulawesi and Borneo, or more precisely, their virtual absence. There is no special link at the generic level; at the specific level there are just four unique links (Fig. 2). Of these four, two are unique in the sense that the distribution patterns are *only* Sulawesi plus Borneo (in the other two cases the species occur more widely, but outside the other areas immediately surrounding Sulawesi).

Of the two species confined to just Borneo plus Sulawesi, one has been investigated cladistically, *Eurema tominia* Vollenhoven. Yata (1988) originally suggested that this butterfly has dispersed westwards, from Sulawesi into Borneo, not the reverse; Yata (1990) still entertains this possible explanation, amongst others. This example is the only case of sympatry in the *tilaha* species-group (to which *tominia* belongs), and it is interesting that Holloway (1973) describes hybrid zones between the two taxa in Borneo. A third possibility, that the distribution of this species is a result of recent vicariance, in line with Katili's geological model of recent separation of Borneo and Sulawesi, is not supported by the mass of other evidence (Holloway 1987; Vane-Wright 1990).

The second unique Borneo-Sulawesi species link is a lycaenid, *Nacaduba normani* Eliot. According to Pan and Morishita (1990), although only first described in 1969, it can be a significant pest of cocoa. These authors point out that there are few records of *N. normani* and in the past it has been confused with a related widespread species, *N. berenice* Herrich-Schäffer. Thus, the distribution of *N. normani* may ultimately prove to be wider than currently thought.

Evidence for a special link to the south, with the Lesser Sunda Islands and Sulawesi only uniquely sharing three species, looks weak. However, there is one piece of evidence from a cladistic analysis that supports such a linkage (see below).

Cladistic Analyses

The cladistic relationships of the species of the milkweed butterfly genus *Idea* Fabricius compare well with the predominant generic and specific patterns. In Fig. 3, based on a diagram by Holloway (1991), the cladistic analysis of Kitching *et al.* (1987) has been transformed into an area cladogram. Sulawesi occurs twice, firstly as a high-level endemic suggestive of relatively long isolation for the island, and secondly as sister to the lowest-ranking group, which thus suggests a recent link between Sulawesi, the Moluccas and western New Guinea. At the highest generic, or most ancient level, however, the pattern is an Asian one. Thus the analysis of *Idea* is consistent with the results based on analysing all genera and species.

Holloway (1991) has also treated a number of Malay Archipelago moth genera, for which he has carried out cladistic analyses, in a similar fashion (Fig. 4). In *Narosa* Walker we can see a partly similar outcome, with Sulawesi sister to New Guinea at a low rank, but with a high-ranking Asian pattern. *Erconholda* Kiriakoff, in contrast, reflects the other anticipated outcome, this time a low-ranking Philippine/Sulawesi link, but also with the high-ranking Asian pattern. This Philippine/Asia link is repeated in the upper part of the *Besida* Walker diagram. The bottom line, however, gives a tantalising suggestion at a still higher level, of an eastern link, or an east–west vicariance event. But, like the butterfly genera such as *Elodina* and *Cephrenes* already mentioned, more work is required to validate this pattern before process explanations are called for.

Ackery and Vane-Wright (1984) carried out a cladistic analysis of all milkweed butterflies, a group rich in species endemic to the Malay Archipelago. Apart from *Idea* (new analysis commented on above), only two other species-groups, both belonging to the genus *Parantica* Moore, shed any light on Sulawesi biogeography. The four endemics of the *P. kuekenthali*-group, from northern and central Sulawesi, form a larger monophyletic group with *P. albata* Zinken (Sumatra, Java and Bali), *P. timorica* Grose-Smith (Timor), *P. tityoides* Hagen (Sumatra) and *P. pseudomelaneus* Moore (Java). The Lompobatang endemic (*P. sulewattan* Fruhstorfer: extreme south-western Sulawesi) forms a separate monophyletic group with *P. philo* Grose-Smith (Sumbawa) and *P. wegneri* Nieuwenhuis (Flores). Such patterns are consistent with Michaux's (1991) view that western-central Sulawesi may be related geologically to southern

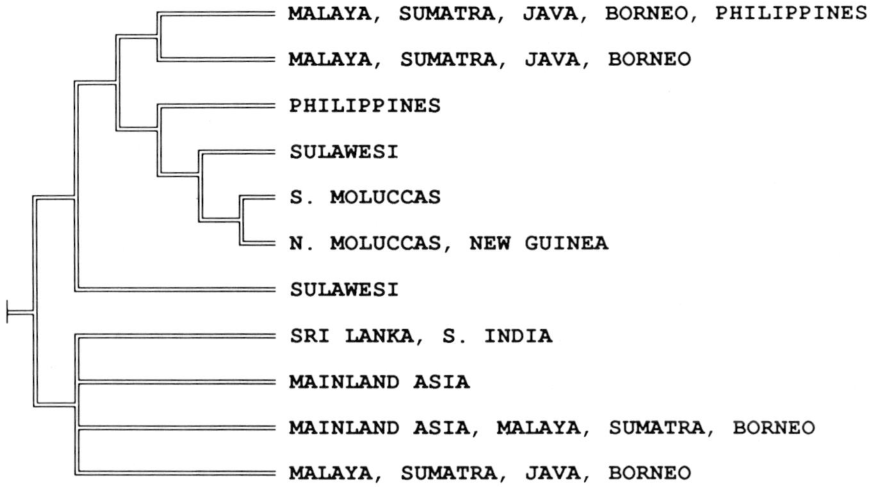

Fig. 3. Area cladogram redrawn from Holloway (1991), based on cladogram for *Idea* developed by Kitching *et al.* (1987). See text for explanation.

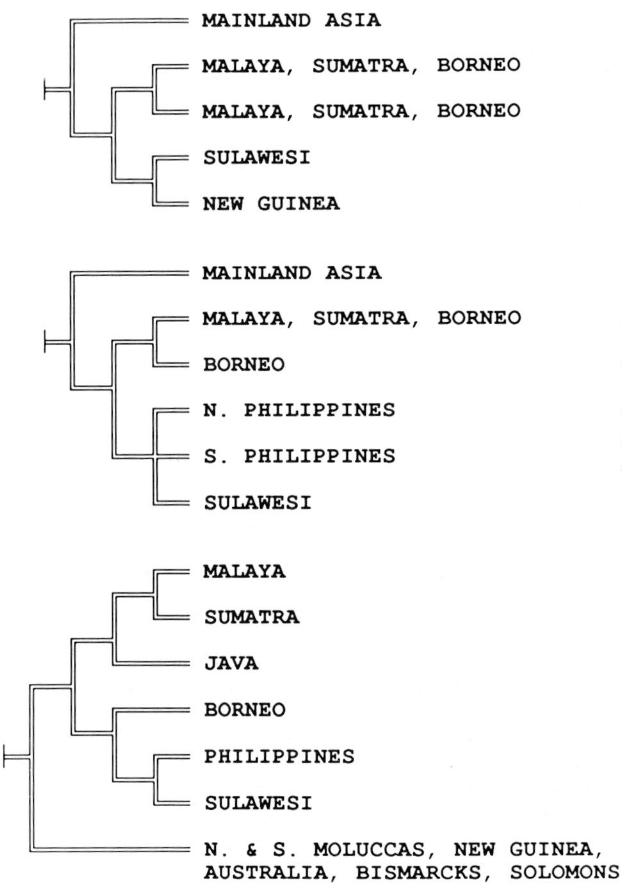

Fig. 4. Area cladograms for three genera of moths, redrawn and based on original cladistic analyses of Holloway (1991). Top: *Narosa*; centre: *Erconholda*; bottom: *Besida*. See text for explanation.

Sumatra, Java and Bali, whereas the south-western tip of Sulawesi may have had a separate southern origin (presumably from within the areas now collectively designated as the Lesser Sundas).

Discussion

The Ages of Butterflies

In relation to earth history, reliable knowledge of the ages of butterfly taxa would be of major significance for interpretation of distribution patterns. In the present context, two particular dates are of special significance: the time of fusion of the Australian and Asian plates, and the time(s) of separation of western Sulawesi from the Asian plate. According to Audley-Charles (1981, 1987), the former is dated at about 15 mya, while Katili (1978) placed the (most recent) separation of western Sulawesi from Asia (Borneo) as little as 2 mya.

Unfortunately, we have scant knowledge of the age of the Lepidoptera. According to Whalley (1986), the oldest *confirmed* lepidopterous fossils are Lower Jurassic (about 180 mya) while modern families (including major lineages of the butterflies) were differentiated by about 65–50 mya (Palaeocene Eocene). In my experience, most entomologists confronted with insect fossils of Oligocene age (*c.* 30 my) or younger consider them to fit comfortably into modern genera (sometimes even species). On this slender basis I consider it likely that most (if not all) modern butterfly genera would have been established by the time the east Sulawesi fusion event took place, and that many present day species lineages would also have been established.

Do Distinct Australian or Papuan Faunal Elements Exist in the Butterflies?

I have already argued above, in considering the significance of the 27 species-level links between Sulawesi and the Moluccas, that there is no convincing Papuan or Australian element in the Sulawesi butterfly fauna. Clearly, if it were the case that no butterfly genus or other higher group is characteristic of these eastern regions, then their 'absence' from Sulawesi would be trivial.

Over 380 butterfly species have been recorded from the continent of Australia, of which 46% are endemic, and a further 29% do not occur west of the Moluccas (Common and Waterhouse 1981). One of the most characteristic elements is the Hypocistini (Nymphalidae: Satyrinae), with many genera and species in Australia and the Papuan region, extending west to Waigeo and Misoöl. Another such element is the Trapezitinae (Hesperiidae), a group of more than 60 species mostly found in Australia, but with a few occurring in New Guinea, Bismarck Archipelago, Aru and Kai islands. At a lower taxonomic level, a number of major genera, such as the lycaenid *Philiris* Röber, are characteristic of the Papuan region, extending into northern Australia and the Moluccas but not Sulawesi. One of the most spectacular of such groups are the true birdwings, *Troides* subgenus *Ornithoptera* Boisduval. Smiles (1982), in a cladistic analysis of the nymphalid genus *Polyura* Billberg, demonstrated strong divisions of Australian region and Asian species; the *Polyura* found in the Moluccas belong to an Australian group, whereas the Sulawesi representatives belong to the single Asian group.

From this it can be appreciated that there are several butterfly groups, including species-groups, genera and higher taxa, defining Australia and the Australian region as an area of endemism. There is no convincing example of any such butterfly group extending west to include Sulawesi, and Sulawesi only.

Butterflies, Bats and Bugs

Butterflies, notably species of the family Papilionidae, exhibit a wide range of faunal connections within Wallacea. However, within this complex situation there are several dominant distribution patterns, as described above. Do these dominant patterns have any generality? Analyses of Sulawesi bat species and aquatic bug genera suggest they may have.

Detailed information on the known distribution of 59 species of bats native to Sulawesi is given by Musser (1987). His data do not permit the separation of Palawan from Borneo (1+7) or the northern and southern Moluccas (3+4). Apart from this limitation, the bat data can be analysed in the same way as the 470 native Sulawesi butterfly species. The 59 bats fall into 23 groups, of which the endemic and widespread categories account for 16 species. This leaves 43 species distributed amongst 21 potentially informative groups, an average of 2.05 species per pattern. The two strongest components, (1+2+6+7) and (3+4), are supported by six and eight species respectively.

Musser (1987) was of the view that, with the exception of the phalangers, 'the native and endemic mammal fauna . . . [of Sulawesi] . . . had its origin on mainland Asia'. *Phalanger* Storr is a marsupial, and so the three Sulawesi Region species must be taken as a Papuan link (Musser notes that the phalangers, in comparison with other marsupials, are exceptional colonisers of islands in the Papuan region). Thus I interpret the two main patterns evident in the Sulawesi bat species data as directly comparable to the butterflies: an apparently older Asian pattern, linking Sulawesi to the Philippines and the Greater Sundas (and thus to mainland Asia), and a younger pattern connecting Sulawesi species with the Moluccas (and thus to the Papuan region).

Musser (1987: 83) also states that the Sulawesi mammal fauna is 'unlike that of the Philippines . . . and I find no close relationship between the faunas of Sulawesi and the Lesser Sunda Islands'. Musser's whole analysis is based on a total of 122 mammal species native to Sulawesi. Conceivably, the weaker, low-ranking links to the Philippines and perhaps to the Lesser Sundas evident in the butterfly data are only apparent because there are four times as many butterflies as mammals found on Sulawesi. In fact, there is one species in the bat data (*Pteropus griseus* Geoffroy) apparently restricted to Sulawesi plus the Lesser Sundas, and at least three species restricted to Sulawesi plus the Philippines, suggesting good concordance between the bat and butterfly data.

Polhemus and Polhemus (1990) have analysed the water bug genera (Heteroptera) of Sulawesi. Of the 42 genera currently known from the island, all but one can be found in Borneo and/or the Philippines, with most occurring widely in South-East Asia, including Sundaland, and many not occurring in New Guinea or Australia. The one exception is an undescribed marine form, known only from Sulawesi and the northern Moluccas. These patterns in the bug genera closely parallel the patterns seen amongst the butterflies. Polhemus and Polhemus comment that 'the present aquatic Heteroptera fauna of Celebes contains very few typically Papuan elements'.

Geology and Areas of Endemism

The comparison among Sulawesi bats, bugs and butterflies is reassuring, insofar as there is evidence of similar dominant patterns for the fauna. It is also reassuring, or challenging, to check if areas of endemism used in the analysis correspond to current geological theories. The paper by Michaux (1991) offers a number of valuable insights.

Area 1. *Palawan*. Considered by Michaux to be a separate micro-continent geologically linked to northern Borneo (Sarawak). Cracraft (1990) does not find evidence from birds to link Palawan specifically to Borneo, and the same appears to be true for butterflies (Vane-Wright 1990); the biogeographic link appears to be with the whole of Sundaland or an even wider area including Indo-China, Burma and the Lesser Sundas (Holloway and Jardine 1968). In this context it may be significant to note that only one Sulawesi butterfly species is restricted, in the surrounding regions, to Palawan plus Borneo.

Area 2. *Mindanao Region*. A complex history for the Philippines is likely, but Michaux points out an implication of the work of McCabe and Cole (1989) to suggest that Mindanao, parts of south and east Borneo, and western Sulawesi 'form a single tectonic unit'. This would also appear to include much of Java and part of Sumatra (Michaux 1991: fig. 2). Altogether, this is broadly consistent with the strong (1+2+6+7) pattern evident in the present analysis. Biogeographic arguments for recognising the Mindanao region as an area of endemism in its own right are summarised by Vane-Wright (1990).

Area 3. *Northern Moluccas*. Michaux (1991) separates this region from Obi and the southern Moluccas, and relates it to a disrupted archipelago to the north of and including part of northern New Guinea. Audley-Charles (1987) tentatively accepted the affinity of the northern Moluccas to New Guinea, but also included Obi as part of the group. Biogeographically, the position of Obi appears ambiguous, with links to both the north and south. In this context it is very interesting to note the distinctive milkweed butterfly *Euploea morosa* Butler, which is endemic to the northern Moluccas (including Obi), Misoöl and Biak, but is not found on the New Guinea mainland (Ackery and Vane-Wright 1984).

Area 4. *Southern Moluccas*. According to Michaux, part of the southern Moluccas, including Obi and Buru, is linked geologically to the Sula, Banggai and Butung island groups. The major island of Seram he places with parts of the eastern Lesser Sundas chain, including Kai, Tanimbar and part of Timor, and links them to Australia. Thus area 4 in the present biogeographic analysis would appear to be a geological hybrid. Furthermore, I have excluded the Sula-Banggai and Butung groups by including them within the Sulawesi Region. Refined biogeographical analyses may shed more light on this complex problem. However, previous work (see Ackery and Vane-Wright 1984) provides some evidence to include Obi with the northern Moluccas (see above), and little evidence to link Seram with Australia rather than New Guinea.

Region 5. *Lesser Sunda Islands.* Michaux groups together the western Lesser Sunda Islands, from Lombok east to part of Timor, apparently by default. The biogeographical arguments for regarding this region as a single area of endemism are not strong. It may be notable that the present analysis reveals only one genus and three species shared between Sulawesi and the Lesser Sunda Islands, to the exclusion of all other regions under consideration.

Region 6. *Java.* Michaux includes Bali with Java, and links this region to Kalimantan, western Sulawesi and south-western Sumatra. This would appear to fit with the recognition of Java plus Bali as an area of endemism, and the existence of a definite pattern linking Sulawesi with Java and Borneo to the exclusion the Philippines proper (6+7: 10 species; 1+6+7: 4 genera, 9 species).

Area 7. *Borneo.* According to Michaux's summary, Borneo is composite, with separate geological links to Palawan plus China, western Sulawesi plus Java and Sumatra, and Indo-China plus the Malay Peninsula. Future biogeographical work on the butterflies of Borneo may be able to detect these complex origins. For the purpose of the present analysis, there are so few unique links between Borneo and Sulawesi that treating Borneo as a single area does not affect the results.

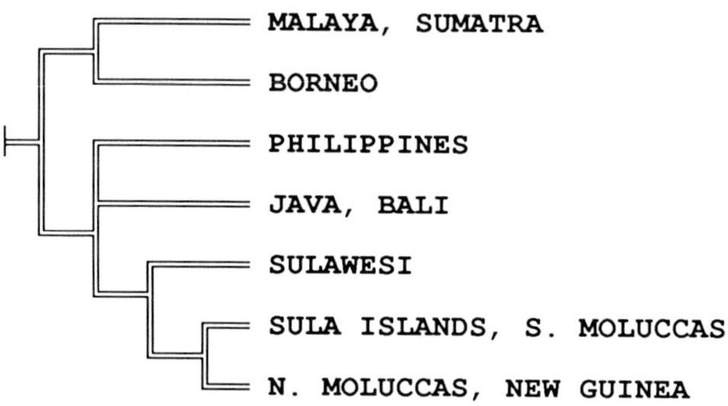

Fig. 5. Area cladogram (redrawn from Humphries 1990) produced by application of consensus tree methods to cladograms for *Idea* (Kitching *et al* 1987) and *Narosa* (Holloway 1987). See text for explanation.

Geological Models and Biogeographical Lines

In Table 1, a number of predictions were made. The present analysis points to an overwhelming Asian linkage for older patterns, and indicates Asian, Papuan (Moluccan), Philippine and Lesser Sundanian links amongst the younger, species-level patterns. None of the eight sets of predictions listed in Table 1 is fully consistent with such an outcome. However, the Michaux system, in which Sulawesi comprises three terranes and was involved in collision with the Australian plate about 15 mya, would fit well if the Australian terrane did not bring any terrestrial biota directly to the Sulawesi Region. This might be so if the plate margin was submerged throughout most, or even all of the approach phase, and then emerged during the collision as virgin land bridging Sulawesi to the Moluccas and New Guinea, or providing 'stepping stones' between these major islands.

Humphries (1990) has analysed cladograms for two genera of Lepidoptera already discussed above, *Idea* (from Kitching *et al.* 1987; Fig. 3) and *Narosa* (from Holloway 1987; Fig. 4), using principles of cladistic biogeography (Humphries and Parenti 1986; Humphries *et al.* 1988). He found that component analysis and Nelson and Majority Rule consensus trees gave an identical area cladogram (Fig. 5). This pattern is consistent with major features of the present analysis of all butterfly genera and species. The younger, more recent patterns linking Sulawesi to the Moluccas and the Papuan region are immediately apparent, but only as an integral part of the older pattern linking Sulawesi to the Philippines, Greater Sunda Islands and Asia. The older pattern ('1+2+6+7') appears to be a biogeographical equivalent of a paraphyletic group. To accept a fixed boundary, such as the Wallace line, between Sulawesi and any of the islands to the west disrupts this pattern by failing to recognise that an asymmetric hierarchy of area relationships is involved.

Conclusions

The original question, raised in the title of this paper, is do the western edges of the Australian Plate and the Australian Region coincide? With respect to any supposed western 'boundary line' for the Australian Region, it is concluded that the original terrestrial biota of Sulawesi was Asian, and that collision with the Australian Plate 15 mya brought few (if any) Australian Region butterflies directly to Sulawesi. This would readily be understandable if the terranes at the Australian Plate margin were submerged throughout most of their approach (Holloway 1987; Kitching *et al.* 1987; see also above).

If a submerged eastern terrane has been uplifted during the collision, then dry land formed from the Australian Plate reaches further westwards than any original indigenous Australian Region terrestrial biota. A significant increase in the land area of Sulawesi over the last 15 my could offer an explanation for the island's relatively low number of species. This, together with a long independent history for western Sulawesi as an offshore Asian island separate not only from Borneo but also from the rest of the Greater Sundas and the Philippines, could also explain the high level of species endemism. A number of groups, most notably the entire nymphalid genus *Lohora*, appear to have undergone minor radiations within Sulawesi (a phenomenon analysed for the Lepidoptera generally by Holloway 1991). The paucity of unique links with Borneo is reflected in the outstanding Makassar Strait discontinuity of the 1863 Wallace Line (Holloway and Jardine 1968; Duffels and de Boer 1990). Finally, dispersal by way of the relatively new lands of Wallacea, notably within Sulawesi itself, and between Sulawesi, the Banggai/Sula archipelago and the Moluccas, may be responsible for the observed lack of a sharp division within Wallacea of Asian and Austral fauna and flora. On the basis of the geological models, a sharp distinction might have been expected (cf. Hutchinson 1989).

The butterflies, both ancient and modern, can but do not necessarily transcend the Wallace lines. Thus the western edges of the Australian Plate and the Australian Region do not, and probably never did, coincide. Indeed, we must reject the notion of a distinct Australian region, at least with respect to 'lines' or 'boundaries' running through Wallacea, while not doubting the reality of distinct Australian biota.

Acknowledgments

My thanks are due to LIPI (Indonesian Institute of Sciences), the Trustees of the British Museum (Natural History), the Royal Entomological Society of London, and Drs W. J. Knight and L. A. Mound, for enabling me to visit Sulawesi during 1985, as a member of Project Wallace (Results of Project Wallace No. 126). My thanks are also due to many colleagues and friends who have helped collate and interpret data on the butterflies of Sulawesi, including Phillip Ackery, Anthony Bedford Russell, Alan Cassidy, Hideyuki Chiba, Rienk de Jong, Wolfgang Dierl, Hans Duffels, John Eliot, Tomoo Fujioka, Jane Goode, Willem Hogenes, Jeremy Holloway, Chris Humphries, Bob Smiles, Campbell Smith, Trig Treadaway and Jason Weintraub. I particularly wish to acknowledge two colleagues for their forbearance: Laurence Mound, who has waited so long to see material results from my obsession with Sulawesi and its butterflies, and Chris Humphries, for indulgence of my idiosyncratic approach to biogeography. Colin Patterson drew my attention to Willis' 'age and area' hypothesis. Dr Holloway kindly allowed me to draw freely on his unpublished paper on moth speciation patterns in the Malay Archipelago. The area analysis used to examine the distributions of Sulawesi butterflies emerged during discussions with Dr de Jong, to whom much credit is due. Ian Kitching very kindly read an earlier draft of this paper, on my behalf, at the 1990 Canberra 'Hennig' meeting.

References

Ackery, P. R., and Vane-Wright, R. I. (1984). 'Milkweed Butterflies.' (Cornell University Press: New York.)

Audley-Charles, M. G. (1981). Geological history of the region of Wallace's line. In 'Wallace's Line and Plate Tectonics'. (Ed. T. C. Whitmore.) pp. 24–35. (Clarendon Press: Oxford.)

Audley-Charles, M. G. (1987). Dispersal of Gondwanaland: relevance to evolution of the angiosperms. In 'Biogeographical Evolution of the Malay Archipelago'. (Ed. T. C. Whitmore.) pp. 5–25. (Clarendon Press: Oxford.)

van Balgooy, M. M. J. (1987). A plant geographical analysis of Sulawesi. In 'Biogeographical Evolution of the Malay Archipelago'. (Ed. T. C. Whitmore.) pp. 94–102. (Clarendon Press: Oxford.)

Common, I. F. B., and Waterhouse, D. F. (1981). 'Butterflies of Australia.' (Revised edn.) (Angus & Robertson: London & Sydney.)

Corbet, A. S., and Pendlebury, H. M. (1978). 'The Butterflies of the Malay Peninsula.' 3rd edn. (Ed. J. N. Eliot.) (Malayan Nature Society: Kuala Lumpur.)

Cracraft, J. (1990). From Malaysia to New Guinea: evolutionary biogeography within a complex continent-island arc contact zone. Proceedings 19th International Ornithological Congress, 1988. Vol. 2, 2581–93.

Croizat, L. (1952). 'Manual of Phytogeography.' (Junk: The Hague.)

Dickerson, R. E. (1928). Distribution of life in the Philippines. *Monographs. Philippine Bureau of Science* (Manila) **21**, 322 pp.

Duffels, J. P., and de Boer, A. J. (1990). Areas of endemism and composite areas in East Malesia. In 'The Plant Diversity of Malesia'. (Eds P. Baas *et al.*) pp. 249–72. (Kluwer Academic Publishers: Amsterdam.)

George, W. (1981). Wallace and his line. In 'Wallace's Line and Plate Tectonics'. (Ed. T. C. Whitmore.) pp. 3–8. (Clarendon Press: Oxford.)

Hamilton, W. (1979). Tectonics of the Indonesian region. *United States Geological Survey Professional Paper* **1078,** ix + 345 pp.

Holloway, J. D. (1973). The taxonomy of four groups of butterflies (Lepidoptera) in relation to general patterns of butterfly distribution in the Indo-Australian area. *Transactions of the Royal Entomological Society London* **125,** 125–76.

Holloway, J. D. (1987). Lepidoptera patterns involving Sulawesi: what do they indicate of past geography? In 'Biogeographical Evolution of the Malay Archipelago'. (Ed. T. C. Whitmore.) pp. 103–18. (Clarendon Press: Oxford.)

Holloway, J. D. (1991). Patterns of moth speciation in the Indo-Australian archipelago. In 'The Unity of Evolutionary Biology'. (Ed. E. C. Dudley.) (Proceedings of the IV International Congress on Systematics and Evolutionary Biology.) (Dioscorides Press: Portland, Oregon.)

Holloway, J. D., and Jardine, N. (1968). Two approaches to zoogeography: a study based on the distributions of butterflies, birds and bats in the Indo-Australian area. *Proceedings of the Linnean Society of London* **179,** 153–88.

Humphries, C. J. (1990). The importance of Wallacea to biogeographical thinking. In 'Insects and the Rain Forests of South East Asia (Wallacea)'. (Eds W. J. Knight and J. D. Holloway.) pp. 7–18. (Royal Entomological Society of London: London.)

Humphries, C. J., and Parenti, L. R. (1986). 'Cladistic Biogeography.' (Clarendon Press: Oxford.)

Humphries, C. J., Ladiges, P. Y., Roos, M., and Zandee, M. (1988). Cladistic biogeography. In 'Analytical Biogeography'. (Eds A. A. Myers and P. S. Giller.) pp. 371–404. (Chapman & Hall: London.)

Hutchison, C. S. (1989). 'Geological Evolution of South-east Asia.' (Clarendon Press: Oxford.)

Katili, J. A. (1978). Past and present geotectonic position of Sulawesi, Indonesia. *Tectonophysics* **26,** 289–322.

Kitching, K. J., Vane-Wright, R. I., and Ackery, P. R. (1987). The cladistics of Ideas. *Cladistics* **3,** 14–34.

MacArthur, R. H., and Wilson, E. O. (1967). 'The Theory of Island Biogeography.' (Princeton University Press: New Jersey.)

McCabe, R., and Cole, J. (1989). Speculations on the late Mesozoic and Cenozoic evolution of the Southeast Asia margin. In 'The Evolution of the Pacific Ocean Margins'. (Ed. Z. Ben-Avraham.) pp. 143–60. (Clarendon Press: Oxford.)

Michaux, B. (1990). Generalized tracks and geology. *Systematic Zoology* **34,** 390–8.

Michaux, B. (1991). Distributional patterns and tectonic development in Indonesia: Wallace reinterpreted. *Australian Systematic Botany* **4,** 25–36.

Musser, G. G. (1987). The mammals of Sulawesi. In 'Biogeographical Evolution of the Malay Archipelago'. (Ed. T. C. Whitmore.) pp. 73–93. (Clarendon Press: Oxford.)

Pan, T-S., and Morishita, K. (1990). Notes on *Nacduba normani* Eliot in Sabah, north Borneo (Lepidoptera, Lycaenidae). *Tyô to Ga* **41,** 149–54.

Polhemus, J. T., and Polhemus, D. A. (1990). Zoogeography of the aquatic Heteroptera of Celebes: regional relationships versus insular endemism. In 'Insects and the Rain Forests of South East Asia (Wallacea)'. (Eds W. J. Knight and J. D. Holloway.) pp. 73–88. (Royal Entomological Society of London: London.)

Rosen, B. R. (1988). From fossils to earth history: applied historical biogeography. In 'Analytical Biogeography'. (Eds A. A. Myers and P. S. Giller.) pp. 437–81. (Chapman & Hall: London.)

Simpson, G. G. (1977). Too many lines: the limits of the Oriental and Australian zoogeographic regions. *Proceedings of the American Philisophical Society* **121,** 107–20.

Smiles, R. L. (1982). The taxonomy and phylogeny of the genus *Polyura* Billberg (Lepidoptera: Nymphalidae). *Bulletin of the British Museum (Natural History)*, Entomology **44,** 115–237.

Vane-Wright, R. I. (1990). The Philippines — key to the biogeography of Wallacea? In 'Insects and the Rain Forests of South East Asia (Wallacea)'. (Eds W. J. Knight and J. D. Holloway.) pp. 19–34. (Royal Entomological Society of London: London.)

Vane-Wright, R. I., de Jong, R., Ackery, P. R., Cassidy, A. C., Eliot, J. N., Goode, J. H., Smiles, R. S., and Smith, C. R. (in press) 'The Butterflies of Sulawesi: Endemism and Biogeography of a Critical Island Fauna.' (The Natural History Museum: London.)

Wallace, A. R. (1867). On the Pieridae of the Indian and Australian regions. *Transactions of the Entomological Society of London* (3) **4,** 301–415.

Whalley, P. E. S. (1986). A review of the current fossil evidence of Lepidoptera in the Mesozoic. *Biological Journal of the Linnean Society* **28,** 253–71.

Willis, J. C. (1922). 'Age and Area: a Study in Geographical Distribution and Origin of Species.' (with additional chapters by de Vries, H., Guppy, H. B., Reid, E. M., and Small, J.) (Cambridge University Press: Cambridge.)

Yata, O. (1988). A revision of the Old World species of the genus *Eurema* Hübner (Lepidoptera, Pieridae). Ph.D. Thesis, Kyushu University, Fukuoka, Japan.

Yata, O. (1990). Cladistic biogeography of *Eurema* butterflies belonging to the subgenus *Terias* (Lepidoptera: Pieridae), with particular reference to Wallacea. In 'Insects and the Rain Forests of South East Asia (Wallacea)'. (Eds W. J. Knight and J. D. Holloway.) pp. 43–8. (Royal Entomological Society of London: London.)

Manuscript received 9 January 1991, accepted 12 March 1991

3 September 1991 | *Aust. Syst. Bot.*, **4**, 199–210

Indo-Pacific Coral Biogeography: a Case Study from the *Acropora selago* Group

Carden C. Wallace[A], *John M. Pandolfi*[B], *Amanda Young*[A] *and Jacqueline Wolstenholme*[A]

[A] Museum of Tropical Queensland, 70–84 Flinders Street, Townsville, Qld 4810, Australia.
[B] Australian Institute of Marine Science, PMB No. 3, Townsville MC, Qld 4810, Australia.

Abstract

We develop species-level biogeographic hypotheses for *Acropora*, the largest extant coral genus and the dominant scleractinian coral of Indo-Pacific reefs, based on morphometric and phylogenetic analyses of the *Acropora selago* group. Fourteen morphometric characters differentiated species from this group with an accuracy of 95%. When the Tukey test was administered, 11 of these characters displayed non-overlapping subsets. The most resolved phylogenetic tree resulted from an analysis based on both morphometric and qualitative characters. Cladistic–biogeographic analysis using this tree and areas derived from species-distribution patterns showed that species with the greatest degree of endemism within the *A. selago* group possess the most derived character states, while the most primitive species (*A. yongei*) is the most widespread.

Within the range of the group, four areas are recognised: (1) the Red Sea, (2) western to central Indian Ocean, (3) eastern Indian Ocean and (4) western to central Pacific Ocean. Species ranges overlap in a stepwise fashion from west to east. Areas adjacent to one another are biogeographically more closely related than non-adjacent areas. Whilst we offer a scenario for the history of distribution patterns of the *A. selago* group, we propose that biogeographical hypotheses based on *Acropora* be tested using a number of different species-groups. Similar distributional ranges for other *Acropora* species-groups, as well as separate distributional ranges for further groups, suggest that *Acropora* will provide an appropriate taxon to examine the biogeography of the tropical marine realm.

Introduction

Reef building scleractinian corals are distributed throughout the tropical and subtropical regions of the world. Distribution patterns are usually expressed in terms of numbers of genera or subgeneric groups present in different regions. The most obvious feature of these patterns is a radial decrease in numbers from a high in the Indo-West Pacific, with smaller centres in the West Indian Ocean and Caribbean (e.g. see Stehli and Wells 1971). Rosen (1988) summarises 13 hypotheses regarding the origins of coral distribution patterns. We note among these the papers of McCoy and Heck (1976, 1983), who attempt to break the run of untestable hypotheses about coral distributions. Hoeksema (1989) reconsiders these hypotheses with species of Fungiidae and concludes that patterns may be explained by a combination of dispersal, vicariance and refugiality. Common to these works is a conclusion that biogeographic analysis will be better facilitated when more species-level data are available. Towards this end,

1030-1887/91/010199$05.00

we are embarking upon a species-level study of the phylogeny and distribution of the coral genus *Acropora*. This genus is distinguished from all other coral genera by its possession of an axial corallite. Because this feature is shared by all of its species, and by no other scleractinian genera, we consider *Acropora* to be monophyletic. *Acropora* is the largest extant coral genus (Wells 1987), occurs in all three oceans (Fig. 1) and is the most obvious component of most Indo-Pacific reefs. Despite this, it has often been omitted from taxonomic revisions, or summarised in non-taxonomic terms in distributional studies. It has always presented problems of identification, because of bewildering within- and between-species variability combined with absence of some skeletal elements used traditionally in coral taxonomy (Wells 1956; Wallace and Dallwitz 1981).

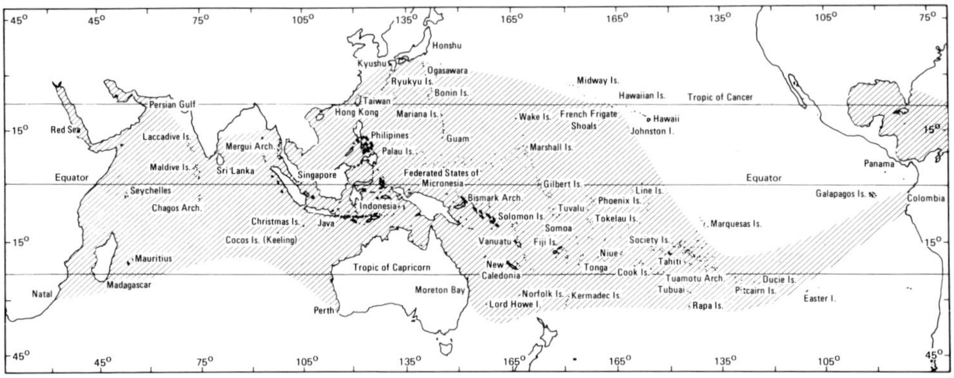

Fig. 1. Distribution of *Acropora* worldwide.

Several factors have contributed to making *Acropora* more accessible to phylogenetic and biogeographic study in recent years. Firstly, the genus has been revised twice (Wallace 1978; Veron and Wallace 1984), species limits being interpreted by observations of variability in the field as well as by microscopic characters. Although Veron and Wallace (1984) presented no phylogenetic analysis, they identified 13 species-groups in which species were thought to be more closely related than species from different groups. Secondly, the mode of reproduction and a mass-spawning phenomenon have been discovered and investigated (e.g. Bothwell 1981; Harrison *et al.* 1984; Wallace 1985; Harrison and Wallace 1990), providing further sources of phylogenetic characters. In particular the sperm has provided ultrastructural characters and nucleotide sequences that vary between species (Harrison 1985, 1988; McMillan and Miller 1988, 1989, 1990).

We report here an investigation of phylogenetic relationships and biogeography of the *Acropora selago* group of Veron and Wallace (1984). The phylogenetic analysis was based on a series of morphometric characters measured from field populations, providing morphological characters independent from those used by Veron and Wallace (1984) to separate *Acropora* species. The morphometric characters, along with some qualitative characters of the *A. selago* group, are used to present the phylogeny.

The objective of the biogeographic study was to develop hypotheses about the relationships amongst areas within the known distribution of the *Acropora selago* group. Our future aim is to test the biogeographic hypotheses generated from this study with a data set drawn from a larger set of *Acropora* species groups.

Methods

Samples comprised 10 colonies each of five species of the *Acropora selago* group (*A. selago*, *A. tenuis*, *A. yongei*, *A donei* and *A. dendrum*) and an outgroup species, *A. millepora*. These colonies were collected from a single reef (Britomart Reef on the Great Barrier Reef) to minimise interpopulation variability. For the two rarest species, *A. donei* and *A. dendrum*, some specimens were obtained from nearby reefs. We did not consider this to present a major problem, since populations of *Acropora* are thought to recruit predominantly externally: fertilisation is external and a precompetency time of 4–7 days is the norm (Harrison and Wallace 1990). Thus, these coral populations may be continuous over scales larger than a single reef.

A further species, *A. striata*, is regarded by CCW as belonging to the *A. selago* group. As this species does not occur on the Great Barrier Reef, it was not included in the morphometric and phylogenetic analyses, but its distribution was plotted for comparison with the other *A. selago* group-species.

In the process of examining specimens for this study, a taxonomic decision had to be made for *Acropora yongei*. Wallace (1978) included this species in *A. haimei*, an apparently broadly distributed species. This decision was revised by Veron and Wallace (1984) who described *A. yongei* as a new species with a known distribution of only the Great Barrier Reef and the Philippines. Subsequent examination of specimens collected by Y. Loya in the Red Sea, as well as Sheppard's interpretation (1987) indicate that they are probably the same species throughout a range extending from the Red Sea and W. Indian Ocean to Johnson Atoll in the Central Pacific. Thus, the two species are herein combined as *A. yongei*, a name which may prove to be a junior synonym of *A. haimei*.

Morphometric Analyses

Fourteen morphological characters were measured to determine whether species that were previously described on the basis of qualitative variation could be differentiated by quantitative morphometrics. Nine describe the branching of the coral ($n = 20$ per colony). These were length of the first order branch (*sensu* Cheetham *et al.* 1980), length of the second order branch, top, middle and bottom diameter of the first-order branch, middle diameter of the second-order branch, the number of radial corallites per mm of first-order branch, the angle the first-order branch makes with its parent branch, and the average distance from the axial corallite to its five nearest neighbours. Five characters were measured from the axial corallites ($n = 40$ per colony). These were the inside and outside diameters of the axial corallites, the length of axial septa, the axial corallite wall thickness and the exertness of the axial corallite. In the analyses presented here, only colony means for each character were used. Thus in analyses of all species, $n = 60$, and in analyses performed only on the *A. selago* group, $n = 50$. Data on the radial corallites are currently being gathered but are not included in the present study.

Two discriminant function analyses (DFA) were performed on two data sets, 'ingroup plus outgroup' and 'ingroup only'. The first analysis, a canonical DFA, was used to investigate species separation and gave Mahalanobis distances (Mahalanobis 1930) between species. Group centroids representing the class means on each canonical variable were plotted for each species for the first two canonical variates.

Because the species seemed to be well separated in the canonical DFA, a second 'classification' DFA was done on the first data set to determine whether the suite of morphometric characters selected for measurement was useful in classifying (in the sense of diagnosing) species of the *A. selago* group and *A. millepora*. Thus we were interested in how our pre-assigned species were classified on the basis of the new morphometric characters. The morphometric analyses were conducted using the SAS software package (SAS 1989).

Phylogenetic Analyses

To investigate whether the measured quantitative characters were sufficiently different among taxa to be useful as phylogenetic characters, we conducted a multivariate analysis of variance (MANOVA) and a Tukey multiple comparisons test on the entire 60 species' means. Quantitative characters were coded using a modification of the Tukey test as suggested in Farris (1990). Each character was weighted equally, regardless of range (ie. number of character states) (Farris 1990). As an outgroup we used *A. millepora*, which occurs in one of the two species-groups placed by Veron and Wallace (1984) closest to the *A. selago* group.

We present phylogenetic results from a data matrix of quantitative character states obtained from the Tukey multiple comparison procedure, and qualitative characters obtained from Veron and Wallace (1984). The phylogenetic analyses used Hennig86 (Farris 1988). Three analyses were made, one using the quantitative data only (characters 0 through 10, Appendix 1), one using the qualitative data only (characters 11 through 21, Appendix 1), and one using a combined data set. The ie option in Hennig86 was used. In the case of the quantitative characters, the nelsen option was used to build a consensus tree. Multistate characters (characters 17, 20 and 21, see Appendix 1) were left unordered (non-additive).

Biogeographic Analyses

Distribution patterns of the *A. selago* group and of *A. millepora* were compiled by interpreting presence/absence data in distributional papers (Hoffmeister 1925; Wells 1954; Rosen 1971, 1979; Jones *et al.* 1972; Pillai *et al.* 1973; Pillai and Scheer 1976; Pillai 1977; Faure 1982; Randall 1983; Scheer and Pillai 1983; Hamilton and Brakel 1984; Sheppard 1985, 1987; Sheppard and Sheppard 1985; Latypov 1986; Maragos and Jokiel 1986; Sheppard and Salm 1988; Veron 1988, 1990*a*, 1990*b*; Veron and Marsh 1988; Best *et al.* 1989; Veron and Hodgson 1989), as well as from specimens lodged in the Museum of Tropical

Queensland, unpublished uncollected site records from Lombok, Indonesia, 1989 by CCW, and records in an unpublished manuscript by the late Jan Verwey on corals collected from the Bay of Batavia (now Jakarta). Taxonomic agreement between species cited in the literature and the study species was interpreted in terms of present and past usage and synonymies. When the interpretation was doubtful, the record was registered by a question mark on the distribution map. The compilation represents a summary of records available to date and is part of an ongoing study to which many further records are to be added before distributional maps can be said to be complete.

To derive an hypothesis of biogeographic area relationships we produced an area cladogram based on the distribution patterns and our phylogenetic analysis for the *A. selago* group. This area cladogram was analysed under Assumptions 0 (Zandee and Roos 1987), 1 and 2 (Nelson and Platnick 1981) using COMPONENT (Release 1·5, Page 1989) to obtain reduced area cladograms.

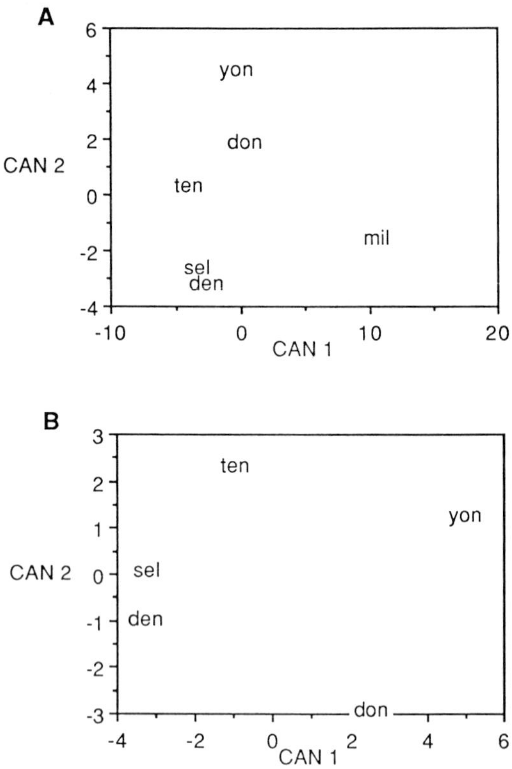

Fig. 2. (*A*) Plot of the group centroids of the discriminant function analysis for all species of the *A. selago* group and for *A. millepora*, the outgroup. *Acropora millepora* plots well away from the other species, especially along canonical variate 1. *Acropora selago* and *A. dendrum* plot very close together. (*B*) Plot for the group centroids for only species of the *A. selago* group. The ingroup species are more separated in the absence of the outgroup. The *A. selago*/*A. dendrum* pair is still apparent, with the other species plotting well away from each other. ten, *A. tenuis*; sel, *A. selago*; yon, *A. yongei*; den, *A. dendrum*; mil, *A. millepora*; don, *A. donei*. (Note variations in scale.)

Results

Morphometrics

In the first canonical DFA in which both the ingroup and the outgroup were included, the first canonical variate explained 67%, the first two canonical variates 88%, and the first three canonical variates 94%, of the total sample variance. In a plot of the group centroids of all species for canonical variates 1 and 2 (Fig. 2*a*), *Acropora millepora*, the outgroup, plots well away from the ingroup along canonical variate 1. *Acropora selago* and *A. dendrum* plot close to one another. Mahalanobis distances were also least between these two species and greatest between *A. millepora* and the ingroup species. When the outgroup *A. millepora* is excluded from the analysis greater separation occurs among the ingroup species (Fig. 2*b*) than occurs when *A. millepora* is included in the analysis (Fig. 2*a*). *Acropora selago* and *A. dendrum* still plot closely, and their Mahalanobis distances are the shortest of any between species. The other three species plot well away from each other and from the *A. selago*/*A. dendrum* pair.

Out of 60 possible misclassifications, the classificatory DFA only misclassified three colonies. Two *A. selago* colonies were misclassified as *A. dendrum* and one *A. dendrum* colony was misclassified as *A. selago*. This represents a correct classification rate of 95%.

Phylogenetic Analysis

The MANOVA on the entire 60 species means showed a significant species effect ($P <$ 0·0001 using Wilk's Criterion) and significant differences in population means ($P <$ 0·0001) for all but one variable among the six species. Only the length of the second-order branch was not signficantly different among species. Thus 13 of the 14 morphometric characters were subsequently analysed to determine character states. When we used the Tukey test on these 13 characters, only six contained wholly non-overlapping homogenous subsets. A further five characters, however, showed little enough overlap to be useful as characters with one, two or three of the species scored as unknown (i.e. missing data).

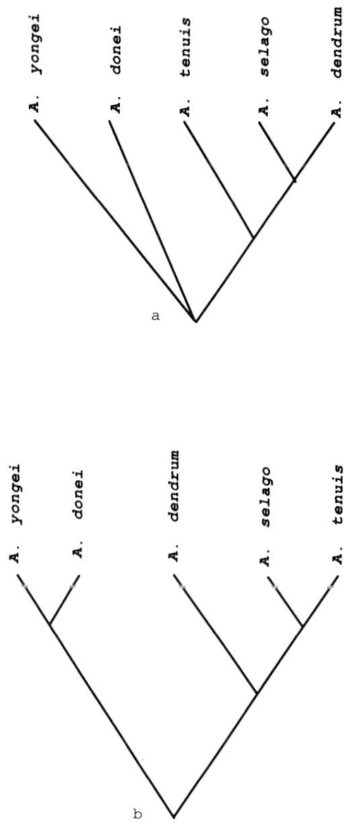

Fig. 3. (*a*) Phylogenetic analysis using only quantitative characters. (*b*) Phylogenetic analysis using only qualitative characters. See Appendix 1 for characters and data matrix.

In the first phylogenetic analysis using only the quantitative data, two trees were obtained with a tree length of 23 and consistency index of 0·91. Fig. 3*a* shows the consensus tree, which contains a trichotomy between *A. yongei*, *A. donei* and a resolved branch comprising *A. tenuis* as a sister-group to the sister-taxa *A. selago* and *A. dendrum*.

In the second analysis using only the qualitative characters, one tree was obtained with a length of 14 and consistency index of 1·00 (Fig. 3*b*). *Acropora dendrum* is here the sister-taxon to the group *A. tenuis* and *A. selago*, and *A. yongei* and *A. donei* are resolved as sister-species.

Fig. 4 shows the tree obtained from the combined analysis. *Acropora tenuis*, *A. selago*, and *A. dendrum* form a clade as in the cladogram based on the analysis using morphometric data (Fig. 3*a*), and the *A. yongei*/*A. donei* branch is resolved as in the analysis using only the qualitative characters (Fig. 3*b*). The tree length for the combined analysis is 38 and the consistency index is 0·92.

Biogeographic Analysis

All taxa of the *Acropora selago* group occur in the western and central Pacific Ocean (Fig. 5). *Acropora dendrum* occurs only in this region, while *A. selago* and *A. donei* occur in this region and the eastern Indian Ocean. *Acropora tenuis* and *A. yongei* occur in these regions and also the central to western Indian Ocean. *Acropora yongei*, the most widespread species, also occurs in the Red Sea.

On the basis of these distribution patterns we recognise four geographic areas (Fig. 6): (*a*) The Red Sea, (*b*) western to central Indian Ocean, (*c*) eastern Indian Ocean and (*d*) western to central Pacific Ocean.

The areas occupied by the species are displayed directly above the names of the terminal taxa in the cladogram in Fig. 4. COMPONENT gave reduced area cladograms under Assumptions 0 (Zandee and Roos 1987), 1 and 2 (Nelson and Platnick 1981) (Fig. 7). Assumption 1 generated 15 trees from the four areas: i.e. all possible trees for four areas were found. Assumption 0 gave just one tree (Fig. 7*a*) whereas Assumption 2 gave two trees (Fig. 7*a*, *b*).

Discussion

The discriminant function analyses show that morphometric characters analysed are useful in differentiating species within the *A. selago* group. Whilst *A. selago* and *A. dendrum* are phenetically similar (Fig. 2), colonies were only misclassified in very low numbers. None of the other four species was misclassified. Thus the qualitative species evaluations previously made (Veron and Wallace 1984) are corroborated by quantitative morphometric analysis and the present morphometric characters appear to represent taxonomic characters in so far as they differentiate among species.

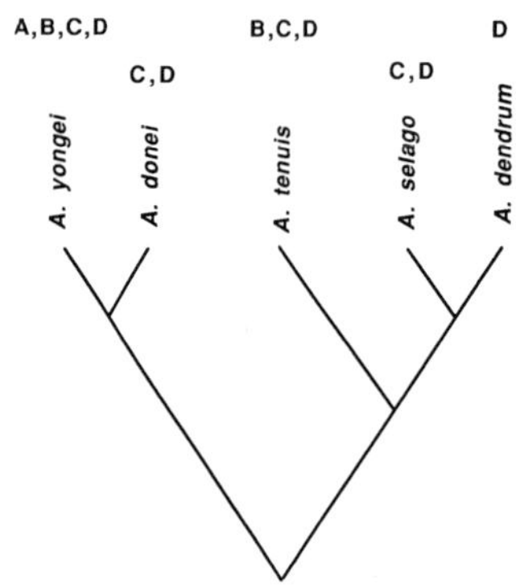

Fig. 4. Phylogenetic tree based on both quantitative and qualitative characters. Biogeographic areas defined in text are shown above terminal taxa. See Appendix 1 for characters and data matrix. A, Red Sea; B, western to central Indian Ocean; C, eastern Indian Ocean; D, western to central Pacific Ocean.

The results of the MANOVA suggest that 13 of the 14 quantitative characters were sufficiently different among species of the *Acropora selago* group to justify using them as characters in a phylogenetic analysis. The Tukey multiple comparisons test, however, resulted in only 11 characters with enough species subsets to be useful in this phylogenetic analysis.

A resolved tree was obtained from a combination of quantitative and qualitative characters. It appears both sets of information are essential for deciphering relationships within this subgroup of *Acropora*. We have yet to add to our quantitative characters another set of morphometric characters, derived from the radial corallite skeleton.

From the relationships between the phylogenetic tree and the distribution of species (from Figs 4 and 5), two patterns are evident: (1) taxa which show the greatest degree of endemism display character states that are the most derived; (2) species ranges overlap in a west to east stepwise fashion with the closest biogeographic relationships occurring between adjacent areas.

Whereas only one area (D, eastern to central Pacific) has an endemic species, the broadest areas are occupied by taxa displaying the least derived character states. *Acropora tenuis* and *A. yongei* have the most cosmopolitan distributions. *Acropora selago* and *A. donei* are more restricted in their distributions than the above taxa, and *A. dendrum* is the most restricted biogeographically. Thus on each branch of the cladogram, taxa possessing more derived character states appear to be more restricted in their distributions. A similar pattern has been found in some other non-*Acropora* coral species (Pandolfi, unpublished data) and is taken as evidence against the centre-of-origin hypothesis proposed in Stehli and Wells (1971).

Assumption 0 gave one reduced area cladogram (Area Cladogram I, Fig. 7*a*), Assumption 1 gave all 15 possible reduced area cladograms for 4 areas, and Assumption 2 gave Area Cladogram I (Fig. 7*a*) and an additional cladogram (Area Cladogram II; Fig. 7*b*).

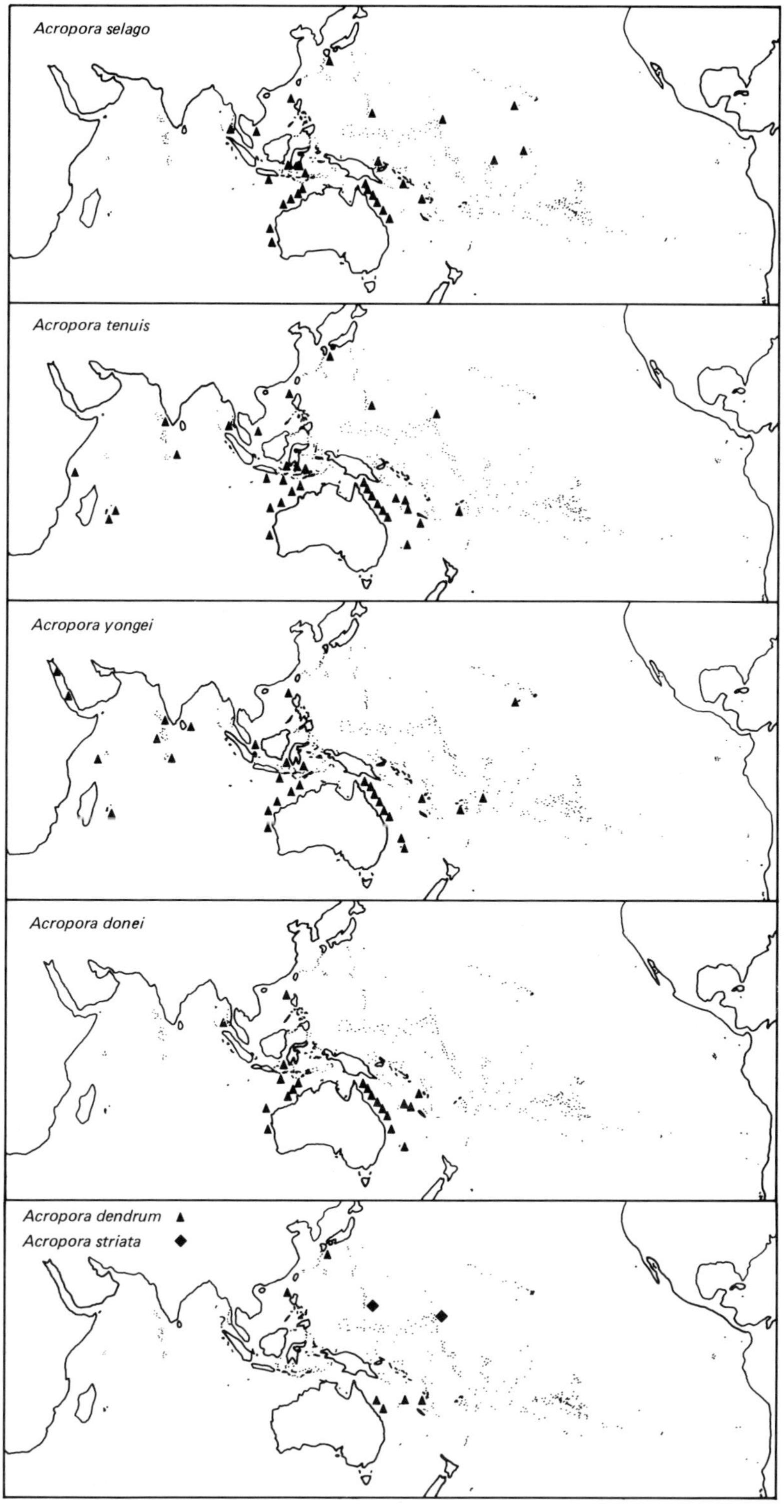

Fig. 5. Distribution of species from the *A. selago* group. (See text for basis of site records.)

To measure the degree of parsimony of the 15 possible area cladograms under Assumption 1, we calculated the number of items of error, the number of extra terms and components required to reconcile the original area cladogram with the reduced area cladogram (Nelson and Platnick 1981) using the FIT option in COMPONENT. We found that Area Cladogram I (Fig. 7*a*) required 6 fewer items of error than Area Cladogram II (Fig. 7*b*). In fact, the former singularly displayed the fewest items of error of any of the 15 possible trees, and is the most acceptable reduced area cladogram.

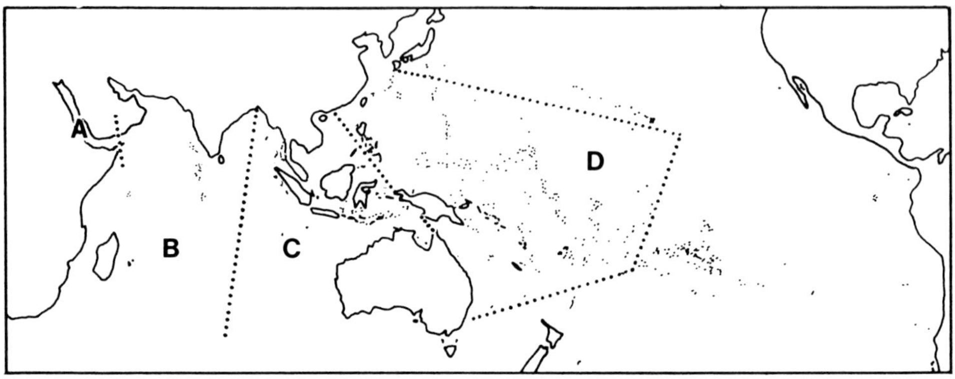

Fig. 6. Biogeographic areas recognised within the distribution patterns of the *Acropora selago* group.

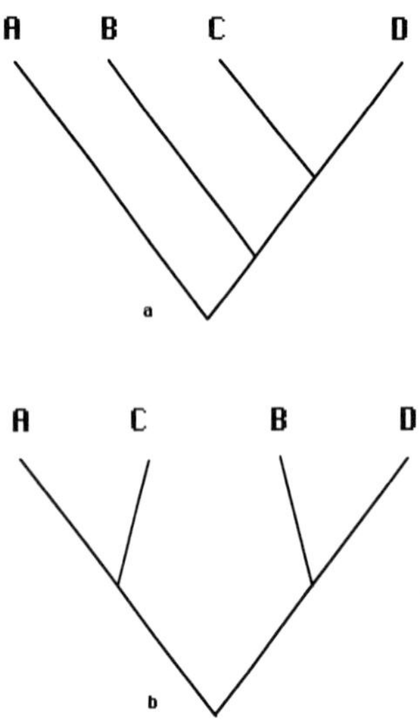

Fig. 7. Reduced four-area statement of relationships of areas A, B, C and D given the phylogeny presented in Fig. 4. Area cladogram (*a*) is the only possibility under Assumption 0. Under Assumption 2 area cladograms (*a*) and (*b*) are possible. Under Assumption 1 all 15 possible trees were obtained (not shown).

When considered under Assumption 2, we believe Area Cladogram I to be more parsimonious than Area Cladogram II. Reduced Area Cladogram I (Fig. 7*a*) shows stepwise separation of a component B + C + D from area A (the Red Sea), then component C + D from area B (western to central Indian Ocean), then of area D (western to central Pacific Ocean) from area C (eastern Indian Ocean). Reduced Area Cladogram II shows the biogeographic closest relationships between non-adjacent areas (Red Sea more closely related to eastern Indian Ocean than to western to central Indian Ocean; western to central Pacific Ocean more closely related to western to central Indian Ocean than to eastern Indian Ocean). It thus requires a number of explanations (dispersals, and/or extinctions) to interpret the species composition of related areas. We also note here that Area Cladogram I shares congruence with the geographical provinces outlined for another group of scleractinians, the fungiids (Hoeksema 1989).

A scenario based on Area Cladogram I for the history of the *A. selago* group is: an ancestral taxon (or taxa) occupied a broad distribution from the Red Sea to the central Pacific. Vicariance

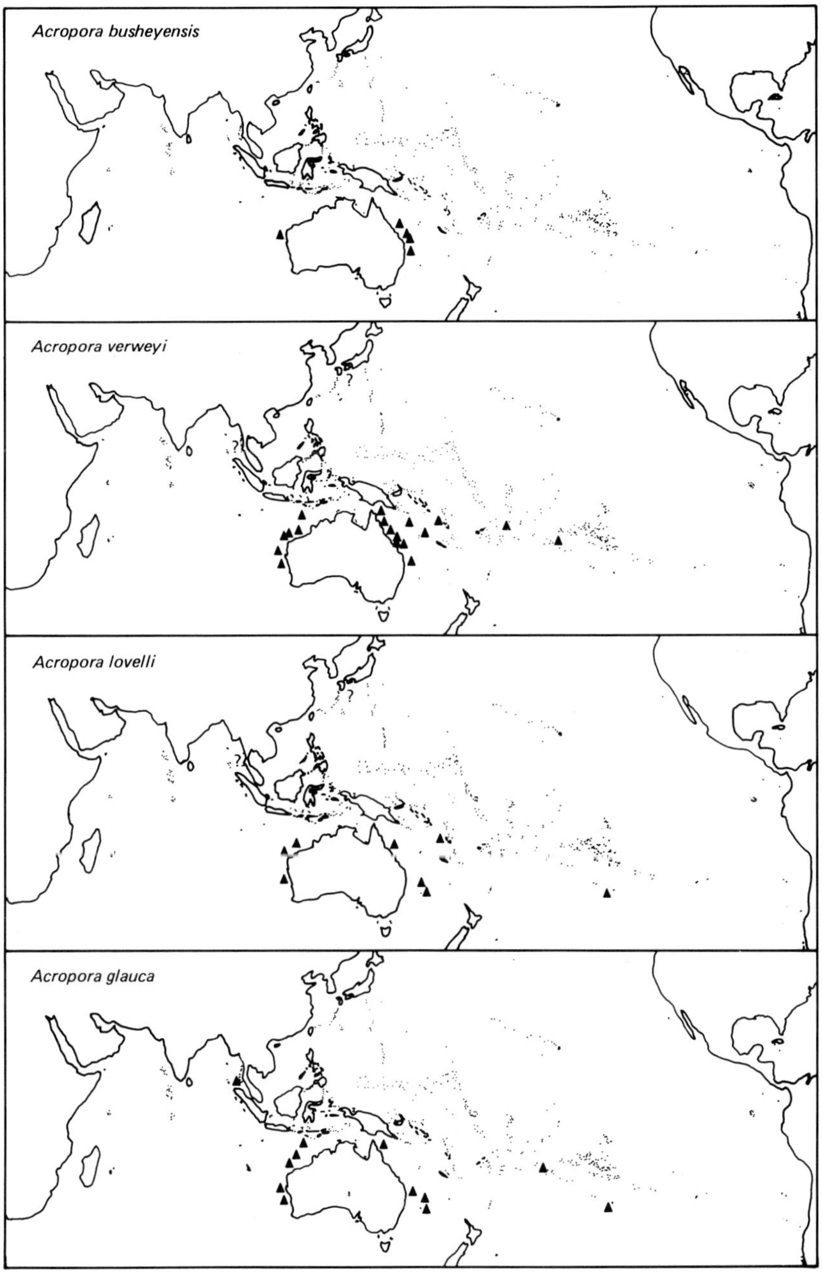

Fig. 8. Distribution of another *Acropora* subgroup, the *A. lovelli* group, for comparison with distribution patterns of the *A. selago* group. (See text for basis of side records.)

events led to the breaking up of this distribution at the mouth of the Red Sea, then west of the Western Australian continental shelf, and then at the Indonesian arc. Speciation events are always to the east of the vicariance line. Under this scenario, speciation events may or may not have been followed by dispersal to broaden the range of the taxa, but no dispersal or extinction need be invoked to explain the area cladogram.

Because this study is based on only a single *Acropora* species-group, we do not further speculate on possible physical, oceanographic, or biological explanations for our scenario. Similar analyses on further *Acropora* species-groups from Veron and Wallace (1984), with subsequent comparative analyses of area cladograms, are needed. Many of these *Acropora* species-groups comprise four or more species and occur within the set of areas A to D derived in this paper, thus they should readily lend themselves to component and consensus analysis of cladistic biogeography (Humphries *et al.* 1988). Other *Acropora* species-groups, however, fall outside the areas A to D posed here. For example, when we plotted distribution records for species of the *Acropora loveli* group, we found them to have a mainly southern tropical to Austral distribution (Fig. 8) and no members present west of the Western Australian continental shelf. In addition, the Caribbean coral fauna includes three species of *Acropora* that do not occur in the Pacific. We are confident that as further groups are included and further areas come under scrutiny, *Acropora* will prove to be an appropriate taxon to examine the biogeographic history of the tropical marine realm.

Acknowledgments

This research was funded by the Australian Research Council Grants Committee. We wish to thank Glen De'ath for fruitful discussions on statistical inference, J. Benzie for critically reading the manuscript and Rob Raven for helping us to run COMPONENT. We acknowledge researchers whose site specimens have been used, especially G. Paulay, D. Itano, S. Mildner and Y. Loya and thank them for their patience in waiting for full identification lists. E. Rowe is thanked for cartography, and Rhondda Leggett for typing the manuscript. This is Contribution no. 529 from the Australian Institute of Marine Science, and no. 1 from the Museum of Tropical Queensland.

References

Best, M. B., Hoeksema, B. W., Moka, W., Moll, H., Sukarsono, and Saturna, I. N. (1989). Recent scleractinian coral species collected during the Snellius II Expedition in Eastern Indonesia. *Netherlands Journal of Sea Research* **23,** 107–15.

Bothwell, A. M. (1981). Fragmentation, a means of asexual reproduction and dispersal in the coral genus *Acropora* (Scleractinia: Astrocoeniina, Acroporidae) — a preliminary report. Proceedings of the 4th International Coral Reef Symposium, Manila. Vol. 2, 137–44.

Cheetham, A. H., Hayek, L. C., and Thompson, E. (1980). Branching structure in arborescent animals: Models of relative growth. *Journal of Theoretical Biology* **85,** 335–69.

Farris, J. S. (1988). 'Hennig86 (Version 1·5) Reference.' Software and manual distributed by the author. (41 Admiral Street, Port Jefferson Station 11776: New York.)

Farris, J. S. (1990). Phenetics in camouflage. *Cladistics* **6,** 91–100.

Faure, G. (1982). Recherche sur les peuplements de Scleractinaires des récifs coralliens de l'Archipel des Mascareignes. These Université d'Aix-Marseille.

Hamilton, H. G. H., and Brakel, W. H. (1984). Structure and coral fauna of East African reefs. *Bulletin Marine Science* **34,** 248–66.

Harrison, P. L. (1985). Sexual characteristics of scleractinian corals: systematic and evolutionary implications. Proceedings of the 5th International Coral Reef Congress, Tahiti, Vol. 4, 337–42.

Harrison, P. L. (1988). Comparative ultrastructure of scleractinian spermatozoa and its evolutionary implications. Ph.D. Thesis, James Cook University, Townsville.

Harrison, P. L., Babcock, R. C., Bull, G. D., Oliver, J. K., Wallace, C. C., and Willis, B. L. (1984). Mass spawning in tropical reef corals. *Science* **223,** 1186–9.

Harrison, P. L., and Wallace, C. C. (1990). Reproduction, dispersal and recruitment of scleractinian corals. In 'Ecosystems of the World: Coral Reefs'. (Ed. Z. Dubinsky.) pp. 133–207. (Elsevier.)

Hoeksema, B. W. (1989). Taxonomy, phylogeny and biogeography of mushroom corals (Scleractinia: Fungiidae). *Zoologische Verhandclingen* **254,** 1–295.

Hoffmeister, J. E. (1925). Some corals from American Samoa and the Fiji Islands. Papers from the Department Marine Biology, Carnegie Institution, Washington No. 22, 1–90, 23 pls.

Humphries, C. J., Ladiges, P. V., Roos, M., and Zandee, M. (1988). Cladistic biogeography. In 'Analytical Biogeography'. (Eds A. A. Myers and P. S. Giller.) pp. 371–404. (Chapman and Hall: London, New York.)

Jones, R. S., Randall, R. H., Cheng, Y., Kami, H. T., and Mak, S. (1972). A marine biological survey of southern Taiwan with emphasis on corals and fishes. Special Publication No. 1. (Institute of Oceanography, College of Sciences, National Taiwan University: Taipei, Taiwan.)

Latypov, Y. Y. (1986). Coral communities of the Namsu Islands (Gulf of Siam, South China Sea). *Marine Ecology Progress Series* **29,** 261–70.

Mahalanobis, P. C. (1930). On tests and measures of group divergence. *Journal and Proceedings of The Asiatic Society of Bengal* **26,** 541–88.

Maragos, J. E., and Jokiel, P. L. (1986). Reef corals of Johnston Atoll: one of the world's most isolated reefs. *Coral Reefs* **4,** 141–50.

McCoy, E. D., and Heck, K. L. Jr (1976). Biogeography of corals, seagrasses and mangroves: an alternative to the center of origin concept. *Systematic Zoology* **25,** 201–10.

McCoy, E. D., and Heck, K. L. Jr (1983). Centers of origin revisited. *Paleobiology* **9,** 17–19.

McMillan, J., and Miller, D. J. (1988). Restriction analysis and DNA hybridization applied to the resolution of *Acropora nobilis* from *A. formosa*. Proceedings of the 6th International Coral Reef Symposium, Townsville. Vol. 2, 775–7.

McMillan, J., and Miller, D. J. (1989). Nucleotide sequences of highly repetitive DNA from scleractinian corals. *Gene* **83,** 185–6.

McMillan, J., and Miller, D. J. (1990). Highly repeated DNA sequences in the scleractinian coral genus *Acropora*: evaluation of cloned repeats as taxonomic probes. *Marine Biology* **104,** 483–7.

Nelson, G., and Platnick, N. (1981). 'Systematics and Biogeography/Cladistics and Vicariance'. (Columbia University Press: New York.)

Page, R. D. M. (1989). 'COMPONENT User's Manual'. (Release 1·5). (University of Auckland: Auckland.)

Pillai, C. S. G. (1977). The structure, formation and species diversity of south Indian reefs. Proceedings 3rd International Coral Reef Symposium, Miami. Vol. No. 1, 47–53.

Pillai, C. S. G., and Scheer, G. (1976). Report on the stony corals from the Maldive Archipelago. Results of the Xarifa Expedition 1957/58. *Zoologica* **43,** 1–83, pls 1–32.

Pillai, C. S. G., Vine, P. J., and Scheer, G. (1973). Notes on a collection of corals from the Seychelles. *Zoologische Jahrbucher. Abteilung sur Systemtik, Okologie und Geographie der Tiere* **100,** 451–65.

Randall, R. H. (1983). 'Guide to the Coastal Resources of Guam: Vol. 2. The Corals.' (University of Guam Press.)

Rosen, B. R. (1971). Annotated check list and bibliography of corals of the Chagos Archipelago (including the recent collection from Diego Garcia) with remarks on their distribution. *Atoll Research Bulletin* **149,** 67–88.

Rosen, B. R. (1979). Checklist of recent coral records from Aldabra (Indian Ocean). *Atoll Research Bulletin* **233,** 1–26.

Rosen, B. R. (1988). Progress, problems and pattern in the biogeography of reef corals and other tropical marine organisms. *Helgolander Meeresunters* **42,** 269–301.

SAS Institute Inc. (1989). 'SAS/STAT User's Guide, Version 6.' 4th Edn, Vol. 1, (Cary, NC: SAS Institute Inc.)

Scheer, G., and Pillai, C. S. G. (1983). Report on the stony corals from the Red Sea. *Zoologica* **133,** 1–198, 41 pls.

Sheppard, C. R. C. (1985). Reefs and coral assemblages of Saudi Arabia. 2. Fringing reefs in the southern region Jeddah to Jizan. *Fauna of Saudi Arabia* **7,** 37–58.

Sheppard, C. R. C. (1987). Coral species of the Indian Ocean and adjacent seas: a synonymized compilation and some regional distributional patterns. *Atoll Research Bulletin* **307,** 1–32.

Sheppard, C. R. C., and Salm, R. V. (1988). Reef and coral communities of Oman, with a description of a new coral species (Order Scleractinia, genus *Acanthastrea*). *Journal Natural History* **22,** 263–79.

Sheppard, C. R. C., and Sheppard, A. L. S. (1985). Reefs and coral assemblages of Saudi Arabia. 1. The Central Red Sea at Yaubu at Sinaiyah. *Fauna of Saudi Arabia* **7,** 17–36.

Stehli, F. G., and Wells, J. W. (1971). Diversity and age patterns in hermatypic corals. *Systematic Zoology* **20,** 115–26.

Veron, J. E. N. (1988). Comparisons between the hermatypic corals of the southern Ryuku Islands of Japan and the Great Barrier Reef of Australia. *Galaxea* **7,** 211–31.

Veron, J. E. N. (1990*a*). Checklist of the hermatypic corals of Vanuatu. *Pacific Science* **44,** 51–70.

Veron, J. E. N. (1990*b*). Re-examination of the reef corals of Cocos (Keeling) Atoll. *Records Western Australian Museum* **14,** 553–81.

Veron, J. E. N., and Hodgson, G. (1989). Annotated checklist of the hermatypic corals of the Philippines. *Pacific Science* **43,** 234–87.

Veron, J. E. N., and Marsh, L. M. (1988). Hermatypic corals of Western Australia: records and annotated species list. *Records Western Australian Museum, Supplement* **29,** 1–136.

Veron, J. E. N., and Wallace, C. C. (1984). Scleractinia of Eastern Australia. Part V. Family Acroporidae. *Australian Institute of Marine Science Monograph Series* **6,** 1–485.

Wallace, C. C. (1978). The coral genus *Acropora* (Scleractinia: Astrocoeniina: Acroporidae) in the central and southern Great Barrier Reef province. *Memoirs of the Queensland Museum* **18,** 237–319, pls 43–103.

Wallace, C. C. (1985). Reproduction, recruitment and fragmentation in nine sympatric species of the coral genus *Acropora*. *Marine Biology* **88,** 217–33.

Wallace, C. C., and Dallwitz, M. B. (1981). Writing coral identification keys that work. Proceedings of the 4th International Coral Reef Symposium, Manila. Vol. 2, 187–90.

Wells, J. W. (1954). Recent corals of the Marshall Islands. *Geological Survey Professional Papers* **260–I,** 385–486, pls 94–185.

Wells, J. W. (1956). Scleractinia. In 'Treatise on Invertebrate Paleontology, F. Coelenterata'. (Ed. R. C. Moore.) F328–440. (University of Kansas Press.)

Wells, J. W. (1987). Notes on Indo-Pacific Scleractinian corals. Part II. A new species of *Acropora* from Australia. *Pacific Science* **39,** 338–9.

Zandee, M., and Roos, M. C. (1987). Component-compatibility in historical biogeography. *Cladistics* **3,** 305–32.

Appendix 1.
Characters and character states for three analyses: quantitative characters only, qualitative characters only, and all characters

0 Length of 1st order branches: range 18·9–28·6 mm
1 Top diameter of 1st-order branch: range 2·9–4·6 mm
2 Middle diameter of 1st-order branch: range 5·3–10·3 mm
3 Bottom diameter of 1st-order branch: range 5·7–11·1 mm
4 Middle diameter of 2nd-order branch: range 7·1–12·3 mm
5 Number of radials/cm on 1st-order branch: range 1·6–7·0
6 Angle of 1st-order branch: range 29·0–51·5 degrees
7 Distance to axial corallite nearest neighbour: range 13·7–27·4 mm
8 Inner axial corallite diameter: range 0·8–1·4 mm
9 Outer axial corallite diameter: range 1·8–3·7 mm
10 Axial corallite exertness: range 1·1–1.9 mm
11 Colony symmetry: symmetrical; not symmetrical
12 Colony growth: determinate; indeterminate
13 Recognisable attachment site in mature individual: present; absent
14 Upper colony surface: planar; not planar
15 Colony branching: arborescent; corymbose or caespito-corymbose
16 Contribution of radial corallites to branch thickness: $> 50\%$; $< 50\%$
17 Radial corallite shape: predominantly lipped without upper wall; predominantly lipped with some upper wall development
18 Coenosteal homogeneity: same on and between radial corallites; different between radials *versus* on radials
19 Radial corallite walls: costate or 'broken costate'; dense arrangement of spinules
20 Axial corallite walls: costate or 'broken costate'; reticulate with spinules; dense arrangement of spinules
21 Spinules: mostly single-tipped and slender; mostly multiple-tipped; mostly blunt-tipped

A. tenuis	0?1?1212121 0100101000?
A. selago	122222?223? 01001010000
A. yongei	?1111200010 10110111010
A. dendrum	??2?22?2232 01001120122
A. millepora	00000001101 01001101000
A. donei	?1111101012 01000111011

Manuscript received 3 September 1990, accepted 25 February 1991

3 September 1991 *Aust. Syst. Bot.* 1991, **4**, 211-27

Patterns of Diversification Within Continental Biotas: Hierarchical Congruence Among the Areas of Endemism of Australian Vertebrates

Joel Cracraft

Departments of Anatomy and Cell Biology and of Biological Sciences, University of Illinois, P.O. Box 6998, Chicago, Illinois 60680, U.S.A.

Abstract

This paper investigates the extent to which there is hierarchical information about area-relationships contained in the distributions of Australian vertebrates. Distributions of genera, their included species, and their included subspecies (when appropriate) were coded present/absent for 14 areas of endemism. The data were then analysed cladistically and most parsimonious area-cladograms constructed; bootstrap consensus trees were used to assess the strength of the cladistic signal. Distributions of birds and mammals were found to be nearly identical in their hierarchical pattern, and snakes shared the same general pattern of area-relationships. Frogs and lizards exhibited greater differences but were still congruent in some respects. This congruence calls for a general explanation. It is proposed that either the pattern can be explained in terms of (1) true area-relationships (vicariance), in which case biogeographic noise, including redundant distributions and widespread taxa do not disturb historical signal; (2) constraints on processes or events that lead to congruent histories of cosmopolitanism; or (3) a combination of both.

Introduction

Speciation produces increases in diversity and is largely a process (pattern over time) resulting from differentiation (change in form) across space. Consequently, the key to understanding patterns and processes of diversification within continental biotas will come from having a detailed picture of the spatial history of their component taxa. Over the last 15 years systematists have reached general agreement that knowledge of biogeographic history is best obtained by searching for congruence among area-relationships (Platnick and Nelson 1978; Rosen 1978, 1979). Evidence for the latter comes from corroborated cladistic hypotheses for clades having endemic taxa in the same areas of endemism. While sharing this primary goal of biogeographic research, historical biogeographers have nevertheless engaged in considerable debate about the details of method, an indication that the complexities of organisms' distributional histories are often not easily resolved. Although this discourse is extensive (Nelson and Platnick 1978, 1981; Platnick 1981; Nelson 1984; Platnick and Nelson 1984, 1988; Humphries and Parenti 1986; Wiley 1987, 1988*a*, 1988*b*; Zandee and Roos 1987; Craw 1988; Cracraft 1988; Craw and Page 1988; Kluge 1988; Page 1988, 1989*a*, 1990; Brooks 1990), many of the important points of difference concern the treatment of phenomena that are considered capable of confounding estimates of area-relationships, namely widespread taxa, taxa missing from areas, and redundant distributions.

Since inferences about the relationships of a given area cannot be made unless representatives of a clade are actually present in it, the problem of missing taxa is a problem of missing information and can be ignored. Thus, only widespread taxa and redundant distributions can be expected to provide confounding information about area-relationships. Several associations are possible between area-relationships and the presence of widespread and redundant distributions. Two areas might share a taxon for which there is as yet no evidence

1030-1887/91/010211$10.00

it has differentiated. In this case, the two populations would provide false information about the cladistic relationships of those areas if one of the populations had dispersed to an area unrelated to the first. On the other hand, the distributions of these two populations might be consistent with, but provide no positive evidence for, a relationship of the two areas if there were dispersal between two related areas, or if the populations had not yet differentiated following vicariance. Whether a clade is considered to have species with 'redundant' distributions is a consequence of the hierarchical level at which that clade is delineated at the inception of the analysis and how narrowly or broadly an investigator individuates areas of endemism (for example, when is 'Australia' for one species actually equivalent to 'Australia' for another in the same clade?). To the extent that redundant distributions can be said to be biogeographically significant (that is, the two areas are judged to be the 'same', but only in the sense that parallelisms or reversals represent the 'same' character), they almost always will come about via dispersal and thus provide false information about area-relationships. But redundant distributions that appear to be noise within one clade may turn out to be signal when intercladal comparisons are made.

Thus, 'false' area-relationships (biogeographic 'noise') arise either through dispersal across barriers or, in the case of widespread species, through a failure to differentiate following vicariance (the 'peril of plesiomorphy' of Nelson and Platnick 1978). Some dispersal events, however, might produce distributions that are consistent (not incongruent) with true area-relationships ('signal'), as do some instances of widespread species that have failed to differentiate. Of interest is whether those events that produce a pattern of biogeographic noise are sufficiently frequent, when compared to those events producing a pattern congruent with biogeographic signal (i.e. cladistic relationships of endemic taxa plus those widespread distributions just noted), that it becomes difficult, if not impossible, to recover the pattern of area-relationships. If long-distance dispersal and misleading plesiomorphic distributions are not frequent events relative to the time-scale of vicariance and subsequent differentiation of populations, then the historical area-relationships of biotas should be recoverable via pattern analysis of present species' distributions, assuming some minimal cladistic information. This proposition is perhaps not significantly different from that which governed Croizat's attempts (see Croizat *et al.* 1974) to decipher biogeographic pattern. Thus, as long as those factors producing biogeographic noise are more or less randomly distributed with respect to areas of endemism, then signal would be expected to be revealed by examination of a sufficiently large sample of distributional data.

The approach just outlined addresses the general question of the relationships of areas of endemism but is not based on an analysis of congruence among two or more clades whose cladistic relationships have been the subject of study. It instead asks two different questions: (1) What cladistic structure exists in raw distributional data? And (2) If such structure exists, what is its historical significance? These two questions also constitute an attempt to discover the relative influence of those factors, noted above, that confound cladistic signal. Within the context of this agenda, however, it must be stressed that such an approach does *not* constitute a general method of biogeographic analysis. It cannot be informative about area-relationships if dispersal and plesiomorphic distributions contribute significant biogeographic noise (i.e. result in spurious associations among areas of endemism). The most satisfactory general method yet proposed is component analysis under assumption 2 (Nelson and Platnick 1981; Humphries and Parenti 1986; Page 1989*a*, 1990), which should be the starting point when fundamental area-cladograms — those produced by cladistic analysis of individual monophyletic groups — are to be compared.

Methods

Data

This paper analyses the native terrestrial vertebrate fauna of mainland Australia, with primary emphasis being placed on avian biogeographic patterns. Bird distributions were taken from Blakers *et al.* (1984). Waterbirds and birds of prey were omitted, as were taxa that were uninformative regarding area-relationships (see below). Distributions of mammals were taken from Strahan (1983), and those of frogs, snakes, and lizards were from Cogger (1983). Michael Tyler (personal communication) kindly provided updated information on the distributions of Australian frogs.

Areas of Endemism

The distribution of each taxon was coded with respect to 14 areas of endemism. These areas (Fig. 1) are based primarily on the distributions of birds, although the distributions of many species in other terrestrial vertebrate groups are also largely congruent with those of birds (Heatwole 1987; this paper). Most of the areas are well defined and have been discussed by various workers (see Keast 1961; Ford 1971, 1978, 1987*a*, 1987*b*; Cracraft 1986). Vegetation descriptions presented below pertain to potential, natural conditions prior to European settlement and follow the *Atlas of Australian Resources: Vol. 6 Vegetation* (AUSLIG 1990).

(*1*) *Kimberley Plateau*

The Kimberley area of endemism is an elevated region of woodland and low woodland having species isolated to the south-west by the Fitzroy River drainage and the Great Sandy Desert, to the south by the northern interior desert, and to the east from Arnhem Land by the lowlands of the Victoria and Daly River valleys (Ford 1978).

(*2*) *Arnhem Land*

Species in the woodland and low woodland habitats of Arnhem Land are isolated from the Kimberley Plateau to the west by the Victoria and Daly River valleys (Ford 1978), to the south by the lowlands of the northern interior desert, and to the south-east by drier, more open habitats around the Gulf of Carpentaria.

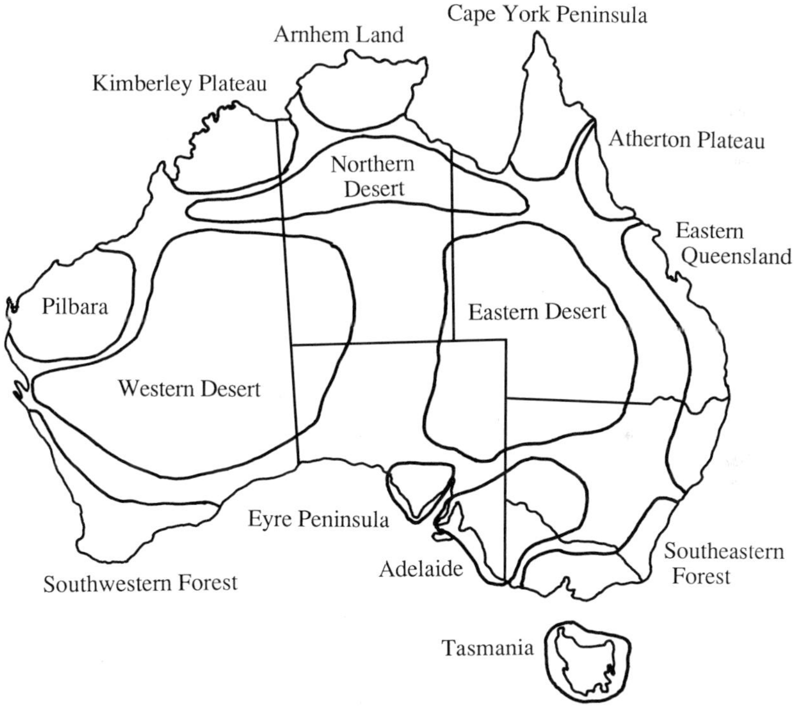

Fig. 1. Areas of endemism based on avian distributional patterns. See text.

(*3*) *Cape York Peninsula*

Species endemic to this area inhabit medium closed forest (rainforest), open forest, and woodland habitats. The area is bounded to the south-east by the uplands of the Atherton Plateau and to the south and south-west by drier, more open habitats.

(*4*) *Atherton Plateau*

Endemics on the Atherton Tableland generally inhabit moist closed-forest (rainforest) habitats and have their distributional limits circumscribed by the lowlands to the north, by drier, more open habitats to the west, and by open woodland in the region of the Burdekin River valley (the Burdekin Gap; Ford 1986, 1987*a*, 1987*b*).

(*5*) *Eastern Queensland*

The boundaries of this area of endemism are somewhat poorly defined. Species generally have their northern limits at the Burdekin Gap and those inhabiting wet-forest habitats are limited westward by drier, more-open woodland habitats. Distributional limits to the south are more ambiguous but many are found in the area of the Hunter River valley where more-open habitats are found; other species extend their ranges into the South-eastern Forest area of endemism. Many species tolerant of drier habitats extend their distributions westward into the Great Dividing Range for various distances, hence the western boundary of this area of endemism is especially poorly defined.

(*6*) *South-eastern Forest*

Species in this area of endemism inhabit tall open forest, open forest, or drier woodland habitats (AUSLIG 1990). The area is bounded to the north by the Hunter River valley, to the north-west (inland) by the Great Dividing Range or drier, more open habitats, and at its western boundary by lowlands vegetated by mallee.

(*7*) *Tasmania*

Species in this area of endemism inhabit the tall open, closed, and open forests of Tasmania.

(*8*) *South-western Forest*

Species in this area are found in open forests and woodland habitats of the south-western corner of Australia. The area is limited to the east by the low shrubland of the Nullarbor Plain and to the north by drier habitats including open scrub and tall shrubland. Species localised along the south-western coastline, and which extend northward to or near the area of Shark Bay, are also included in this area.

(*9*) *Pilbara*

The uplands of the Pilbara (Hammersley Plateau) are vegetated primarily with tall shrubland and tall open shrubland. Species endemic in this area have their distributions limited to the north by the lowlands of the Great Sandy Desert, to the west by the Gibson Desert, and to the south by lowlands.

(*10*) *Western Desert*

This is an ill-defined area of endemism that extends from south and west of the Pilbara east to the Simpson Desert (although some species are limited by the Central Highlands). It roughly conforms to the distribution of mulga vegetation (e.g. AUSLIG 1990), but some of the species are distributed to the north in tall open shrubland where the boundary of the area of endemism is ambiguous.

(*11*) *Eastern Desert*

This area of endemism lies to the east of the Central Highlands and the drier regions of the Simpson Desert and extends to the uplands of the Great Dividing Range. The southern distributions of species are limited by the Darling River or areas of mallee vegetation.

(*12*) *Northern Desert*

The Northern Desert area of endemism is the most poorly defined of all the areas. It consists of mostly low open woodland and lies between the southern edges of the Kimberley Plateau and Arnhem Land areas of endemism and the northern boundaries of the Eastern and Western Desert areas. The area extends eastward to the base of the Gulf of Carpentaria.

(*13*) *Eyre Peninsula*

The distributions of taxa endemic to the Eyre Peninsula are limited to the north and east by the more arid Eyrean barrier and the Flinders Ranges and to the west by the drier area of the Nullarbor Plain (although some species may well have distributions extending to the South-western Forest area of endemism; Ford 1971).

(*14*) *Adelaide*

This area of endemism includes those species restricted to the mallee vegetation east of the Flinders and Mount Lofty Ranges, south of the Darling River valley, and west of the Great Dividing Range.

Some species that extend their distributions westward through the Mount Lofty Ranges to the Yorke Peninsula but do not extend around Spencer Gulf to the Eyre Peninsula are also included.

Methods of Analysis

Taxa were coded for their absence or presence (0, 1) in each of the areas of endemism. Three taxonomic levels were coded hierarchically: the genus, species within the genus and, if information was available, subspecies within each species (subspecies distributions were known only for birds and mammals). Monotypic genera or species were included in the analysis only if they had a distribution in two or more areas. If the evidence suggested that the subspecies boundaries were arbitrary subdivisions of continuous, probably clinal variation, the subspecies was ignored and only the species was coded.

Other systematists have attempted to examine the hierarchical structure of raw distributional data; Rosen (1988), for example, proposed a method termed 'parsimony analysis of endemicity' (PAE). While sharing some similarities with PAE, the methods of this study do differ substantially. First, this study uses pre-defined areas of endemism, whereas Rosen (1988) used point sample localities. Thus, some degree of historical and biogeographic non-randomness is already assumed in the data based on the original empirical distributions. Second, Rosen coded only one taxonomic level — genera — whereas this study imposes some hierarchical structure on the raw data by coding subspecies (when possible), species, and generic distributions.

Hypotheses of area-relationships were generated using the Branch and Bound algorithm of Phylogenetic Analysis Using Parsimony, version 3.0L (PAUP; Swofford 1990). As an aid to evaluating and describing the strength of the cladistic signal in the data, bootstrap consensus trees were produced from 100 replicate trees using various heuristic branch-swapping routines contained in PAUP.

An obvious aspect of understanding the historical relationships of areas of endemism is having knowledge of the 'root' of the area-cladogram. Conventionally, that root is chosen to reflect the root of the consensus (however determined) of all the taxic cladograms. One difficulty is that, in large geographic regions such as continents, relatively few of the clades whose cladistic relationships have been sufficiently analysed have endemics in all, or perhaps even most, areas of endemism. Given this lack of knowledge, it becomes especially problematic to specify the root of the area-cladogram when undertaking an analysis such as is being done here. Short of having cladistic information based on taxic interrelationships, there are several ways in which a root might be assigned. The first two — arbitrarily designating an area as the root or adopting a form of mid-point rooting — are unsatisfactory for obvious reasons. The third — including a hypothetical 'outarea' coded all zeroes — while not entirely satisfactory, is perhaps the best, non-arbitrary means of rooting the trees in order to provide an unbiased comparison of area-cladograms across the major groups of vertebrates. The only assumption being made is one of ingroup monophyly: all the Australian species are more closely related to each other than they are to species outside of Australia. Given that so many genera and species-groups of vertebrates are endemic to Australia, this assumption may generally hold, but there are many obvious exceptions. For example, numerous taxa in Cape York seemingly have sister-group relationships to taxa in New Guinea. Yet, cases such as this may not have an effect on the overall relationships specified by the cladogram inasmuch as New Guinea (or any other area) could simply be added to its appropriate place on the cladogram and the outarea retained. For these reasons, area-cladograms in this analysis will include an 'outarea' as the root of the tree. Rosen (1988: 458) suggested that areas with few taxa be omitted because they will be viewed as being primitive by parsimony analysis, and as a consequence be placed near the root of the tree. This will certainly be the case in some instances but, depending on the pattern of shared taxic-relationships, it need not necessarily follow. There are several explanations for areas having fewer taxa relative to others. One of these is historical, namely that the areas are truly primitive in the sense that they are the sister-area of a larger area that was subsequently subdivided repeatedly. If such is the case, then omitting areas with few taxa precludes us from investigating this history using this method.

Results

The Avian Pattern

Based on 562 codings of genera, species, and subspecies-level taxa, a parsimony analysis yielded a single parsimonious tree of 957 steps (Fig. 2*A*). Three major clusters of areas are recognised. The first includes areas along the northern and eastern margins of the continent. Kimberley and Arnhem Land are united and related in turn to the eastern forest areas. Within the latter the East Queensland and South-eastern Forest are most closely related, and Atherton Plateau has its relationships to these rather than to Cape York.

The second cluster includes the four interior desert areas, which as a group are united with the preceding cluster of areas. The Western Desert and Eastern Desert areas are grouped together and then cluster with the Pilbara relative to the Northern Desert area.

The final cluster includes the areas of the southern margin of the continent. The Eyre Peninsula and Adelaide are sister-areas that are united with the South-western Forest relative to Tasmania.

Using a bootstrap analysis to assess the strength of the cladistic signal just described suggests that some of these relationships are tenuous at best. In general, the bootstrap confirms a strong cladistic signal for the areas of the northern and eastern continental margins but less so for the interior and southern clusters of areas (Fig. 2*B*). Although the Eastern Desert and Western Desert areas are strongly associated, the relationships of the Pilbara and Northern Desert are uncertain. Within the cluster of southern areas, the relationships of the South-western Forest, Eyre Peninsula, and Adelaide areas are confirmed, but the placement of Tasmania is problematic.

Birds

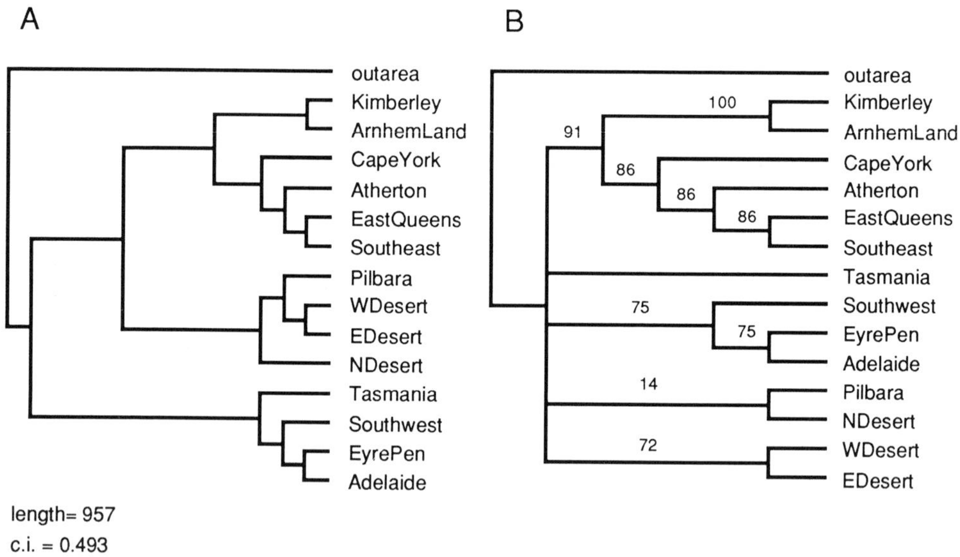

Fig. 2. *A*, most parsimonious tree for 562 avian distributions. *B*, bootstrap solution for avian distributional patterns based on 100 replicates. See text.

Evaluating the Biogeographic Significance of the Avian Pattern

Potential interpretations and tests

The fundamental question regarding the pattern of Fig. 2 is the extent to which it represents the true cladistic relationships of the areas in question versus whether it expresses similarity relationships that reflect shared widespread taxa due to geographic proximity. Two different approaches can be used to help evaluate these alternatives.

First, in a large data-set such as that for birds, the analytical method used here might be expected to find area-relationships that result from a combination of sister-group vicariance, area plesiomorphy, and long-distance dispersal. The problem is to estimate the potential contribution of each. One way to do this is to map this pattern onto a consensus of fundamental area-cladograms that have been derived using component analysis under assumption 2 (Nelson and Platnick 1981; see also Page 1988, 1989*b*). In the case of birds, some cladistic results are available from which fundamental area-cladograms can be constructed, but the data are too few to make a satisfactory evaluation of the pattern of Fig. 2 (see discussion next section).

A central tenet of comparative biology is the concept of congruence of cladistic pattern, and this can be employed to develop a second method for evaluating the biogeographic significance of the avian pattern. That method entails repeating the analysis using independent distributional data and searching for congruence with the avian pattern. If little or no congruence is observed, the hypothesis that the avian pattern is an accurate representation of the vicariance history of the areas would be called into question. This method will be investigated below in a subsequent section.

Congruence with fundamental area-cladograms of avian taxa

(*i*) *Tregellasia* and *Ptiloris*. Relationships within the flycatcher genus *Tregellasia* and within the riflebirds, genus *Ptiloris*, are informative about the area-relationships of areas of endemism along the eastern coast of Australia (Cracraft 1986). Unfortunately, these two sets of relationships are only informative for three areas within Australia, but both are congruent with each other and with the general hypothesis of Fig. 2: (Cape York (Atherton Plateau + Eastern Queensland)). The sister-group of each of these clades is in New Guinea.

(*ii*) *Poephila, Petrophassa*, and *Malurus*. Three genera contain species having endemics along the northern and eastern continental margin (see Cracraft 1986 for distributions and systematic relationships). The fairy wrens, genus *Malurus*, include numerous species distributed over most of Australia. Their relationships are not fully understood but one clade of three species specifies the following area-relationships (Cracraft 1986): (Cape York (Kimberley Plateau + Arnhem Land)).

A genus of pigeons, *Petrophassa*, and a clade of finches (*Poephila sensu stricto*) provide information on area-relationships but the patterns are complicated by some widespread species (Cracraft 1986). For *Petrophassa*, there are 76 trees for seven areas under assumption 2. A strict consensus of this set produces no resolution but a Nelson consensus tree shows the following pattern: ((Pilbara + Northern Desert)((Kimberley Plateau + Arnhem Land) + (Cape York, Atherton Plateau, Eastern Queensland)). That pattern is congruent with the area-relationships of Fig. 2. Likewise, an analysis of *Poephila* using assumption 2 produces 60 trees with a Nelson consensus tree having the pattern: ((Kimberley Plateau + Arnhem Land)(Cape York, Atherton Plateau, Eastern Queensland)).

(*iii*) *Psophodes*. The whipbirds consist of a number of differentiated taxa having distributions along the southern and eastern margins of the continent as well as in the interior desert (see Cracraft 1986 for a discussion of relationships). Because of several widespread species within the clade, the use of assumption 2 produces no significant resolution. Applying assumption 1, on the other hand, yields a strict consensus tree of the form: (Eastern Desert + Western Desert)((Atherton Plateau, Eastern Queensland, Southeastern Forest) + (Southwestern Forest, Eyre Peninsula, Adelaide))). This pattern exhibits some similarities in lower-level area-relationships with those of Fig. 2, but conflicts for the higher-level area-relationships.

(*iv*) Other fundamental area-cladograms. In general, the cladistic relationships of species within clades of Australian birds are still poorly known. Those that have been studied provide little or no information relative to evaluating the hypothesis of Fig. 2 because of the presence of numerous widespread species or redundant distributions. With few narrowly endemic species in a clade, the large number of possible area-relationships under assumption 2, or even assumption 1, makes comparisons to Fig. 2 extremely tenuous. Within the quail-thrushes (*Cinclosoma*) there is a strong pattern of Eastern–Western Desert sister-species whose relationships are to species having broad distributions along the southern and eastern continental margins (Cracraft 1986, 1989). This set of relationships has the appearance of a conflict with the pattern in Fig. 2 but taxa in the north are lacking.

Parrots in the genus *Platycercus* are broadly distributed in numerous areas of endemism, but most of the species are involved in redundant distributions and/or are widespread. Applying assumption 2 to one hypothesis of their relationships based on mitochondrial DNA restriction fragments (Ovenden *et al.* 1987) yields a Nelson consensus tree that unites Cape York, Atherton Plateau, and Eastern Queensland but places the South-eastern Forest area with Adelaide and Tasmania, which conflicts with the hypothesis of Fig. 2.

Other studies, including those on the babblers *Pomatostomas* (Kocher *et al.* 1989), the scrubwrens *Sericornis* (Christidis *et al.* 1988), poephiline finches related to the species discussed above (Christidis 1986), do not speak directly to area-relationships because of the presence of many widespread species.

Congruence with patterns of other vertebrates

(*i*) *Frogs*. A cladistic analysis of 202 frog taxa produced a single most parsimonious tree (Fig. 3*A*) for the 14 areas of endemism of Fig. 1. Three major clusters of areas are recognised. A northern and eastern group of areas has Kimberley + Arnhem Land as the sister-group to Cape York + Atherton Plateau. As in the avian pattern, this cluster is related to a cluster composed of interior desert areas. The latter has a sister-group relationship between the Pilbara and Western Desert, which are in turn related to the Eastern Desert area. As in the case of birds, the Northern Desert is the sister-area of these three. Finally, a southern margin cluster is also delineated but with relationships very different from those of birds. Most significantly, the Eastern Queensland + South-eastern Forest clade is grouped with the southern areas rather than with the northern and eastern clusters. Moreover, these two eastern areas are related to the Adelaide area, with Tasmania and the South-western Forest area occupying more distant relationships. In this solution, the Eyre Peninsula is at the base of the tree.

A bootstrap analysis (Fig. 3*B*) reveals a strong cladistic signal for a number of these groupings, particularly those involving areas along the northern and eastern margins of the

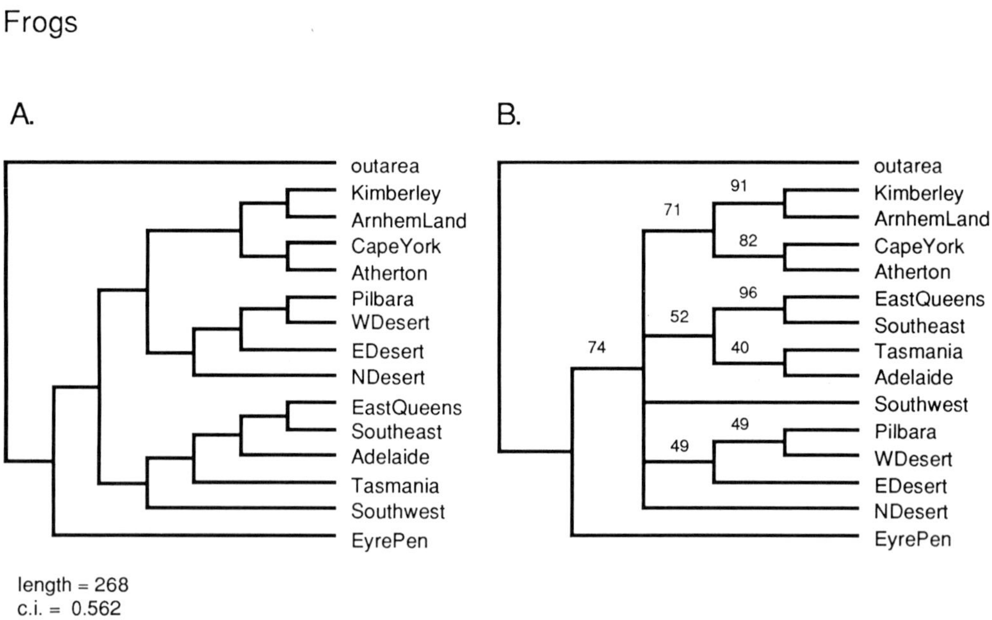

Fig. 3. *A*, most parsimonious tree for 202 frog distributions. *B*, bootstrap solution for frog distributional patterns based on 100 replicates. See text.

continent. The relationship between the northern and north-eastern cluster of areas and the interior desert cluster is not supported in the bootstrap tree. Relationships within the latter cluster are maintained by the bootstrap tree, but support is not especially strong. Finally, except for the relationship between the Eastern Queensland and South-eastern Forest areas, those involving the other areas of the southern margin are poorly supported. Surprisingly, the basal position of the Eyre Peninsula area of endemism is moderately well supported by the bootstrap, but this very possibly reflects the depauperate nature of the frog fauna.

(*ii*) *Lizards*. On the basis of 475 taxa, a cladistic analysis produced a single most parsimonious tree (Fig. 4*A*). Three primary clusters of areas are found. The first unites the sister-group, Kimberley + Arnhem Land, to the four interior desert areas. Within the latter, the Pilbara and Western Desert areas are most closely related, followed successively by the Eastern Desert and Northern Desert areas. Related to this first cluster is a second that includes Cape York and Atherton Plateau. All of these, in turn, are related to the third cluster which includes the eastern margin sister-group, Eastern Queensland + South-eastern Forest, and the southern margin group, ((South-west + Eyre Peninsula) Adelaide)). Tasmania is positioned at the base of the tree.

A bootstrap analysis (Fig. 4*B*) indicates that the Kimberley-Arnhem Land / interior desert cluster is fairly strongly supported except for the ambiguous position of the Northern Desert area. Cape York and Atherton are strongly united but their relationship to the preceding cluster is not strongly supported. Relationships within the third cluster are generally poorly supported, although a grouping of South-western Forest, Eyre Peninsula, and Adelaide receives moderate support. All the areas are strongly separated from Tasmania, which may be basal merely because it has a depauperate lizard fauna.

(*iii*) *Snakes*. A parsimony analysis was undertaken using 149 taxa of snakes. Two equally parsimonious trees were found, differing only in the placement of two areas (Pilbara and Northern Desert). The strict consensus is shown in Fig. 5*A*. One clade has the northern areas, Kimberley and Arnhem Land, as sister-areas and related to a suite of eastern margin areas. Within the latter, Eastern Queensland and South-eastern Forest areas are clustered, and their sister-area is the Atherton Plateau, and their sister-area, in turn, is Cape York. Western Desert and Eastern Desert are united but the relationships of the Pilbara and North Desert areas are with the northern and eastern areas. Finally, the Eyre Peninsula and Adelaide areas are united as the sister-group of the South-western Forest area. Tasmania is placed at the base of the tree, again possibly because it has a depauperate snake fauna.

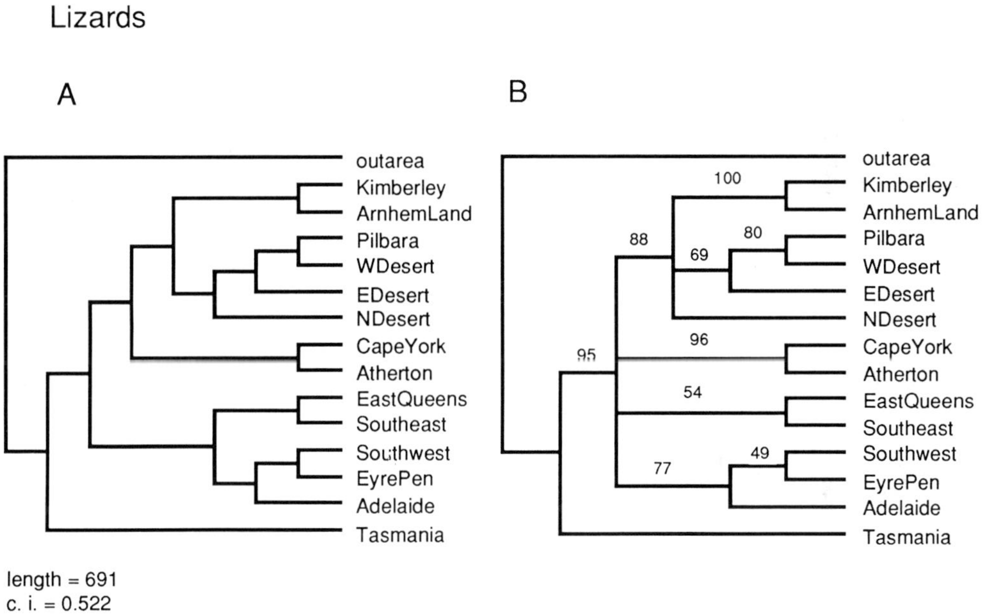

Fig. 4. *A*, most parsimonious tree for 475 lizard distributions. *B*, bootstrap solution for lizard distributional patterns based on 100 replicates. See text.

(*iv*) *Mammals*. A cladistic analysis of 359 taxa yielded two equally parsimonious trees differing only in the placement of Tasmania. The strict consensus tree is shown in Fig. 6 and also represents the topology of the bootstrap tree. There is strong support for a group of areas along the northern and eastern margins of the continent. Kimberley and Arnhem Land are always united as are the Cape York and Atherton Plateau areas and the East Queensland and South-eastern Forest areas; the latter two pairs are sister-groups. A second cluster consists of three southern margin areas having somewhat less strong support: ((Eyre Peninsula + Adelaide) Southwest)). Finally, the interior desert areas cluster together and their relationships are also moderately well supported. The Western Desert and Eastern Desert areas are united and related successively to Pilbara and then to Northern Desert. All three of these clusters, along with Tasmania, have unresolved relationships to one another (Fig. 6).

(*v*) *Comparison among the vertebrate groups*. The patterns found in non-avian vertebrates demonstrate many instances of congruence with the avian pattern of Fig. 2. That congruence is strongest in the mammals. Fig. 7*A* shows a consensus tree — in this case strict, majority-rule, and Nelson trees are all the same — for birds and one of the most parsimonious trees of mammals. The second mammalian tree differs only in the placement of Tasmania, and accordingly that area should be regarded as ambiguously placed on the consensus tree. The bird–mammal tree shows only one instance of incongruence, namely in the

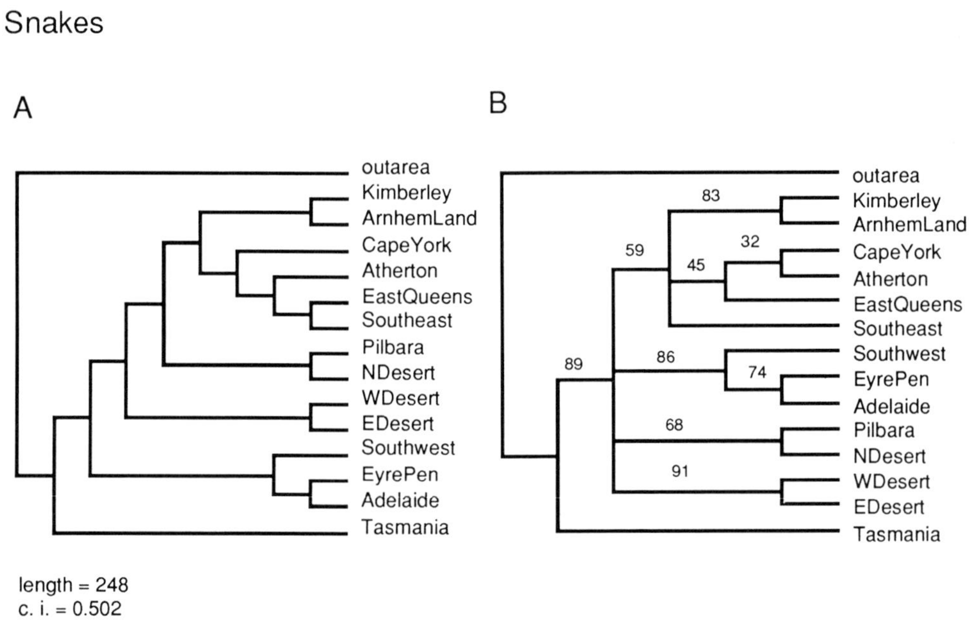

Fig. 5. *A*, most parsimonious tree for 149 snake distributions. *B*, bootstrap solution for snake distributional patterns based on 100 replicates. See text.

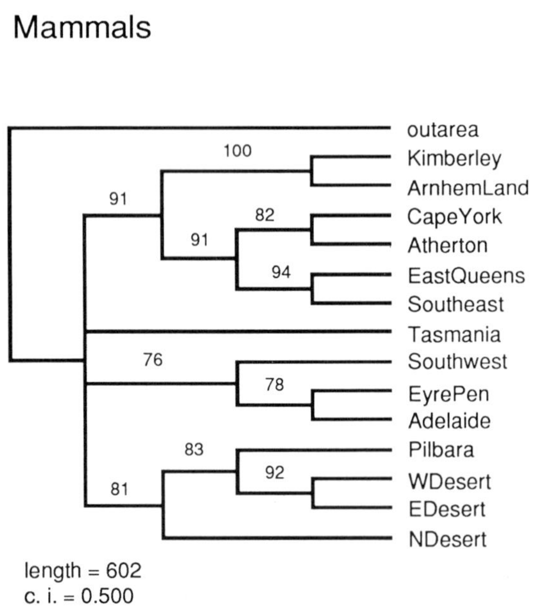

Fig. 6. Strict consensus for the two most parsimonious trees for 359 mammalian distributions. The bootstrap solution based on 100 replicates has the same topology and its values are also shown. See text.

relationship of Atherton either to East Queensland + South-eastern Forest (birds) or to Cape York (mammals). Both birds and mammals possess three groups of related areas: those of the northern and eastern continental margins, interior desert and, finally, those of the southern margin. The first two are united relative to the third.

The distribution patterns of snakes also exhibit overall congruence with birds and mammals, although in details they differ. Fig. 7*B* shows a strict consensus of all three groups. The areas of the northern and eastern continental margin are united and have very similar interrelationships across the three groups. The interior desert areas also cluster with the northern and eastern clade although, with snakes added, relationships within the desert areas are now unresolved (except for the Western and Eastern Desert areas of endemism). Finally, relationships among the areas of the southern margin are fully congruent among the three groups, and the ambiguous placement of Tasmania is now more clearly recognised.

The remaining two groups of vertebrates, lizards and frogs, exhibit instances of congruence with the three groups just discussed but they also show two major differences (see Figs

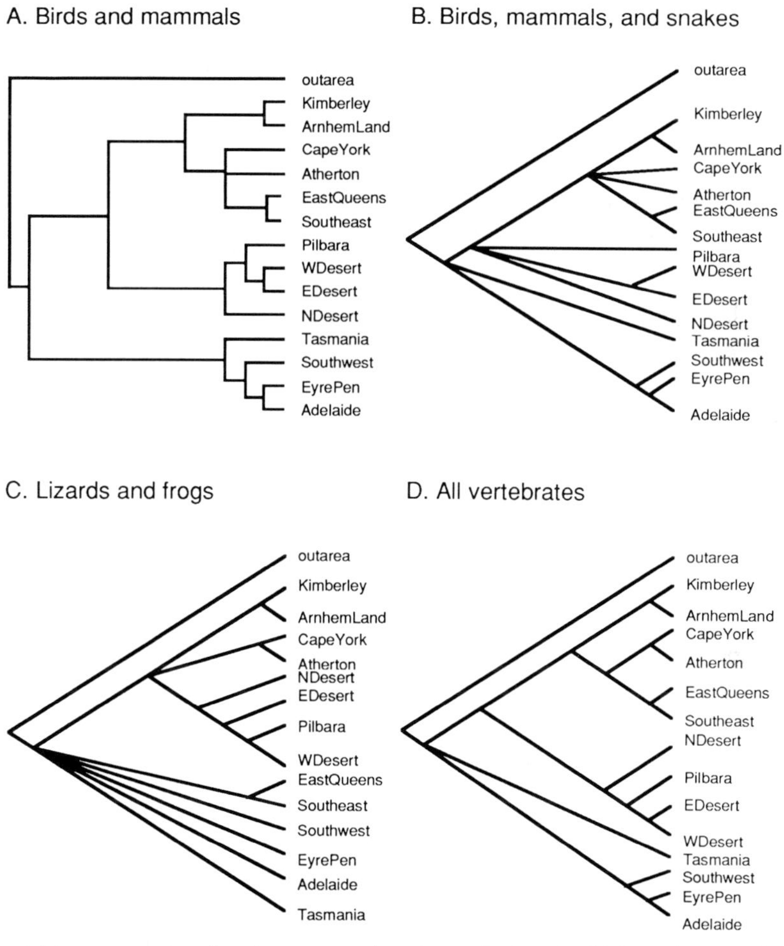

Fig. 7. *A*, strict consensus tree for the most parsimonious solutions for birds and mammals. *B*, strict consensus tree for the most parsimonious solutions for birds, mammals, and snakes. *C*, strict consensus tree for the most parsimonious solutions for lizards and frogs. *D*, Nelson (and majority-rule) consensus tree for most parsimonious solutions for all five vertebrate groups. See text.

3–4 and the strict consensus of Fig. 7*C*). First, the Eastern Queensland–South-eastern Forest clade is still recognised but is no longer closely related to the areas of the northern and north-eastern continental margins. And, second, frogs and lizards show a different pattern of congruent relationships within the interior desert areas, but the relationships of this cluster are still with the four northern and north-eastern areas. Because the five groups of vertebrates can be partitioned into two distinct patterns of area-relationships, a strict consensus of all five yields very little resolution, with only two pairs of sister-areas being held in common: Kimberley + Arnhem Land and Eastern Queensland + Southeastern Forest. A Nelson, or a majority-rule, consensus tree such as shown in Fig. 7*D* has considerable resolution only because it summarises the patterns shown by three groups (birds, mammals, and snakes) instead of two. Thus, the two alternatives shown in Figs 7*B* and 7*C* are perhaps more appropriate for future investigations of area-relationships.

Discussion and Conclusions

What is the Biogeographic Significance of the Avian Pattern?

The data of this study presumably carry information about the cladistic relationships of areas but they also contain noise produced by plesiomorphic distributions, widespread taxa, and long-distance dispersal. What is remarkable, perhaps, is that despite many potential problems with the data — inadequate understanding of systematics at the subspecies, species, and generic levels, ambiguities in our knowledge about distributions, ambiguities in delineating areas of endemism and assigning taxa to those areas, and so forth — rather strong hierarchical patterns emerge from the different analyses. In particular, the strong congruence in the bird and mammal patterns (Fig. 7*A*) and in those of birds, mammals, and snakes (Fig. 7*B*) requires a general explanation. Moreover, even the considerable congruence of lizard and frog distributions (Fig. 7*C*) is surely not coincidental.

Possible explanations for these patterns seem few in number. At the one extreme, they might represent the true vicariant relationships of the areas in question, depicting the sequential subdivision of a once cosmopolitan Australian vertebrate biota. At the other extreme, it might be that none of the relationships represent vicariant relationships but simply the overwhelming influence of plesiomorphy (failure to differentiate), episodes of dispersion, or rampant long-distance dispersal. The answer probably includes elements of both extremes, but with historical patterns of vicariance playing the larger role. Not only do the majority of the sister-areas share vicariant sister-taxa, many of the taxa that are widespread in those sister-areas are also disjunct and found nowhere else and thus very possibly reflect the vicariance history even though they have apparently failed to differentiate (although certainly a number of these taxa will eventually be shown to be distinct when investigated further). Proximity of two areas is by itself not an obvious predictor of area-relationships in the vertebrate data, inasmuch as many contiguous areas of endemism are not generally seen to be sister-groups.

Patterns of vertebrate distributions appear to have a strong hierarchical structure. If these area-relationships do not represent past patterns of vicariance, then the observed congruence among groups of vertebrates implies some very strong historical constraints on recent biotic patterns of dispersion within the Australian vertebrate fauna. These results would thus be interpreted as indicating that the process of biotas becoming more cosmopolitan is historically structured, that biotic dispersion does not take place randomly.

At the extreme, one might deny a major influence for history altogether and propose that the congruence in area-relationships represents 'ecological similarity'. Yet, such an explanation probably raises a false dichotomy — that is, between ecology versus history. We expect, more often than not, that closely related taxa will share ecological attributes, but this should not be taken to mean that ecological similarity itself is the causal dynamic producing a pattern of relationships among areas of endemism. These patterns arise as a consequence of the distributions of very many taxa having rather distinct ecological characteristics. Moreover, closely related areas need not be ecologically similar. Finally, 'ecological similarity' is itself often viewed in terms of biotic similarity which makes any discussion of 'history versus ecology' semantically tenuous. History and ecology may well be inseparable: temporal and spatial variation in the abiotic environment — geomorphological complexity, moisture and temperature regimes, and so on — influence not only patterns of biotic dispersion but also patterns of vicariance.

How Much Noise is Present in Distributional Data?

The results of this study suggest the conclusion that plesiomorphic distributions and those of widespread taxa may not contribute a sufficient amount of false information about area-relationships to overwhelm historical signal. At some hierarchical level, plesiomorphic distributions should be congruent with historically determined area-relationships. Therefore, as long as widespread taxa do not numerically dominate a biota and as long as their distributions do not reflect the effects of random dispersion among areas, their contribution to false area-relationships will be minimised. In fact, within any continental biota, most taxa

— subspecies, species, and to some extent genera — are relatively narrowly endemic. For example, in the bird data-set, the 204 subspecies taxa were distributed, on average, in only two (mean = 1·94) areas of endemism (50·5% were endemic to a single area). Furthermore, 86·8% were endemic in three or fewer areas and no subspecies was found in more than six areas. For 300 species-level taxa, their distributions encompassed a mean of 3·5 areas (23·0% were endemic to a single area). The distribution of areas occupied by species taxa is strongly skewed: 41·7% of the species occupy no more than two areas, 70·0% occupy no more than four, and 90·3% occupy no more than six areas. On average, genera (N = 68) occupy 7·79 areas, with 53% of the genera occupying seven areas or less. In light of these data, it becomes less likely that widespread taxa will obscure patterns of area-relationships especially at the level of subspecies or species, which comprised the vast amount of the informative taxa in this study. One conjecture that these congruent patterns do contradict is the hypothesis that long-distance dispersal is an important process in the assembly of continental biotas. Rampant long-distance dispersal, if present, would destroy any general pattern of congruence.

These are possible arguments as to why biogeographic noise, or false area-relationships, might not be contributing significantly to the patterns of Figs 2–7. Yet, as shown by the bootstrap analyses, many of the postulated area-relationships are very poorly supported. We might expect that as support for a biogeographic component decreases, so too might the adverse influence of plesiomorphic distributions and widespread taxa increase.

What Do These Biogeographic Patterns Say About the Diversification of the Australian Vertebrate Fauna?

That terrestrial vertebrates share many of the same areas of endemism has been recognised for some time, and those areas have been interpreted as centres of differentiation, especially within the context of a Pleistocene refugium hypothesis (Keast 1961, 1962, 1981*a*, 1981*b*; Ford 1987*a*, 1987*b*; Heatwole 1987). Relatively little consideration has been paid, on the other hand, to understanding the historical relationships of these areas (Cracraft 1986).

To the extent that the congruent patterns discussed in this paper represent accurate estimates of the relationships of Australian areas of endemism, they can be used to evaluate several previous hypotheses regarding the diversification of the Australian vertebrate fauna. However, given the paucity of our knowledge about the cladistic relationships of vertebrates in general, it is probably premature to place too much significance on many of the details of the patterns discussed earlier, although some of the broader patterns having relatively strong support are worthy of discussion.

Four general themes have dominated previous conceptions of biotic diversification in Australia:

(1) That it has developed as a result of numerous waves of colonisation from the north and subsequent radiation within the continent;
(2) That two major biotas were developed, one in the arid interior and one in the more moist, peripheral areas of the continent;
(3) That the biota of the arid interior arose via numerous independent colonisations from marginally placed centres of diversity and radiation; and finally
(4) That the speciation events producing the modern biota, especially that occupying the peripheral areas, are predominately Pleistocene in age.

Because much of the evidence used to support these conjectures was derived from numerous studies of taxonomically restricted groups of species, emphasis has consequently been placed on idiosyncratic events rather than broader patterns. When one thus tries to explain distributional patterns of individual groups, they often seem to fit one or more of these thematic conjectures. And so support for them continues to grow. Example 1: many species groups have relatives in New Guinea, Timor, or elsewhere, hence each represents a colonisation and subsequent radiation. Example 2: many groups are restricted to the arid interior, many to the periphery, hence phenetically at least there seems to be two distinct biotas. Example 3: evidence indicates that Australia has become more arid over time, evolutionary modifications for life in arid environments seem to be derived, and many groups in the arid interior have relatives in peripheral areas, hence one can conclude that many of these interior forms were derived from the geographically closest population in the more moist periphery. And, example 4: there is evidence of arid cycles in the Pleistocene that apparently led to contractions of moist habitats into smaller refuges, and significant species-level differentiation presumably took place at this time.

Like all grand generalisations, there is no doubt some truth lurking about in them; but if the areas of endemism have a hierarchical history roughly similar to those in Fig. 7, then these conjectures will have to be modified significantly, or perhaps abandoned, at least as generalisations about the diversification of the vertebrate fauna. Consider each of the generalisations just noted.

Under the centres of origin paradigm (e.g. Darlington 1957), diversification of continental biotas results from radiations following numerous colonisations of ancestral species. These colonisations can take place either as long-distance dispersal events or as faunal dispersions across corridors or land bridges. Long-distance dispersal events do occur, and perhaps some small proportion of these lead to subsequent differentiation, but long-distance dispersal cannot establish the complexly hierarchical patterns of Fig. 7. These reflect congruent histories of numerous faunal elements, hence ancestral, undifferentiated biotas move more or less as a unit (at least within a temporal scale of climatic and geological change), much as we infer to have happened during climatic cycles of the Pleistocene. Australia has established regional biotas with other areas at various times in the past, and through biotic migrations Australia may have donated as many faunal elements to these areas as it has received. A key problem in deciphering Australia's biotic history will be to search for sets of general area-cladograms that specify different histories for the same areas of endemism (e.g. Cracraft 1988). Such patterns might be predicted to reflect temporally separated ancestral biotas that underwent vicariance in the same areas but as a consequence of vicariance events of different ages.

The hierarchical patterns of Fig. 7 support the presence of regionalism in the diversification of areas of endemism within Australia. They also are suggestive of the notion of an historically cohesive interior biota but, with the general exception of an Eastern Desert + Western Desert relationship, the affinities of the Pilbara and Northern Desert areas remain ambiguous. What is strongly supported, furthermore, is an historical disjunction within the peripheral areas themselves. The Kimberley Plateau, Arnhem Land, Cape York and Atherton Plateau always are associated, whereas the South-western Forest, Adelaide, and Eyre Peninsula areas are almost always clustered. In birds, mammals, and snakes, the Eastern Queensland and South-eastern Forest areas cluster with areas to the north, whereas in frogs and lizards they have relationships to the southern group of areas. In all five groups of vertebrates, moreover, the evidence suggests that the interior desert areas have their historical relationships, as a regional fauna, to the north and north-east rather than to the areas along the southern margin of the continent (although there are obviously numerous individual groups within the interior having relationships to taxa in these latter areas and thus the true history of the interior is no doubt far more complex than described here). Conceiving of biotic diversification with Australia in terms of an interior–peripheral dichotomy, therefore, oversimplifies the historical complexities of these patterns.

Many previous efforts to understand the biogeographic history of the fauna of the arid interior proposed that the latter was derived from multiple invasions from peripherally contiguous areas. Thus, faunal elements in the Western Desert area of endemism are thought to originate from the South-western Forest area, Eastern Desert faunal elements from the South-eastern Forest area or from the Adelaide area, and so forth. The fact that many, possibly most, species within interior areas have their sister-species in other arid areas suggests a vicariance perspective will help us understand the history of this complex biota. The strong association between the Eastern Desert and Western Desert areas argues for a vicariant relationship, and this hypothesis also receives support from cladistic studies of endemics (reviewed earlier). Affinities of the Pilbara and Northern Desert areas, as already noted, are more ambiguous and may reflect shared plesiomorphy or widespread taxa as much as they do true cladistic pattern.

Congruence across vertebrate groups also suggests that these areas have had a deep history, not one of Pleistocene contemporaneity. These hierarchical relationships imply that the areas, their vicariance events, and the taxa themselves are of different ages. Supporters of the refuge hypothesis have assumed these ages to be relatively recent, rarely older than the Pleistocene. Unfortunately, there is little evidence one way or the other. Many different types of vicariance events can be conceived of as being potentially significant, and these events vary greatly in age. L. R. Maxson and her colleagues have used a molecular clock calibration based on immunological distance of albumins to estimate the dates of divergence for various species-pairs of frogs having vicariant distributions along the southern margin of the continent (Maxson and Roberts 1984; Roberts and Maxson 1985*a*, 1985*b*, 1988). Those dates range between 5·0 and 27·6 million years, with most of them over 8·0 million years. Even assuming a 10%, or even a 20%, error in the correlation between immunological distance and date of divergence, we would still conclude that these vicariance events are far older than the Pleistocene. And given that speciation patterns in frogs parallel those in other groups of vertebrates suggests that these areas of endemism may be older than previously thought. This is not to say, of course, that many species-pairs of vertebrates might not be recent vicariants resulting from Pleistocene shifts in climate and vegetation. In fact, it is probable that many of these areas are characterised by biotas of different ages, much as we apparently see in frogs; if this is the case, then analyses such as done in this paper will have difficulty in revealing such patterns. Instead, it will be necessary to use component analysis to look for repeated patterns of redundant relationships. At this time, however, there is a pressing need, first, to define areas of endemism more narrowly than has been done in this paper because only after areas of endemism are precisely delineated will we know what questions to ask, and second, to undertake many more cladistic studies of clades possessing narrow endemics. After all this is done, we may well find that the complexities of real-world data exceed the ability of our current methodologies to make complete sense of them.

Acknowledgments

I thank Drs H. G. Cogger and C. Gans for facilitating my access to some of the data used in this paper, and especially to Dr M. Tyler who provided corrections and updated information about frog distributions. Dr G. Nelson provided his usual trenchant comments and good-natured encouragement. This paper was greatly improved by the comments of two anonymous referees, and I extend my thanks to whomever they may be. I am grateful to Drs C. J. Humphries and P. Ladiges for extending an invitation to participate in the biogeography symposium at the Willi Hennig meetings. Finally, this research was supported, in part, by grants from the National Science Foundation (BSR-8805957 and BSR-9007652).

References

AUSLIG (Australian Surveying and Land Information Group). (1990). 'Atlas of Australian Resources (3rd ser.). Vol. 6. Vegetation.' (Canberra.)

Blakers, M., Davies, S. J. J. F., and Reilly, P. N. (1984). 'The Atlas of Australian Birds'. (Melbourne University Press: Melbourne.)

Brooks, D. R. (1990). Parsimony analysis in historical biogeography and coevolution: methodological and theoretical update. *Systematic Zoology* **39**, 14–30.

Christidis, L. (1986). Chromosomal evolution within the family Estrildidae (Aves) I. The Poephilae. *Genetica* **71**, 81–97.

Christidis, L., Schodde, R., and Baverstock, P. R. (1988). Genetic and morphological differentiation and phylogeny in the Australo-Papuan scrubwrens (*Sericornis*, Acanthizidae). *Auk* **105**, 616-29.

Cogger, H. G. (1983). 'Reptiles and Amphibians of Australia', 3rd Edition. (A. H. & A. W. Reed: Sydney.)

Cracraft, J. (1986). Origin and evolution of continental biotas: speciation and historical congruence within the Australian avifauna. *Evolution* **40**, 977-96.

Cracraft, J. (1988). Deep-history biogeography: retrieving the historical pattern of evolving continental biotas. *Systematic Zoology* **37**, 221-36.

Cracraft, J. (1989). Speciation and its ontology: the empirical consequences of alternative species concepts for understanding patterns and processes of differentiation. In 'Speciation and Its Consequences'. (Eds Endler, J. A. and Otte, D.) pp.28-59. (Sinauer Associates: Sunderland, Massachusetts.)

Craw, R. (1988). Panbiogeography: method and synthesis in biogeography. In 'Analytical Biogeography'. (Eds A. A. Myers and P. S. Giller.) pp. 405-35. (Chapman and Hall: London.)

Craw, R., and Page, R. (1988). Panbiogeography: method and metaphor in the new biogeography. In 'Evolutionary Processes and Metaphors'. (Eds M.-W. Ho and S. W. Fox.) pp. 163-89. (John Wiley: London.)

Croizat, L., Nelson, G., and Rosen, D. E. (1974). Centers of origin and related concepts. *Systematic Zoology* **23**, 265-87.

Darlington, P. J. Jr. (1957). 'Zoogeography The Geographical Distribution of Animals'. (John Wiley and Sons: New York.)

Ford, J. (1971). Distribution and taxonomy of southern birds in the Great Victoria Desert. *Emu* **71**, 27–36.

Ford, J. (1978). Geographical isolation and morphological and habitat differentiation between birds of the Kimberley and the Northern Territory. *Emu* **78**, 25–35.

Ford, J. (1986). Avian hybridization and allopatry in the region of the Einasleigh Uplands and Burdekin-Lynd Divide, north-eastern Queensland. *Emu* **86**, 87–110.

Ford, J. (1987*a*). Minor isolates and minor geographical barriers in avian speciation in continental Australia. *Emu* **87**, 90–102.

Ford, J. (1987*b*). Hybrid zones in Australian birds. *Emu* **87**, 158-78.

Heatwole, H. (1987). Major components and distributions of the terrestrial fauna. In 'Fauna of Australia. Vol. 1A.' pp. 101-35. (Australian Government Publication Service: Canberra.)

Humphries, C. J., and Parenti, L. R. (1986). 'Cladistic Biogeography.' (Oxford University Press: Oxford.)

Keast, A. (1961). Bird speciation on the Australian continent. *Bulletin of the Museum of Comparative Zoology* **123**, 303–495.

Keast, A. (1962). Vertebrate speciation in Australia: some comparisons between birds, marsupials, and reptiles. In 'The Evolution of Living Organisms'. (Ed. G. W. Leeper.) pp. 380–407. (Melbourne University Press: Melbourne.)

Keast, A. (1981*a*). The evolutionary biogeography of Australian birds. In 'Ecological Biogeography of Australia'. (Ed. A. Keast.) pp. 1587–1635. (Dr W. Junk: The Hague.)

Keast, A. (1981*b*). Distributional patterns, regional biotas, and adaptations in the Australia biota: a synthesis. In 'Ecological Biogeography of Australia'. (Ed. A. Keast.) pp. 1891–1997. (Dr W. Junk: The Hague.)

Kluge, A. G. (1988). Parsimony in vicariance biogeography: a quantitative method and a Greater Antillean example. *Systematic Zoology* **37**, 315-28.

Kocher, T. D., Thomas, W. K., Meyer, A., Edwards, S. V., Paabo, S., Villablanca, F. X., and Wilson, A. C. (1989). Dynamics of mitochondrial DNA evolution in animals: amplification and sequencing with conserved primers. *Proceedings of the National Academy of Sciences of the USA* **86**, 6196-200.

Maxson, L. R., and Roberts, J. D. (1984). Albumin and Australian frogs: molecular data a challenge to speciation model. *Science* **225**, 957-8.

Nelson, G. J. (1984). Cladistics and biogeography. In 'Cladistics: Perspectives on the Reconstruction of Evolutionary History'. (Eds T. Duncan and T. F. Stuessy). pp. 273-93. (Columbia University Press: New York.)

Nelson, G. J., and Platnick, N. I. (78). The perils of plesiomorphy: widespread taxa, dispersal, and phenetic biogeography. *Systematic Zoology* **27**, 474-7.

Nelson, G. J., and Platnick, N. I. (1981). 'Systematics and Biogeography: Cladistics and Vicariance.' (Columbia University Press: New York.)

Ovenden, J. R., Mackinlay, A. G., and Crozier, R. H. (1987). Systematics and mitochondrial genome evolution of Australian rosellas (Aves: Platycercidae). *Molecular Biology and Evolution* **4**, 526-43.

Page, R. D. M. (1988). Quantitative cladistic biogeography: constructing and comparing area cladograms. *Systematic Zoology* **37**, 254-70.

Page, R. D. M. (1989*a*). Comments on component-compatability in historical biogeography. *Cladistics* **5**, 167-82.

Page, R. D. M. (1989*b*). Component User's Manual. Release 1.5 MSDOS computer program. (Auckland, New Zealand).

Page, R. D. M. (1990). Component analysis: a valiant failure? *Cladistics* **6**, 119-36.

Platnick, N. I. (1981). Widespread taxa and biogeographic congruence. *Advances in Cladistics* **1**, 223-7.

Platnick, N. I., and Nelson, G. J. (1978). A method of analysis for historical biogeography. *Systematic Zoology* **27**, 1-16.

Platnick, N. I., and Nelson, G. J. (1984). Composite areas in vicariance biogeography. *Systematic Zoology* **33**, 328-35.

Platnick, N. I., and Nelson, G. J. (1988). Spanning-tree biogeography: shortcut, detour, or dead-end? *Systematic Zoology* **37**, 410-19.

Roberts, J. D., and Maxson, L. R. (1985*a*). The biogeography of southern Australian frogs: molecular data reject multiple invasion and Pleistocene divergence models. In 'Biology of Australasian Frogs'. (Eds G. Grigg, R. Shine and H. Ehmann.) pp. 83-9. (Royal Zoological Society of New South Wales: Sydney.)

Roberts, J. D., and Maxson, L. R. (1985*b*). Tertiary speciation models in Australian anurans: molecular data challenge Pleistocene scenario. *Evolution* **39**, 325-34.

Roberts, J. D., and Maxson, L. R. (1988). Speciation processes in southeastern Australian anurans: a critique. *Proceedings of the Academy of Natural Sciences of Philadelphia* **140**, 295-307.

Rosen, B. R. (1988). From fossils to earth history: applied historical biogeography. In 'Analytical Biogeography'. (Eds A. A. Myers and P. S. Giller.) pp. 437-81. (Chapman and Hall: London.)

Rosen, D. E. (1978). Vicariant patterns and historical explanation in biogeography. *Systematic Zoology* **27**, 159-88.

Rosen, D. E. (1979). Fishes from the uplands and intermontane basins of Guatemala: revisionary studies and comparative geography. *Bulletin of the American Museum of Natural History* **162**, 267–376.

Strahan, R. (Ed.) (1983). 'Complete Book of Australian Mammals.' (Angus & Robertson: Sydney.)

Swofford, D. L. (1990). 'PAUP: Phylogenetic Analysis Using Parsimony.' MacIntosh Version 3.0L. (Illinois Natural History Survey: Champaign, Illinois.)

Wiley, E. O. (1987). Methods in vicariance biogeography. In 'Systematics and Evolution: a Matter of Diversity'. (Ed. P. Hovenkamp.) pp. 283–306. (Utrecht University: Utrecht, Netherlands.)

Wiley, E. O. (1988*a*). Parsimony analysis and vicariance biogeography. *Systematic Zoology* **37**, 271-90.

Wiley, E. O. (1988*b*). Vicariance biogeography. *Annual Review of Ecology and Systematics* **19**, 513-42.

Zandee, M., and Roos, M. C. (1987). Component-compatibility in historical biogeography. *Cladistics* **3**, 305-32.

Manuscript received 27 November 1990, accepted 19 April 1991